OF THE ELEMENTS

III	IV	V	VI	VII	VIII
					2,2 **He** 4.00
2,3 5 **B** 10.81	2,4 6 **C** 12.01	2,5 7 **N** 14.01	2,6 8 **O** 16.00	2,7 9 **F** 19.00	2,8 10 **Ne** 20.18
2,8,3 13 **Al** 26.98	2,8,4 14 **Si** 28.09	2,8,5 15 **P** 30.97	2,8,6 16 **S** 32.06	2,8,7 17 **Cl** 35.45	2,8,8 18 **Ar** 39.95

				III	IV	V	VI	VII	VIII
2,8,15,2 27 **Co** 58.93	2,8,16,2 28 **Ni** 58.71	2,8,18,1 29 **Cu** 63.55	2,8,18,2 30 **Zn** 65.37	2,8,18,3 31 **Ga** 69.72	2,8,18,4 32 **Ge** 72.59	2,8,18,5 33 **As** 74.92	2,8,18,6 34 **Se** 78.96	2,8,18,7 35 **Br** 79.90	2,8,18,8 36 **Kr** 83.80
2,8,18,16,1 45 **Rh** 102.91	2,8,18,18 46 **Pd** 106.4	2,8,18,18,1 47 **Ag** 107.87	2,8,18,18,2 48 **Cd** 112.40	2,8,18,18,3 49 **In** 114.82	2,8,18,18,4 50 **Sn** 118.69	2,8,18,18,5 51 **Sb** 121.75	2,8,18,18,6 52 **Te** 127.60	2,8,18,18,7 53 **I** 126.90	2,8,18,18,8 54 **Xe** 131.30
2,8,18,32,15,2 77 **Ir** 192.2	2,8,18,32,17,1 78 **Pt** 195.09	2,8,18,32,18,1 79 **Au** 196.97	2,8,18,32,18,2 80 **Hg** 200.59	2,8,18,32,18,3 81 **Tl** 204.37	2,8,18,32,18,4 82 **Pb** 207.19	2,8,18,32,18,5 83 **Bi** 208.98	2,8,18,32,18,6 84 **Po** (210)	2,8,18,32,18,7 85 **At** (210)	2,8,18,32,18,8 86 **Rn** (222)

Atomic weights are based on carbon-12;
values in parentheses are for the most stable or the most familiar isotope.
† Symbol is unofficial

2,8,18,24,8,2 62 **Sm** 150.35	2,8,18,25,8,2 63 **Eu** 151.96	2,8,18,25,9,2 64 **Gd** 157.25	2,8,18,27,8,2 65 **Tb** 158.92	2,8,18,28,8,2 66 **Dy** 162.50	2,8,18,29,8,2 67 **Ho** 164.93	2,8,18,30,8,2 68 **Er** 167.26	2,8,18,31,8,2 69 **Tm** 168.93	2,8,18,32,8,2 70 **Yb** 173.04	2,8,18,32,9,2 71 **Lu** 174.97

2,8,18,32,23,9,2 94 **Pu** (242)	2,8,18,32,24,9,2 95 **Am** (243)	2,8,18,32,25,9,2 96 **Cm** (247)	2,8,18,32,26,9,2 97 **Bk** (247)	2,8,18,32,27,9,2 98 **Cf** (249)	2,8,18,32,28,9,2 99 **Es** (254)	2,8,18,32,29,9,2 100 **Fm** (253)	2,8,18,32,30,9,2 101 **Md** (256)	2,8,18,32,31,9,2 102 **No** (254?)	2,8,18,32,32,9,2 103 **Lr** (257)

CHEMISTRY
AND THE
LIVING ORGANISM

Molly M. Bloomfield

John Wiley & Sons, Inc.

New York • Santa Barbara • London • Sydney • Toronto

This book was printed and bound by Halliday Lithograph Corp.
It was set in Helvetica Light by Ruttle, Shaw & Wetherill, Inc.
Text & cover design by Nicholas A. Bernini/Blaise Zito Assoc., Inc.
Eugene Patti was the copy editor.
The drawings were designed and executed by John Balbalis
with the assistance of the Wiley Illustration Department.
Picture research was done by Stella Kupferberg.
Marion Palen was the production manager.

Library of Congress Cataloging in Publication Data:

Bloomfield, Molly M.   1944–
   Chemistry and the living organism.

   Includes bibliographies and index.
    1. Chemistry.   2. Biological chemistry.
I. Title.
QD33.B672   540   76-26573
ISBN 0-471-08255-4

Printed in the United States of America

10 9 8 7 6 5 4 3 2 1

To Stefan

Preface

Chemistry courses for nonscience majors have undergone remarkable changes in recent years. Abstract chemical theories, with their rigorous mathematical requirements, have given way to applied approaches presenting the principles of chemistry in contexts familiar to the student. This book adopts such a student-oriented approach to serve the needs of students in the allied health sciences and other disciplines.

The ideas for this textbook developed during several years of teaching students whose principal academic interests lay outside the field of chemistry. These students approached their study of chemistry with a great deal of apprehension, perceiving it to be a subject taught in a strange language and having little apparent relevance to their immediate personal needs. This text was written to address the fundamental concerns of such students, while presenting an accurate and comprehensive coverage of the basic principles of chemistry.

To achieve this goal, I have used a conversational writing style more characteristic of recreational reading than of technical texts. Scientific vocabulary appears only when it is needed, and new terminology is clearly defined. Furthermore, the complex mathematical manipulations required in chemistry courses that prepare students for advanced scientific work are largely absent from this book. This is not to say that there is no math! Mathematical calculations are of fundamental importance in many practical uses of chemistry (such as calculating the correct amount of drug for a prescribed dose), and such basic computations are carefully and fully described. In short, my main objective has been to provide a complete and accurate introduction to the basic principles of chemistry in a style that is easy to understand and enjoyable to read.

This textbook includes the student as an integral part of the subject matter. By presenting the principles of chemistry in the context of their clinical and biological applications, I have continually emphasized the relevance of this material to the student's personal and professional life. Each chapter begins with a short story vividly illustrating some impact of the subject matter on the human body. The student then learns about atoms, molecules, chemical reactions, equilibrium, and other basic concepts all in the context of the living organism. The principal motivating factor is the student's growing awareness, chapter by chapter, of the ways in which the human body is directly influenced by its chemical environment, both internal and external.

This book is intended for survey courses for students seeking a basic understanding of chemical principles, and especially for students interested in the biological applications of such principles. It is particularly appropriate for students in such fields as the allied health sciences, physical education, and home economics.

The book consists of four sections: The "Introduction," in which the discussion of a PKU child furnishes a rationale for the study of chemistry; "A Chemical Background," in which the basic vocabulary of chemistry is introduced; "The Elements Necessary for Life," in which the functions of the elements critical to living systems are examined; and "The Compounds of Life," in which the large molecules of living organisms and the interactions among these molecules are discussed.

A list of learning objectives appears at the beginning of each chapter. These will help the student identify the important concepts to be covered in the chapter, and may serve as a study guide for later review. An extensive set of questions and problems, based upon the learning objectives, is at the end of each chapter. Also included after each chapter is a reading list that will help students locate additional information about the topics being discussed. Two appendixes carefully detail the mathematics required for this book.

A **student's study guide** is available for this textbook. The study guide contains section-by-section summaries for each chapter, a list of important terms for each section, and an extensive set of self-test questions and answers.

A **laboratory manual** written by Joseph Bauer of William Rainey Harper College is also available. The laboratory exercises are keyed to each chapter and reinforce the principles discussed in the text by allowing the student to personally observe the interactions described there. Laboratory report sheets can be detached from the lab manual to facilitate grading.

A **teacher's manual** contains solutions to the problems for each chapter, answers to the laboratory exercises, and a list of chemicals and equipment (as well as some helpful hints) for the laboratory experiments.

I would like to express my appreciation to the students and colleagues who have contributed their ideas to the development of this book. I especially thank Joseph M. Bauer, William Rainey Harper College; John R. Holum, Augsburg College; Gerald E. Humiston, Harcum Junior College; Dennis Lehman, The Loop College; Thomas R. Riggs, University of Michigan; L. C. Smith, Indiana State University; Betty Tarr, Los Angeles Southwest College; Mary Vennos and James L. Wyatt, Essex Community College; Michael A. Wartell, Metropolitan State College; and Charles A. Whittemore, Colorado Women's College, whose critical comments and suggestions were invaluable in the development of the final manuscript. Grateful thanks also go to Gary Carlson and the staff at Wiley for their support and expert advice, and to Karen Bland for her excellent typing. Above all, I thank my husband Stefan, without whose patience, support, advice, and editorial criticism this book would never have been written.

Corvallis, Oregon   Molly M. Bloomfield

CONTENTS

section I
introduction

chapter 1

PKU—A Case for Understanding Chemistry

1.1 Why Study Chemistry?

Chemistry is the study of the way in which substances interact.

Now, that may sound like a pretty general definition for such a specialized field of study. However, the broad scope of this definition is one way of indicating just how thoroughly chemistry pervades each of our lives. For example, you drink water from your tap at home or from a fountain at school without a second's hesitation, but someone has added chemicals to the water to insure its safety. You will seldom need to use an iron, thanks to the development of chemicals that give your clothes a permanent press. Just picture your daily routine: You awake in the morning under sheets either made of synthetic fibers chemically produced in a factory, or made of cotton—which was created through chemical reactions in the blossom of the cotton plant. You don clothes made largely of synthetic materials, brush your teeth with a toothpaste containing fluoride, and eat a breakfast fortified with minerals and synthetic vitamins. You may drive to school in a car powered by the energy released through chemical reactions in the engine, or perhaps you pedal a bicycle—powered by the energy released through chemical reactions in your muscles. And now you're reading this textbook, whose paper was created through a chemical process and whose ink was chemically blended. Literally every facet of your life is closely related to the field of chemistry, whether it be the synthetic chemistry of the test tube and the modern laboratory, or the basic chemistry that pervades all of nature.

Chemistry also affects each of our lives in very personal ways, and scientists are just beginning to understand many of these processes. For example, your physical appearance is governed by chemicals. Chemical substances called hormones help determine your height, your weight, your build, and your sexual characteristics. Your good health is dependent upon chemicals that preserve the food you eat, chemicals that protect you from disease, and chemicals (in the form of food) that supply your body with the energy it needs to function properly. Chemicals influence your behavior and your emotional outlook. Some types of mental illness have been found to be chemically induced, and can be treated chemically. Much of your memory may be chemical; your thoughts and experiences may be stored in your brain in the form of chemical compounds. In essence your entire life is chemical, and a basic knowledge of chemistry will allow you to be more aware of your total self and the way in which you interact with your environment.

1.2 A Case For Understanding Chemistry

To illustrate how a knowledge of chemistry might be helpful in understanding the events surrounding us, let's consider a story about a family—it could have been your family or the family of a friend. Don't concern yourself too much with the exact chemistry in this story now. We will return to this case later in the text when we have developed the basic vocabulary and background necessary to understand the chemical processes we will now be describing.

Figure 1.1 This family illustrates the characteristic genetic distribution of phenylketonuria. The parents are both carriers of the trait. By the laws of probability, each child has a 25% chance of being normal, a 50% chance of being a carrier of the trait, and a 25% chance of having PKU. The son at the right is normal, the two daughters are carriers, and the son in the wheelchair has PKU. (Courtesy Willard R. Centerwall, M.D. from *Phenylketonuria,* Frank L. Lyman, ed., Charles C. Thomas, 1963.)

Billy was brought home from the hospital as a happy, healthy baby. People who came to visit the family commented on how fair his skin and hair appeared compared to the rest of the family. As Billy entered his fourth month, his mother started noticing that he no longer watched his mobile as it turned above his crib, and he rarely returned her smiles. Billy was slow in learning to sit up by himself. But then again, his brother had also been late in doing such things and he had turned out to be a very active young child. As Billy grew older, however, his parents became increasingly concerned about his development and behavior. He had become irritable, and would fly into temper tantrums for no reason at all. Although his parents worked very hard to teach Billy to talk, he was able to master only a few words. Furthermore, his mother noticed a strange musty odor about him when she changed his diapers, and his skin was often inflamed and flaky. As Billy neared the age of three, his parents began to acknowledge that he was retarded. He still didn't walk or talk, and he was becoming uncontrollable; he would throw his toys and sit for hours rocking back and forth. His parents started worrying that they might have to put him in an institution.

Finally, a friend convinced them that the best thing for Billy and for themselves would be to take him to a clinic for diagnosis. At the clinic Billy was sent through a battery of tests, after which the doctor told his parents that Billy had been found to have a disorder called phenylketonuria — PKU for short. Billy's parents were quite visibly upset, and their first question was whether there was any hope for Billy. The doctor replied that nothing could be done to reverse the retardation. When untreated, PKU causes irreversible brain damage, and Billy's IQ was found to be 40. (The average person's IQ is 100, and individuals with an IQ below 70 are considered retarded.) However, the doctor assured them that Billy could be placed on a special diet that would probably improve his behavior and his skin condition. He further explained that PKU was an inherited disease, which meant that each parent must be a carrier of a defective gene. Therefore, there was a 1 out of 4 chance that any other child that they might have would inherit the disease (Figure 1.1). Nevertheless, the doctor did not discourage Billy's parents from having more children. He explained that the disorder was now understood and could be treated, and that if PKU were diagnosed soon after birth, children stricken with the disease could lead normal lives. Since recent laws in most states have made the testing of newborn infants for PKU mandatory, most new cases of PKU were being diagnosed soon after birth.

Billy was taken home, and his parents placed him on the special diet that the doctor had given them. They were pleased to find that he no longer had the musty odor, that his skin and hair color darkened, that his behavior brightened, and that he even began to smile. To her relief, Billy's mother found that she could now manage to care for him.

Billy's sister Susan was born five years later. Before she was brought home from the hospital, a sample of blood was taken from her heel and placed on a piece of filter paper for testing. Three weeks later Susan was brought back to the clinic to have the test repeated. The doctor then met with Susan's parents to tell them the tests had revealed an elevated level of a substance called phenylalanine in Susan's blood. This meant that Susan, like Billy, had PKU. However, unlike Billy, the outlook for Susan was very hopeful. In order to reduce the abnormally high level of phenylalanine in her blood, Susan was immediately put on a special diet consisting of a synthetic preparation that provided all of the essential nutrients, but very little phenylalanine. To her parents' delight, Susan is growing into a healthy, active child with an above average IQ. As she grows into maturity and her brain completes its development, Susan's diet can become more varied. The contrast between Billy and Susan is remarkable, and it resulted entirely from an elevation of one chemical compound in the blood (Figure 1.2).

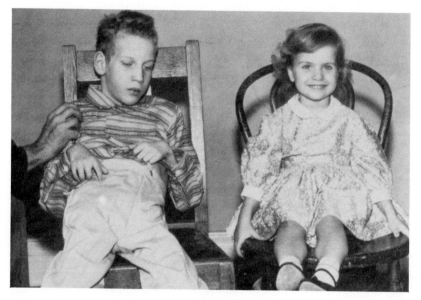

Figure 1.2 Treated and untreated siblings with phenylketonuria. The eleven-year-old boy is severely retarded, while his 2½-year-old sister is normal. Photo by Willard R. Centerwall, M.D. From *Phenylketonuria*, Frank R. Lyman, ed. Charles C. Thomas, 1963.

The physical manifestations of PKU are the result of an upset in the delicate chemical balance that exists within the human body. You are probably aware of instances in which the chemical balance that exists within your own body has been disrupted. Hangovers or muscle cramps are the unpleasant consequences of some common minor disruptions in the body's chemistry. We will be seeing that our bodies are extremely complex organisms that contain many different interlocking chemical systems, all of which must remain in balance to function properly.

The origin of the chemical disruption occurring in PKU is an error in the body's mechanism for converting phenylalanine into another substance called tyrosine. Both of these chemicals, in proper concentration, are required for the normal functioning of the body. It is the abnormally high concentration of phenylalanine and the low concentration of tyrosine that result directly or indirectly in the various abnormalities observed in Billy.

High levels of phenylalanine, and other chemicals produced by the body as a result of this high level, block energy-releasing reactions in the brain of an infant, preventing cells from obtaining the amount of energy necessary for normal functioning. This high level also seems to delay the formation of the protective coating around the brain and spinal column. In addition, large amounts of phenylalanine in the fluid of the brain prevent the normal uptake of other substances by the brain cells, so these cells will not have the normal mix of chemicals from which to construct their essential and permanent components. The brain of an infant at birth is only 25% of its mature weight, and it grows rapidly, reaching 89% of its mature weight by six years of age. During these years of rapid growth, it is critically important for the brain to be surrounded by the correct chemical environment. The altered chemical environment of the brain of a PKU child results in the formation of defective brain cells, explaining the mental retardation of such children.

The chemical tyrosine also plays an important role in the normal functioning of the body. The dark pigment found in hair and skin is formed from tyrosine, and the low level of tyrosine in untreated PKU children leads to the fair hair and skin observed on Billy. Tyrosine is also one of the building blocks of other substances important in the normal functioning of nerves.

If PKU is diagnosed soon after birth, the child can be placed on a special diet that will allow his brain to develop under normal chemical conditions, thus preventing retardation. Otherwise the brain will have grown in an abnormal chemical environment, leading to severe and irreversible mental damage. Even then, however, some of the other clinical manifestations such as skin disease, odor, skin and hair color, seizures, and destructive behavior are reversible, and will improve once the child is placed on the proper diet. This is possible because the body systems responsible for these abnormalities can begin to function normally when they are placed in the correct chemical environment.

There are other examples of similar diseases, which will be discussed in later chapters, but the point of this story is to underscore the critical necessity of the correct chemical balance in the body, from conception throughout life. As researchers become more knowledgeable about the chemistry of living organisms, they will be able to control many more of the diseases that result from chemical irregularities.

1.3 This Textbook

If you had a PKU child in your family you would certainly want to find out as much as you could about the disease, its cause, symptoms, and treatment. To understand the material written about PKU you would first need to learn the vocabulary used in such discussions, and perhaps you would want to read some general books on chemistry, biology, and the human body in order to get more out of the technical papers written about the disease.

This actually is an approach with which you should be quite familiar. You might want to learn how to change the spark plugs on a car, adjust the gears on a bicycle, or make a soufflé. In each case you would have to be familiar with the vocabulary used in the instruction manual, and would need to have at least some general knowledge about cars, bicycles, or kitchens. In the same way, before we can completely understand the many aspects of chemistry affecting our lives, we must first become familiar with the vocabulary used in chemical discussions, and with some of the basic principles and laws that govern the chemical behavior of living organisms. In Section II of this book we will introduce many of the vocabulary terms and fundamental concepts that you will need to know. In Sections III and IV, then, we will use this new vocabulary to discuss the many chemical substances that are essential to life.

Additional Reading

Books
1.   John B. Stanbury *et al.* (ed.), *The Metabolic Basis of Inherited Disease*, McGraw Hill, New York, 1972.

2.   *The Clinical Team Looks at Phenylketonuria*, U. S. Department of Health, Education, and Welfare, Public Health Service, 1972.

Article
1.   "PKU: New Insights into Cause and Effects," *Science News, 103* (150), March 10, 1973.

section II
a chemical background

Chapter 2

Matter and Energy

Learning Objectives

By the time you have finished this chapter, you should be able to:

1. Describe the difference between mass and weight.

2. Convert measurements between the metric and English system.

3. Define the following terms: element, compound, atom, and molecule.

4. State the difference between homogeneous and heterogeneous mixtures, and give three examples of each.

5. Describe the differences between kinetic energy and potential energy, and cite several examples of each.

6. Describe the relationship between the Celsius, Kelvin and Fahrenheit temperature scales.

7. Describe the changes that occur on the molecular level as an ice cube is warmed from $-5°C$ to $110°C$.

8. State, in your own words, five laws of gas behavior, and cite everyday examples of each.

9. Perform calculations using Boyle's law, Charles' law, and the general gas law.

10. Define a "calorie," and calculate the number of calories in a food sample from experimental data.

Jerry and Sue invited their neighbors John and Rosemary to sail with them the morning of July 2 for a cruise to one of the outer islands. The sea was calm, the air was a warm 75°F, and the weather was clear and sunny. They were having a wonderful time until, about 15 miles from their destination, the wind died. Since they were to meet friends for dinner that evening on the island, Jerry decided to start the auxiliary motor. Suddenly an explosion ripped through the boat, and the engine burst into flames. Jerry quickly ordered everyone to don life jackets, and tried unsuccessfully to put out the fire. As the flames spread, Jerry and John tore the seat planking from the boat and threw it overboard for flotation, and everyone jumped into the water. The boat's radio had been destroyed in the explosion, so the four friends could only quietly cling to the planking and hope that their friends would soon mount a search effort.

Their friends did indeed worry when the boat became overdue, and alerted the Coast Guard. A rescue boat was sent out, and the four neighbors were quickly discovered floating near the debris of the burnt boat. Although it had only been four hours since the accident occurred, Rosemary was dead and Jerry died before he reached the hospital. Sue and John were in serious condition, but both survived thanks to the immediate treatment they received.

All four had suffered from hypothermia, a name that denotes the lowering of the body's inner temperature. Hypothermia can be fatal when body temperature drops as few as six degrees below the normal level of 98.6°F. Surprisingly, a person can survive a 40 to 50 degree temperature drop in his hands and feet, but only a small drop in the temperature of the body core can cause death. Because heat flows from a warm region to a colder region, hypothermia can occur under ordinary conditions even when the air temperature is relatively mild. Being wet increases the flow of heat from the body, since water will conduct heat about 240 times better than still air.

Hypothermia begins as soon as the body starts to lose heat faster than it can be produced. The hands and feet are affected first. A drop of only three degrees in body temperature will reduce manual dexterity to the point that one is unable to perform the basic tasks necessary for survival. In addition, as the temperature of the body core is lowered, the brain becomes numb. Individuals become confused and irrational. Such an effect was apparent when Rosemary, after two hours in the water, decided that she was going to swim the 15 miles to shore. Jerry had to swim after her and physically drag her back to the rest of the group. This vigorous activity exhausted both of them, critically depleting their diminishing supplies of energy.

The speed with which hypothermia develops varies considerably with the energy reserves of the individual, and the nature of the survival situation. However, anything that one can do to prevent heat loss is critically important. The proper choices of clothing, insulation, or shelter, and the minimization of muscular activity will all help prevent the onset of hypothermia.

The need for an individual to maintain a stable body temperature in the face of possibly hostile environments is only one example of the complex interactions between living systems and their environments. Any study of the environment and chemistry of the living organism requires an examination of the nature of the various substances involved, the way in which they interact, and the energy changes that occur. As you proceed through this chapter you may suddenly feel bombarded by unfamiliar terms and definitions. In a way, getting through this first chapter is a bit like getting through the first few weeks of a class in conversational French. You receive lists and lists of vocabulary terms and definitions, and are asked to memorize them all. It may seem at first like an impossible task, but as the words are used over and over again, their meanings become second nature. In many cases you will already have an intuitive notion of what various chemical terms and concepts mean, but this may be the first time that you have been asked to learn the precise scientific definition.

Matter

2.1 What Is Matter?

The physical world in which we live is composed of matter. **Matter** is defined as anything that has mass and occupies space. Of course that definition doesn't do you much good unless you know what mass is. You probably have a general feeling about the concept of mass, and would certainly be able to tell which has a greater mass—a brick or a feather. You might recall having heard the Houston Astrodome referred to as a massive structure. The **mass** of an object is a measure of how hard it is to start the object moving, or how hard it is to change its speed or direction once it is moving. For example, a bowling ball is harder to push than a balloon because the bowling ball has a greater mass. The mass of an object is constant no matter where in the universe it is found. Imagine a bowling ball and a balloon both floating around weightlessly in the cabin of an orbiting spacecraft. The balloon would bounce harmlessly off an astronaut's head, while the bowling ball could do substantial damage to his skull. This illustrates the difference between mass and weight; **weight** is a measure of the gravitational attraction or force on an object. For example, since the moon's gravitational pull is roughly one-sixth that of the earth, an astronaut who weighed 180 pounds on the earth would tip his bathroom scales on the moon at only 30 pounds. But, of course, he would have the same mass in either location (Figure 2.1).

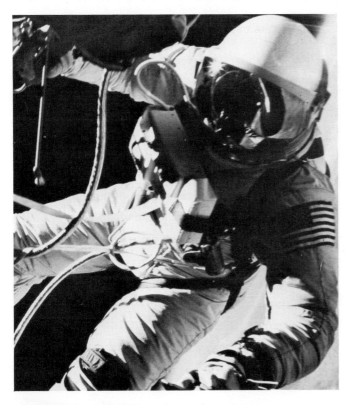

Figure 2.1 The mass of this astronaut has not changed even though he is now weightless in outer space. (Courtesy NASA)

2.2 Measuring Matter

Scientists throughout the world use the metric system for the measurement of matter. The metric system units are the basis of the International System of Units (or SI units), and their use will eventually be mandatory in the United States.

Using the familiar English system of measurements, we have all learned that four quarts make a gallon, 12 inches make a foot, and 16 ounces make a pound. This is a complicated system of measurements to use, since there is no consistent relationship between the number of smaller units needed to make up a larger unit of measure. By contrast, the great advantage of the metric system is that all units of measure are related to their subparts by multiples of 10, and standard prefixes are used to specify the number of multiplications or divisions by 10 that are required. The notion of such a numerical prefix is not new to you—we all know that the prefix tri-, such as found in the words tricycle, tripod, or trio, denotes that the object in question has three of something. In a like manner, the prefix kilo- used in the metric system means one thousand. One kilometer, then, is one thousand meters, and one kilogram is one thousand grams. Table 2.1 shows some of the prefixes used in the metric system. Table 2.2 displays the correspondence between the English system of measurements and the metric system.

Table 2.1   Some Prefixes Used in the Metric System

Prefix	Symbol	Multiple	Example
kilo-	k	$1000 = 10^3$	kilometer, km $=$ 1000 meters
hecto-	h	$100 = 10^2$	hectometer, hm $=$ 100 meters
deka-	dk	$10 = 10^1$	dekameter, dkm $=$ 10 meters
			meter, m (basic unit)
deci-	d	$0.1 = 10^{-1}$	decimeter, dm $=$ 0.1 meter
centi-	c	$0.01 = 10^{-2}$	centimeter, cm $=$ 0.01 meter
milli-	m	$0.001 = 10^{-3}$	millimeter, mm $=$ 0.001 meter
micro-	μ	$0.000001 = 10^{-6}$	micrometer, μm $=$ 0.000001 meter

Table 2.2   Some Relationships between Units in the Metric System
and the English System of Measure

	Metric Units	Conversion Factor	English Units
Length	Meter	1 m = 39.37 in.	Inches
Mass	Kilogram	1 kg = 2.2 lb	Pound
Volume	Liter	1 l = 1.06 qt	Quart

Length

The unit measure of length in the SI system is the **meter.** This unit was
defined in 1790 rather arbitrarily as one ten-millionth of the distance from
the north pole to the equator. Although this arbitrary standard has since
been twice redefined, its length has remained essentially unchanged. One
meter is equivalent to 39.37 inches and is, therefore, just slightly longer
than a yard (Figure 2.2).

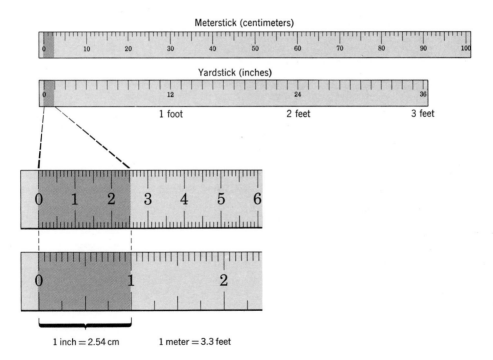

Figure 2.2   A meter is just slightly longer than a
yard, and an inch equals 2.54 centimeters.

Mass

The unit measure of mass in the metric system is the **gram.** The SI standard used to define this mass is a block of platinum metal weighing exactly one kilogram (1000 grams), which is kept in a vault in France by the International Bureau of Weights and Measures. One kilogram is equivalent to 2.2 pounds.

Volume

You will commonly find the volume of an object stated in terms of some unit of length. For example, to calculate the volume of a box we might multiply the length × width × height. The SI unit of volume is the cubic meter, abbreviated m³. This is a fairly large unit of measure. For most practical purposes, therefore, we will use the metric unit of volume — the **liter** — which is only slightly larger than a quart. There are 1000 liters in one cubic meter, and one liter contains 1000 milliliters (abbreviated ml). One milliliter, then, is equivalent to a cubic centimeter (abbreviated cm³ or cc). That is, 1 ml = 1 cc. You will commonly find laboratory measuring devices such as syringes or pipets labeled in either cubic centimeters or milliliters (Figures 2.3 and 2.4).

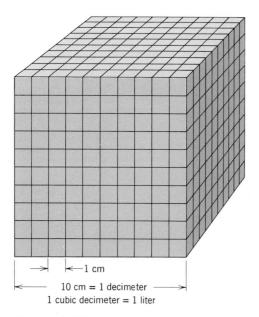

←— 1 cm

10 cm = 1 decimeter

1 cubic decimeter = 1 liter

Figure 2.3 This box has a volume of one liter.
Volume = 10 cm × 10 cm × 10 cm = 1000 cm³
(or cc) = 1000 ml = 1 liter.

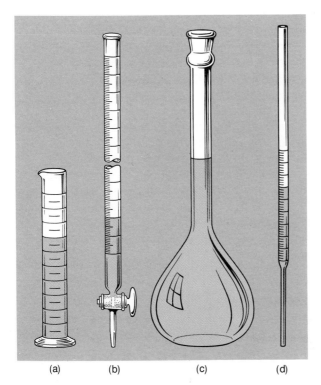

Figure 2.4 This laboratory equipment is commonly used to measure volume: (a) graduated cylinder, (b) buret, (c) volumetric flask, and (d) pipet.

The decimal character of the metric system makes calculations much easier to perform than in the English system. Appendix 2 explains how to perform calculations with the metric system, and describes a method for the conversion of measurements between the English system and the metric system.

2.3 Forces That Act On Matter

You may once have asked what pulls a piece of metal toward a magnet; you were probably told that it was magnetic force. Force may be defined as any action that causes an object to speed up, slow down, or change direction. In this section we will briefly discuss three kinds of forces: gravitational force, electrostatic force, and nuclear force.

Gravitational force is an attraction between two objects that depends upon the masses of the two objects and the distance between them. The attraction increases as the mass of the objects increases, but decreases as the objects move farther apart. It is the gravitational attraction of the earth for a bowling ball that makes the ball so heavy. On a grander scale, the ocean tides are a result of the gravitational forces of the moon and the sun acting on the earth, and the planets of the solar system are held in their orbits about the sun by the gravitational attraction of the sun.

Electrostatic forces exist between particles of matter that bear electric charges. Matter can take on one of two opposite charges, arbitrarily designated as positive and negative. Objects with the same charge repel each other, and objects with opposite charges attract (Figure 2.5). The greater the charges on the objects and the shorter the distance between them, the greater will be the electrostatic force between the objects. If you've noticed your hair being attracted to your comb on a cold, dry day, you have observed the effects of the electrostatic attraction between your hair and the comb. Electrostatic forces are very much larger than gravitational forces (10^{35} times larger, which means 1 followed by 35 zeros — see Appendix 1).

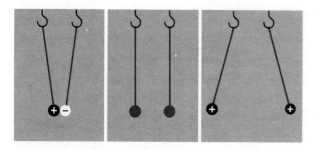

Figure 2.5   Objects with opposite electrical charges attract one another, while those with like electrical charges repel.

Nuclear forces operate between the particles in the center of an atom (called the nucleus), and are presently not completely understood. They operate over extremely short distances and are very large, which is why so much energy is released in a nuclear explosion. We will learn more about this nuclear force in Chapter 3.

2.4 The Atom

Matter is composed of extremely small particles called atoms. The diameter of an atom is about eight-billionths of an inch (0.00000002 cm, or 2×10^{-8} cm — see Appendix 1 if you are confused by this notation). It is very difficult to imagine anything so small. To give you an example, a single page of this textbook is about 500,000 atoms thick.

The existence of atoms was proposed by the Greek philosopher Democritus in the fifth century B.C., but not until John Dalton revived the atomic theory in the mid-nineteenth century was the idea of the atom studied in detail. The last 100 years have seen a tremendous body of knowledge develop about the atom, its structure, and the principles of its behavior, even though no one has actually seen a single atom. Scientists have been able to build elaborate models of atomic structure that fit experimental data very closely. Recently, very fuzzy pictures of heavy atoms have been made with the use of a specially designed electron microscope (Figure 2.6).

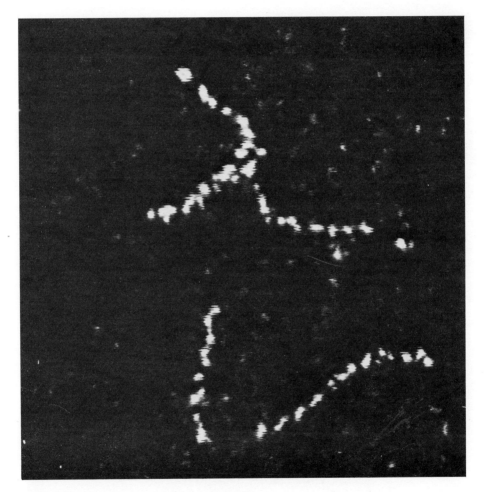

Figure 2.6 Chains of thorium atoms separated by organic molecules were formed on a thin carbon film and photographed with a high resolution scanning electron microscope. The strings of white spots are the chains of thorium atoms. (Courtesy A. V. Crewe. From A. V. Crewe, R. B. Park, J. Biggins, *Science 168*, Figure 4. © 1970 by the American Association for the Advancement of Science)

The Greeks pictured the atom as being indivisible, but the work of many scientists over the last 80 years has shown the atom to be made up of smaller particles. Dozens of subatomic particles have now been identified, but for us only three are important: the **proton,** the **neutron,** and the **electron.** It is the number of these particles and the manner in which they are arranged that give each atom its characteristic chemical properties.

2.5 Elements and Compounds

All living and nonliving matter is composed of elements, either alone or in combination with other elements. An **element** is a pure substance that cannot be broken down into simpler substances by ordinary chemical processes. There are 106 known elements, the heaviest of which are man-made. These man-made elements are produced by bombarding smaller elements with particles that have been accelerated to great speeds in instruments such as cyclotrons or linear accelerators. It is likely that even more elements will be synthesized as scientists continue to refine their techniques.

An **atom** is the smallest unit of an element with the properties of that element. Atoms of some elements remain single, while in other elements the atoms link up with each other. The element copper, for example, is made up of single copper atoms, while the element oxygen is found in the form of two oxygen atoms joined tightly together. When two or more atoms are joined together, a **molecule** is formed. Atoms of more than one element can combine to form a **compound,** which has characteristics that are totally different from those of the elements from which it was formed (Figure 2.7). For instance, two atoms of the element

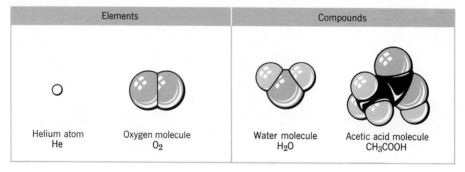

Elements		Compounds	
Helium atom He	Oxygen molecule O_2	Water molecule H_2O	Acetic acid molecule CH_3COOH

Figure 2.7 Elements such as helium exist as single atoms, but other elements such as oxygen exist as diatomic molecules. Compounds, such as water and acetic acid, have molecules containing atoms of more than one element.

hydrogen, which can burn, will react with one atom of the element oxygen, which is the gas we breathe, to form the compound called water. But water, as you know, won't burn, and no living organism can sustain life by breathing water (Table 2.3). (You might have thought that fish breathe water, but they only filter water through their gills in order to remove the oxygen dissolved in the water.)

Table 2.3   Some Elements and Compounds

Elements	Compounds
Aluminum	Albumen — egg white
Copper	Methane — burner gas
Hydrogen	Silicon dioxide — white sand
Mercury	Sodium bicarbonate — baking soda
Oxygen	Sodium chloride — table salt
Sulfur	Water

The table on the inside back cover of the book lists the known elements together with their symbols. The symbol of an element is a shorthand way to denote one atom of that element. The symbol assigned to an element is the first letter in the Latin or Latinized name of that element; for example, H for hydrogen, C for carbon, and O for oxygen. If more than one element has a name starting with the same letter, the first two letters of the Latin name can be used for the symbol, with the first letter capitalized and the second in lower case; for example, Co for cobalt, and Ca for calcium. Note that since chlorine and chromium share the same first two letters, these two elements have the symbols Cl and Cr, respectively. In a few cases, the Latin name of an element is different from the common English name. For instance, the symbol Fe for iron comes from the Latin name ferrum; the symbol Ag for silver comes from the Latin name argentum; and the symbol Pb for lead comes from the Latin name plumbum.

2.6 Mixtures

Elements, and compounds made up of elements, are substances composed of atoms in some definite ratio. For a given compound this ratio will always be the same no matter how the substance is formed. For example, water always contains two atoms of hydrogen for every atom of oxygen. But much of the world around us is made up of mixtures of elements and compounds, and mixtures can be made up in any proportion.

Sugared coffee is a mixture of sugar and coffee, and the proportion of sugar to coffee in a given cup depends upon how sweet one likes his coffee. The air that we breathe, the water that we drink, the ground on which we walk, and the gasoline that we put into our cars are each mixtures of various substances. Mixtures may be **homogeneous,** meaning so uniform that you can't tell one part from another, or **heterogeneous,** meaning nonuniform, allowing one part to be distinguished from another. For example, a well-mixed cup of coffee with sugar in it is a homogeneous mixture; it will have the sugar molecules evenly distributed throughout the coffee, and each sip will taste the same. On the other hand, coffee with sugar in it that has not been stirred is an example of a heterogeneous mixture; you would certainly be able to distinguish the last few sips of coffee from the first (Figure 2.8).

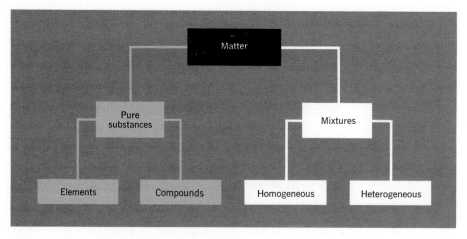

Figure 2.8 Matter can be classified as pure elements and compounds, or heterogeneous or homogeneous mixtures.

2.7 Physical and Chemical Changes

In this book we will be concentrating on two types of changes that matter can undergo: physical change and chemical change. Common examples of substances undergoing physical change include ice melting in a glass of water, wax melting on a candle, sugar dissolving in a cup of coffee, hair drying in the hot air of a hair dryer, and fingernails being trimmed with a nail clipper. A **physical change** is one in which a substance changes form, but retains its chemical identity. For example, water still retains all of its chemical properties whether it is found in the form of ice, water, or steam. Therefore when ice melts to form water, the water has simply undergone a physical change and may be returned to its former state by placing it in a freezer. Similarly, wax melting on a candle undergoes a series of physical changes, first changing to a liquid and then changing back to a solid as it cools. However, the wax retains the same chemical identity throughout all of these physical changes (Figure 2.9).

Chemical changes are occurring when you fry an egg, allow your bicycle to rust in the rain, accelerate your car from a stoplight, digest your dinner, or run a hundred yard dash. In a **chemical change,** the starting materials are consumed, and different substances are formed in their place. Obviously a fried egg has a different appearance and taste from a raw one, and when it cools on your plate it does not return to the raw state. Similarly, if you were to collect the products that come out of the exhaust pipe of a car they would have none of the characteristics of the gasoline that went into the engine. The basic distinction to remember, therefore, is that **a physical change involves only a change of form, while a chemical change involves a basic change in the nature of the substances involved.**

Energy

2.8 What Is Energy?

Energy has become a widely discussed topic, and the "energy crisis" has affected each of our lives. A great deal has been said and written about potential shortages in the world's supply of petroleum, and great controversies have arisen over alternative sources of energy such as electricity and nuclear power. Even though you have felt the effects of our nation's energy supply problems, you may have only a vague notion of precisely what energy is. Some days you may wake up in the morning feeling energetic, ready to get things accomplished—and the tasks you have completed by the end of the day will have resulted from your expenditure of energy. It is precisely the ability to accomplish something—the capacity to do work—that defines **energy.** A rushing stream, a rock poised at the top of a hill, the gasoline in a car, the muscles in an arm,

Figure 2.9 Physical changes involve a change in form. (Mimi Forsyth/Monkmeyer)

Figure 2.9 (continued) Chemical changes involve a change in the basic nature of the substance. (Top left and middle, Mimi Forsyth/Monkmeyer; top right, Hugh Rogers/Monkmeyer; bottom, Gerry Cranham/Rapho-Photo Researchers)

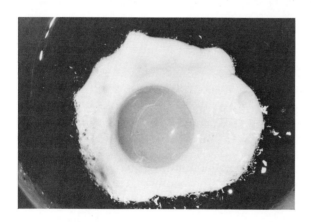

each has the capacity to cause change, to do work on an object; therefore, each is said to possess energy.

Although scientists define energy as the capacity to do work, work can mean many things to many people. Painting a fence, shoveling snow, or chopping logs may be work to some people, but fun and games to others. For this reason, scientists have given the term "work" a mathematical meaning that is independent of an individual's opinion. **Mechanical work** is defined to be the application of a force through a distance. Earlier in the chapter we mentioned that force is a push or pull on an object that causes the object to start moving, or to change speed or direction once it is moving. If you push a car and set it into motion, you have done work on the car. However, if you push a car but are unable to get it to move, you have not done any mechanical work.

One implication of the above definition is that we can't measure the energy contained in an object until it has been transferred from that object to another one. Using the above example, the energy expended by your muscles can be measured only when this energy is transferred from your muscles to the car that you are pushing.

2.9 Kinetic Energy

Energy can take many forms, but it is often convenient to categorize energy as either energy of motion or energy of position. The name applied to energy of motion is **kinetic energy.** A car traveling at 20 mph, a rock hurtling through the air, or a pot of boiling water: all have energy by virtue of motion; that is, each possesses kinetic energy. The moving car could do a great deal of damage if it hit a parked car, the rock could break a window, and the boiling water could cook vegetables. While it is obvious that both the car and the rock have motion and, therefore, possess kinetic energy, you may wonder about the boiling water. The rock and the car are large objects whose motion we can see, but if we had a "super" microscope that could let us see at the level of atoms and molecules, we would realize that all matter is in constant motion. The molecules of water boiling in the pot are moving extremely rapidly and, for this reason, possess a great deal of kinetic energy (Figure 2.10).

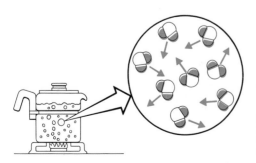

Figure 2.10 We can use the kinetic energy of the rapidly moving molecules in boiling water to do work—such as cooking vegetables.

Kinetic energy can be measured; its magnitude depends upon the mass of the particle and the particle's speed or velocity. Algebraically,

$$\text{Kinetic energy} = \tfrac{1}{2} \times \text{mass} \times (\text{velocity})^2$$
$$\text{K.E.} = \tfrac{1}{2} \times m \times v^2, \quad \text{or} \quad \tfrac{1}{2}\,mv^2$$

Although you may not have thought of it in this way before, you already have a good general feeling for the kinetic energy possessed by an object. For example, you would probably rather have a five-year-old child throw a baseball at your stomach than have a major league pitcher try it. In each case the baseball would have the same mass, but would be traveling at a different speed and, therefore, would have different kinetic energy. The ball thrown by the child would be traveling at a much slower speed; it would have less kinetic energy and would do less damage to your tissues (Figure 2.11). As a second example, would you rather have your parked car hit by a bicycle going 5 mph or a dump truck going 5 mph? In this example, the speed is the same, but the much larger mass of the dump truck gives it a much larger kinetic energy and a much greater capacity to do work—that is, to crumple your fender.

Figure 2.11 Which ball would you rather have hit you? (Right, M. E. Newman/Woodfin Camp).

2.10 Temperature

A baseball, a car, and a dump truck are all objects that we can weigh, whose speed we can determine, and whose kinetic energy we can then calculate. But how can we determine the kinetic energy of a particle that we can't see, such as an atom or a molecule? It is the temperature of a substance that allows us to measure the kinetic energy of its molecules. Scientifically, **temperature** is a measure of average kinetic energy. The term "average kinetic energy" is used because all molecules, like all people, are not alike (Figure 2.12). At any temperature, some molecules will be moving very fast, some very slow, and the majority somewhere in between. But just as we can talk about the average behavior or performance of a group of students, so can we refer to the average behavior of a group of molecules or atoms. As the average kinetic energy of a group of molecules increases, so does the temperature of that substance. From experience, you know that the hotter a pan of water is, the more severely it can burn your hand, or the faster it can cook vegetables. You now understand that the hotter water has a higher average kinetic energy and, therefore, has a greater ability to do work in the form of damaging tissues or cooking vegetables.

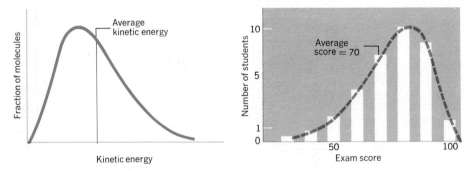

Figure 2.12 At any one temperature each gas molecule will have a specific kinetic energy just as each student will have a specific score on an exam. However, just as there is an average score for the exam, there is an average kinetic energy for the gas molecules, and temperature is a measure of this average kinetic energy.

2.11 Measuring Temperature

Many instruments have been devised to measure the temperature of a substance. The one with which you are undoubtedly most familiar is the mercury thermometer, a thin graduated glass tube with a mercury-containing bulb on the end. As this bulb is heated, the mercury expands and rises in the tube. The particular units marked on the glass tube depend upon the temperature scale used. We are used to seeing temperatures expressed in the **Fahrenheit (°F)** scale, which is associated with the English system of measurement. On this scale, the freezing point of water is 32°F and the boiling point of water is 212°F (a difference of 180 degrees). In the metric system, the **Celsius (°C)** temperature scale is used. On this scale the freezing point of water is 0°C and the boiling point is 100°C (a difference of 100 degrees) (Figure 2.13 and Table 2.4). A third

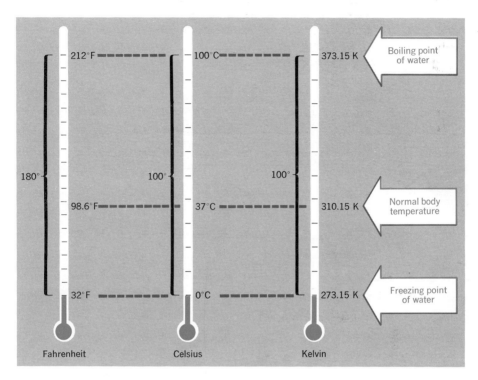

Figure 2.13 Relationships between the Fahrenheit, Celsius, and Kelvin temperature scales.

Table 2.4　A Comparison of Temperatures on the Fahrenheit and Celsius Scales

	Fahrenheit	Celsius
A cold winter day	−5°F	−21°C
Freezing point of water	32°F	0°C
Room temperature	77°F	25°C
Normal body temperature	98.6°F	37°C
A hot summer day	100°F	38°C
Boiling point of water	212°F	100°C

temperature scale is often used by scientists, and is the one recognized in the SI system; this is the **Kelvin (K)** temperature scale. (Note that Kelvin temperatures are written as K, not °K.) The freezing point of water on the Kelvin scale is 273.15 K and the boiling point is 373.15 K. (As with the Celsius scale, this is a difference of 100 degrees.) You may wonder how a number like 273.15 was chosen for the freezing point of water. This number was picked so that zero degrees on the Kelvin scale corresponds to the lowest temperature it is theoretically possible to reach. Zero degrees Kelvin is known as **absolute zero.** Figure 2.14 shows how to make conversions between these temperature scales.

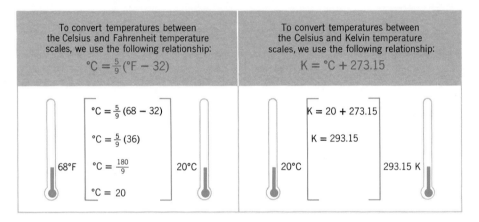

To convert temperatures between the Celsius and Fahrenheit temperature scales, we use the following relationship:

$$°C = \frac{5}{9}(°F - 32)$$

$$°C = \frac{5}{9}(68 - 32)$$

$$°C = \frac{5}{9}(36)$$

68°F $\quad °C = \frac{180}{9} \quad$ 20°C

$$°C = 20$$

To convert temperatures between the Celsius and Kelvin temperature scales, we use the following relationship:

$$K = °C + 273.15$$

$$K = 20 + 273.15$$

$$K = 293.15$$

20°C $\qquad$ 293.15 K

Figure 2.14 Temperature unit conversions.

2.12 Potential Energy

A rock hurtling through the air has kinetic energy; when it strikes a window, it can break it. But a rock perched on the top of a cliff also possesses energy; it has the capacity to do work. If it fell off the cliff onto a passing car, it could do severe damage. The rock on the top of the cliff is said to possess **potential energy,** or energy of position. Such energy is not in use; it is stored, and has the capacity to do work when it is converted to other forms of energy. For example, the water stored behind a dam possesses potential energy. When released in a controlled fashion, it can drive turbines to produce electrical energy or, if released in an uncontrolled fashion by collapse of the dam, the water could display its energy by destroying everything in its path (Figure 2.15). Compounds also have a certain amount of potential energy stored within their molecules. This energy can be released during a chemical reaction and used to do work. For example, the potential energy stored in gasoline molecules is released when gas is burned in your car engine. It does work in the form of driving the pistons that power your car.

Figure 2.15 The water stored behind a dam possesses potential energy which, when released in a controlled fashion, can drive turbines to produce electrical energy (top) or, when released in an uncontrolled fashion, can cause death and destruction. Such a disaster occurred when a dam collapsed on Buffalo Creek in West Virginia in 1972 (bottom). (Top, Courtesy Bureau of Reclamation, Pacific Northwest Region; bottom, Wide World Photos)

Kinetic-Molecular Theory

You may be a bit uncomfortable with the notion that the particles making up all of matter are in constant motion. The kinetic-molecular theory describes, on a molecular level, the motion of chemicals—what occurs when they are warmed or cooled (Figure 2.16), or undergo chemical changes.

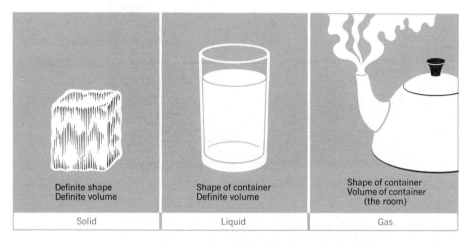

Solid	Liquid	Gas
Definite shape Definite volume	Shape of container Definite volume	Shape of container Volume of container (the room)

Figure 2.16   The three states of matter: water can exist as a solid, liquid, or gas.

2.13 States of Matter: Solid

Matter in the **solid** state consists of particles arrayed in an orderly, closely packed arrangement, restricting their freedom to move from place to place (Figure 2.17). Each particle is attracted to its neighbors, and this explains

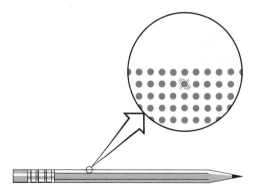

Figure 2.17   A solid consists of closely-packed particles arrayed in an orderly arrangement. Each particle vibrates around a fixed point.

why solids hold together. These particles, however, are not totally motionless. They move back and forth and up and down, or vibrate around a fixed point. This vibration is not enough to destroy the attraction between the particles — the temperature is too low. But if heat is added to the solid, the kinetic energy of the particles increases. Their vibrations become increasingly violent until they reach a point when the solid begins to break apart, or melt. The temperature at which this occurs is called the **melting point** of the solid. At this point the heat that is added does not go into increasing the average kinetic energy of the particles — that is, the temperature — but goes entirely into melting the solid. When the solid is completely melted, the temperature of the substance will again begin to rise as heat is added (Figure 2.18).

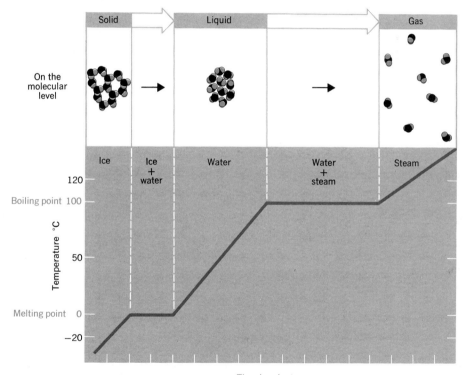

Figure 2.18 This diagram shows the phase changes of water from solid to liquid to gas as heat is added at a constant rate. The temperature at which the ice changes to liquid water is the melting point of water; the temperature at which liquid water changes to steam is the boiling point of water.

2.14 Liquid

The particles in a **liquid** are not held together as tightly or rigidly as in the solid. Although they are fairly close together, they can move from place to place by slipping over one another (Figure 2.19). Thus, although a liquid

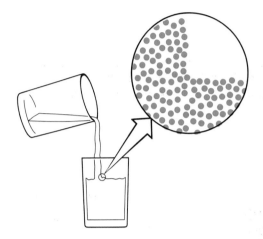

Figure 2.19   A liquid consists of particles which are fairly close together and can move from place to place by slipping over one another.

will maintain a constant volume, it will take on the shape of the container. If particles near the surface of the liquid have enough energy, they can break away from the surface of the liquid. If these particles don't collide with air molecules and return to the liquid, they escape from the liquid completely. This is the process called **evaporation.** Adding more heat to the liquid will increase the kinetic energy of the particles, increasing the rate of evaporation.

Adding enough heat to the liquid can increase the motion of the particles to a point where violent collisions occur within the liquid and bubbles of vapor (gas) are formed. This is the process called **boiling.** The temperature at which boiling occurs under conditions of normal atmospheric pressure is the **normal boiling point** of the substance. At this temperature, additional heat added to the substance goes into pulling the particles apart—that is, changing the substance from a liquid to a gas (Figure 2.18). When all of the substance is a gas, the temperature will again increase as heat is added.

2.15 Gas

Particles in the **gaseous** state are moving very rapidly in a completely random fashion. They move so rapidly that there is a great deal of space between the particles (compared to their size). This explains why it is easier to run through air than through water. Particles of a gas travel in straight lines until they collide with other particles or with the sides of the container. Thus, gas particles will move throughout a container, taking on the shape of the container. The pressure exerted by a gas is nothing more than the collisions of the billions of gas particles against the side of the container (Figure 2.20). When a gas is heated, its particles simply

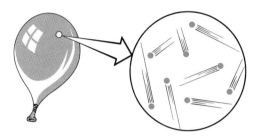

Figure 2.20 A gas consists of widely separated particles which are moving very rapidly in a random, chaotic fashion.

move faster and faster. Theoretically, when a gas is cooled, the particles move slower and slower until a temperature is reached when all motion stops. This temperature, absolute zero, is equal to $-273.15°C$ (or zero degrees K).

2.16 Laws of Gas Behavior

Extensive study has led scientists to formulate laws describing the behavior of gases. Although you may not know the precise mathematical statement of these laws, you certainly are familiar with their general content from everyday experience. For example, an aerosol can when heated will explode, and a tire may blow out when driven at high speeds on a very hot day (increasing the temperature of a gas increases its pressure). When you push down on a bicycle pump, the plunger goes down (increasing the pressure on a gas decreases its volume). When you accidentally burn the cookies you were baking, the whole house soon smells like burnt cookie (a gas will diffuse to all regions of a container). When you open a bottle of carbonated beverage, you hear a hissing sound (the solubility of a gas in a liquid depends upon the pressure).

2.17 Units of Pressure

Before we can discuss the precise mathematical formulations of the laws mentioned above, we need to take a look at how pressure is measured. Pressure is defined as a force per unit of area; two of the units used to measure pressure are as follows:

Atmosphere (atm). One atmosphere of pressure is equivalent to 14.7 pounds per square inch, the pressure exerted by the earth's atmosphere at sea level. One atmosphere of pressure will support a column of mercury 760 mm high (Figure 2.21). **Standard atmospheric pressure** is defined as 1 atm.

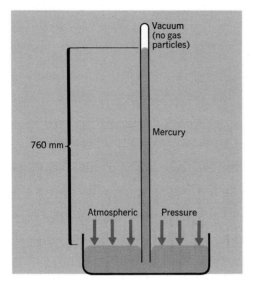

Figure 2.21 The mercury barometer is an instrument used to measure atmospheric pressure. The height of the column will vary from place to place and from time to time as weather conditions affect the atmospheric pressure. At sea level the atmosphere will support a column of mercury about 760 mm high, but at an altitude of three and one-half miles the column would be only about 380 mm high.

Millimeters of mercury (mm Hg). This has long been the most common unit used to express pressure. 1 mm Hg $= \frac{1}{760}$ atm; this is a very small amount of pressure.

2.18 Boyle's Law (The Relationship between Pressure and Volume)

In the seventeenth century, the British chemist Robert Boyle discovered that the volume of a gas varies inversely with its pressure if the temperature is kept constant (Figure 2.22). In other words, if you increase the pressure,

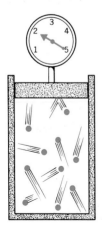
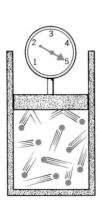

Figure 2.22 Boyle's law states that when the volume of a gas decreases, the pressure will increase (when the temperature of the gas remains constant).

the volume decreases; decrease the pressure and the volume increases. Mathematically,

$$V \propto \frac{1}{P} \quad (T = \text{constant})$$

or

$$PV = k \quad \text{(a constant)}$$

Given two conditions of temperature and pressure, P_1V_1 and P_2V_2:

$$P_1V_1 = k \quad \text{and} \quad P_2V_2 = k$$

Therefore

$$P_1V_1 = P_2V_2$$

Example: _____

Imagine a gas to be contained in a cylinder with a movable piston. If the volume of the gas is 1.5 liters and the pressure is 760 mm Hg, what would be the volume if the pressure were increased to 1140 mm Hg?

1. From the statement of the question we can identify the following:

$$V_1 = 1.5 \text{ liters}$$
$$P_1 = 760 \text{ mm Hg}$$
$$P_2 = 1140 \text{ mm Hg}$$
$$V_2 = \text{ ?}$$

2. Substituting these values into the equation $P_1V_1 = P_2V_2$,

$$760 \text{ mm Hg} \times 1.5 \text{ liters} = 1140 \text{ mm Hg} \times V_2$$

Hence,

$$V_2 = \frac{760 \text{ mm Hg} \times 1.5 \text{ liters}}{1140 \text{ mm Hg}} = 1.0 \text{ liters}$$

An illustration of Boyle's law at work in our bodies is provided by examining how we breathe (Figure 2.23).

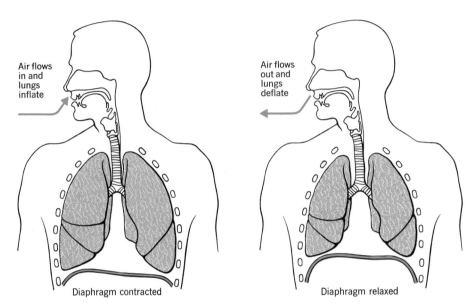

Air flows in and lungs inflate

Air flows out and lungs deflate

Diaphragm contracted

Diaphragm relaxed

Figure 2.23 Breathing is an example of Boyle's law in action. Your lungs are located in the thoracic cavity, surrounded by the ribs and a muscular membrane called the diaphragm.
a) *To inhale:* The diaphragm contracts, flattening out and increasing the volume of the thoracic cavity. This lowers the pressure in the cavity below that of the atmosphere, causing air to flow into the lungs.
b) *To exhale:* The diaphragm relaxes, pushing up into the thoracic cavity. This decreases the volume of the cavity, increasing the pressure above atmospheric pressure and causing air to flow out of the lungs.

2.19 Charles' Law (The Relationship between Volume and Temperature)

In the early nineteenth century, the French physicist Jacques Charles discovered that the volume of a gas varies directly with Kelvin temperature when the pressure is constant. In other words, if you increase the temperature, the volume will increase (Figure 2.24). Mathematically,

$$V \propto T \qquad (P = \text{constant})$$

or

$$\frac{V}{T} = k \quad \text{(a constant)}$$

If we have two values of temperature and volume, $V_1 T_1$ and $V_2 T_2$, then

$$\frac{V_1}{T_1} = k \qquad \text{and} \qquad \frac{V_2}{T_2} = k$$

so that

$$\frac{V_1}{T_1} = \frac{V_2}{T_2}$$

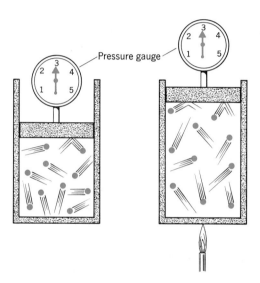

Figure 2.24 Charles' law states that when the temperature of a gas is increased, the volume of the gas will increase (when the pressure remains constant).

Pressure gauge

Example:

Imagine that a gas occupies a volume of 2.0 liters at a temperature of 27°C. To what temperature in degrees Celsius must the gas be cooled to reduce its volume to 1.5 liters?

1. From the question we know that

$$V_1 = 2.0 \text{ liters}$$
$$V_2 = 1.5 \text{ liters}$$
$$T_1 = 27 + 273 = 300 \text{ K}$$
$$T_2 = \quad ?$$

2. Substituting these values into the equation for Charles' law,

$$\frac{2.0 \text{ liters}}{300 \text{ K}} = \frac{1.5 \text{ liters}}{T_2}$$

or $\qquad 2.0 \text{ liters} \times T_2 = 1.5 \text{ liters} \times 300 \text{ K}$

$$T_2 = \frac{1.5 \text{ liters} \times 300 \text{ K}}{2.0 \text{ liters}} = 225 \text{ K} = -48°C$$

2.20 General Gas Law

We can combine the laws stated by Charles and Boyle in a general mathematical equation that will allow us to predict gas behavior given specific conditions of pressure, temperature, and volume:

$$\frac{P_1V_1}{T_1} = \frac{P_2V_2}{T_2}$$

2.21 Graham's Law of Diffusion

In the nineteenth century the Scottish chemist Thomas Graham observed that gases diffuse (intermingle with other particles) at different rates. In particular, he observed that lighter gases diffuse more rapidly than heavier gases at the same temperature.

2.22 Henry's Law (Solubility of Gases)

When you open a bottle of carbonated beverage and listen to the hiss of the escaping carbon dioxide, you are observing a consequence of Henry's law. At the beginning of the nineteenth century the English chemist William Henry discovered that the solubility of a gas in a liquid at a given temperature is directly proportional to the pressure of that gas on the liquid. Carbonated beverages are bottled under high pressures; when you open the cap you reduce the pressure in the bottle, lowering the solubility of the carbon dioxide and permitting the gas to escape. Another example of changes in solubility caused by changes in pressure is the **bends,** a disorder that occurs when deep-sea divers are brought up to the surface too quickly (Figure 2.25).

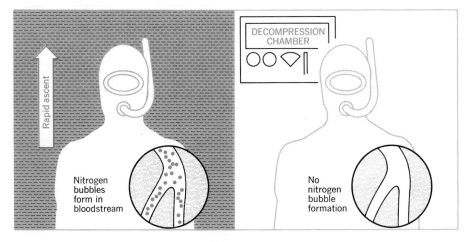

Figure 2.25 Deep sea divers breathe air under pressure, which increases the solubility of the nitrogen in the blood (at normal atmospheric pressure very little nitrogen will dissolve in the blood). If the diver is brought to the surface too quickly, the pressure of the dissolved gas will be greater than atmospheric pressure. The nitrogen in the tissues will then form tiny bubbles which cause severe pain. The only cure for the bends is slow decompression, allowing the nitrogen to escape without forming bubbles.

2.23 Dalton's Law of Partial Pressures

We have stated that the pressure of a gas stems from the collisions between the gas particles and the walls of the container. There are two ways we can increase the frequency of collisions, and thereby increase the pressure. One method is to increase the temperature of the gas, thus increasing the kinetic energy of the gas particles and the number of collisions that occur. The second way is to increase the number of particles of gas in the container.

The pressure exerted by a gas is independent of the type of gas particles, but depends directly on the number of particles of gas that are present. For example, we can double the pressure in a container by adding an equal number of particles of the same gas or a different gas (Figure 2.26). In a mixture of gases, the pressure exerted by each gas is called

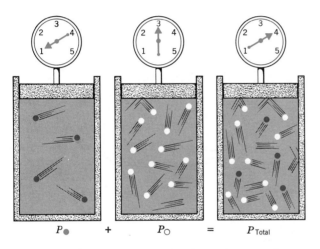

$$P_{\bullet} \qquad + \qquad P_{\bigcirc} \qquad = \qquad P_{\text{Total}}$$

Figure 2.26 Dalton's law states that the total pressure in a container will be equal to the sum of the partial pressures of the gases in that container.

the **partial pressure, P,** of that gas, and depends only upon the number of particles of that gas present. Dalton's law states that the total pressure of a mixture of gases is equal to the sum of the partial pressures of each of the gases in the mixture. For a mixture of four gases, A, B, C and D,

$$P_{\text{total}} = P_A + P_B + P_C + P_D$$

Example: _____

The earth's atmosphere is a mixture of nitrogen, oxygen, argon, and other gases found in small amounts. By Dalton's law then,

$$P_{\text{atmosphere}} = P_{N_2} + P_{O_2} + P_{Ar} + P_{\text{other}}$$

We can use this equation to answer the following question: what is the partial pressure of oxygen at sea level if $P_{N_2} = 593$ mm Hg, $P_{Ar} = 7$ mm Hg, and $P_{\text{other}} = 0.2$ mm Hg?

At sea level atmospheric pressure is 760 mm Hg. Substituting into the above equation, we get

$$760 \text{ mm Hg} = 593 \text{ mm Hg} + P_{O_2} + 7 \text{ mm Hg} + 0.2 \text{ mm Hg}$$

Hence,

$$P_{O_2} = 59.8 \text{ mm Hg}$$

The movement of respiratory gases (oxygen and carbon dioxide) within our bodies is directly related to the partial pressures of these gases, and is described in Figure 2.27.

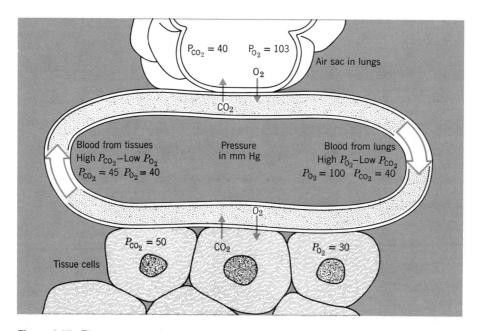

Figure 2.27 The movement of oxygen and carbon dioxide within our bodies depends upon the partial pressure of those gases. The gas will diffuse from a region of higher partial pressure to a region of lower pressure.

In the lungs: Blood entering the lungs from the tissues has been depleted of its oxygen supply and is carrying the waste product carbon dioxide from the cells. Oxygen will diffuse from the lungs to the blood and the carbon dioxide will diffuse from the blood to the lungs.

In the tissues: Tissue cells are constantly using oxygen, so the partial pressure of oxygen in the cells is low. The oxygen carried in the blood from the lungs has a higher partial pressure, so it will diffuse from the blood to the tissues. Carbon dioxide which is produced by the cells will diffuse from the cells to the bloodstream.

Energy Sources

It is convenient to classify the energy around us according to the source of that energy. Such categories include electrical energy, chemical energy, nuclear energy, gravitational energy, heat energy, mechanical energy, and light energy.

2.24 Heat

Heat is a form of energy with which you should be quite familiar. You may never have thought of heat as doing work, but hot gases produced by exploding gasoline push the pistons that power your car, and hot steam turns the turbines in a generating plant to power your electric appliances.

Heat energy is measured in units called calories. A **calorie (cal)** is the amount of energy required to raise the temperature of one gram of water exactly one degree Celsius. Raising the temperature of a cup of water (250 grams) from room temperature (25°C) to boiling (100°C) to make a cup of coffee requires 18,750 calories of energy.

$$\text{calories} = \text{grams of water} \times \text{temperature change (denoted } \Delta t) = g \times \Delta t$$
$$\text{cal} = 250 \text{ g} \times (100° - 25°)$$
$$= 250 \text{ g} \times 75°$$
$$= 18{,}750 \text{ cal}$$

A calorie is such a small unit of energy that it is often more convenient to talk in terms of 1000 calories, called a kilocalorie. A **kilocalorie (kcal)** is the amount of heat energy required to raise the temperature of 1000 g (about 1 quart) of water one degree Celsius. The food Calorie (note the capital C) that you have heard so much about, is actually a kilocalorie. That one ounce bag of potato chips that you had for a snack contains 160 food Calories, or 160 kilocalories. To avoid confusion we will always refer to food energy content in terms of kilocalories.

The instrument that was used to determine the calorie content of those potato chips, or any other substance, is called a **calorimeter,** and is shown in Figure 2.28. The sample to be tested is placed in the inner chamber, and is then burned completely. The energy released from this burning warms the water in the surrounding container, and the number of calories released can be calculated from the rise in temperature.

Example: _____

If the temperature of 1000 grams of water in the calorimeter increases by 50°C when 10 peanuts are burned, then the average calorie content of a peanut would be 5000 cal or 5 kcal.

$$\text{calories} = \text{grams} \times \Delta t$$
$$= 1000 \text{ g} \times 50°$$
$$= 50{,}000 \text{ cal produced by 10 peanuts}$$

$$\frac{50{,}000 \text{ cal}}{10 \text{ peanuts}} = \frac{5000 \text{ cal}}{1 \text{ peanut}} = \frac{5 \text{ kcal}}{1 \text{ peanut}}$$

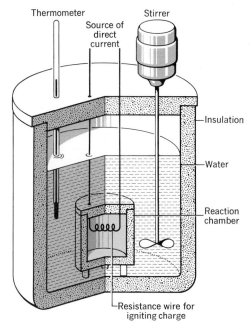

Figure 2.28   A calorimeter consists of a reaction chamber, where the sample is burned, surrounded by water. The water, in turn, is contained in an insulated unit constructed so that the water can be stirred and the temperature change of the water can be measured.

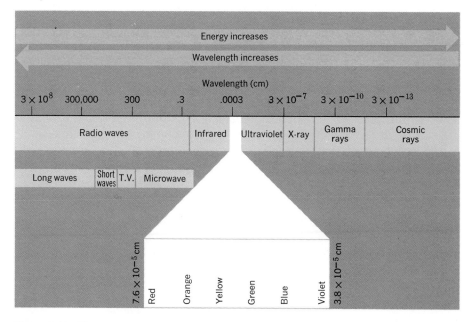

Figure 2.29   The electromagnetic spectrum. Visible light is only a very small part of this spectrum.

2.25 Light

Light represents another form of energy that pervades our lives. The light, or radiant, energy that we see is only a tiny segment of an entire range of electromagnetic energy from low energy radio waves to very high energy X rays and gamma rays. What we see as white light is actually made up of the whole spectrum of colors as displayed by a rainbow. The location of this visible spectrum in the **electromagnetic spectrum** is shown in Figure 2.29.

Light is actually an extremely complex concept; in some respects it has wavelike characteristics and, in other respects, it acts as a particle. However, in this book we will discuss only some of its wavelike properties. Light waves may be compared to ocean waves with their crests and troughs. The **wavelength** of light is the distance from one crest to the next crest, or from trough to trough, or from middle to middle for that matter (Figure 2.30). Each wavelength of light is associated with a particular level

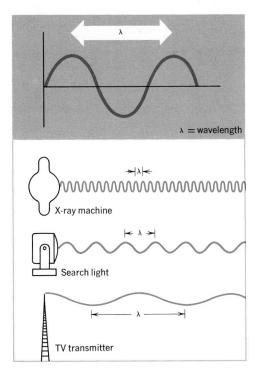

Figure 2.30 Light travels in waves. The wavelength of light can be defined as the distance from crest to crest. Each wavelength is associated with a particular energy: the longer the wavelength, the lower the energy.

of energy; the longer the wavelength, the lower the energy of the light. Look at the visible spectrum shown in Figure 2.29. Visible light with the highest energy is in the purple region, with the shortest wavelengths, while red light possesses the lowest energy. Light energy outside the visible range can have profound effects on living matter. Recent experiments show that radiation of both long and short wavelengths can affect animals on both a physiological and behavioral level. For example, effects of light on the human body include the synthesis of vitamin D and skin pigment, thickening of the skin, absorption of calcium, and the regulation of biological rhythms.

Radiation in the **infrared** range cannot be seen, but can be felt as heat and can be detected by a thermometer. Infrared radiation from the sun warms us, and light fixtures that emit this radiation are used to keep food warm in restaurants. Although infrared radiation doesn't affect ordinary photographic film, special film can detect the infrared radiation given off by warm objects. Such film has been used to map thermal pollution from power plants and the density of vegetation in certain regions (Figure 2.31).

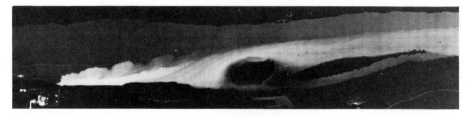

Figure 2.31 Infrared-sensitive film was used to record the heat discharge from a power plant on the Connecticut River. The power plant and an oil tanker with its hot engine room is at the lower left, and the glowing cloud running down the bank is the warm water being discharged from the cooling system of the power plant. (Courtesy Environmental Analysis Department-HRB-Singer, Inc.)

Two species of snakes have infrared sensors that can detect low density infrared radiation given off by their prey. Using these sensors, they can pinpoint and capture their prey in the dark, and can locate hiding places with comfortable living temperatures.

Man has made significant use of long wavelength, low energy radiation to transport radio and television signals, and has used energy in the radar range to track objects in the air and under the water. **Microwave** radiation is becoming increasingly popular as a means of rapid cooking. Microwaves interact with food by setting the molecules within the food into motion, causing the food to heat up uniformly. This can be contrasted with conventional cooking, in which the molecules on the exterior of the food are heated first, and the heat is then transferred gradually to the molecules in the center of the food. Because microwave ovens uniformly heat all of the food at the same time, cooking time is greatly reduced.

Ultraviolet light has shorter wavelengths and higher energy than visible light. It is penetrating radiation that can be quite harmful to living tissue, causing changes in certain biological systems or causing irreparable damage to cells. Some wavelengths of ultraviolet light are very effective in killing bacteria, and are used in sterilizing units. Surprisingly, ultraviolet light is also quite important to our bodies; it penetrates the skin and causes the production of vitamin D, a compound necessary for the prevention of rickets. Unfortunately, too much ultraviolet exposure can produce the burns we call sunburn, and the possibility of skin cancer.

Most of the ultraviolet radiation coming to us from the sun is absorbed by the upper atmosphere of the earth. Our skin also possesses a protective mechanism against overexposure to ultraviolet radiation. Upon exposure to this radiation the skin will produce a dark pigment that absorbs the excess ultraviolet light; it is this increase in protective pigment that we call a suntan.

X rays and **gamma rays** have progressively shorter wavelengths, higher energies, and greater penetrating power. The shortest X rays can pass through steel castings, and gamma rays can pass through a 10-inch thick lead plate. Since X rays will not pass as easily through bones and teeth as they will through tissue, they are a very useful diagnostic tool for the medical and dental professions. However, penetration of cells by high energy radiation such as X rays or gamma rays can disrupt normal chemical processes, causing cells to grow abnormally and to die. The effects of radiation are cumulative over time, so dosages and repeated exposure to X rays must be carefully controlled. Cancer cells are more sensitive to radiation than normal cells, so X rays and gamma rays can be used to treat certain forms of cancer. Medical radiation equipment is designed to control the exact degree of penetration of the radiation and to concentrate the radiation on the cancerous area. Still, some painful burns result when cancer is treated with high energy radiation.

Gamma rays, with their high energy and penetrating power, are extremely harmful to tissues. In the next chapter we will see that gamma rays are one form of radiation emitted by radioactive material, and are the source of the extreme danger of such radioactivity.

2.26 Conservation of Energy*

Energy can be converted from one form to another. For example, electrical energy is converted into heat energy in our furnaces, stoves, and toasters, and into light and heat in our lamps. Chemical energy is converted into heat energy when gas is burned in stoves or furnaces, or into mechanical energy when gasoline is burned in car engines. Although energy can be converted from one form to another, it must all be accounted for. This is a statement of the **First Law of Thermodynamics,** which says that energy can neither be created nor destroyed, but only changed in form. In other words, the total amount of energy at the end of a reaction or process must

---

* This section, "Conservation of Energy" and Section 2.27, "Entropy" are optional, and may be skipped without loss of continuity.

equal the total amount of energy at the beginning. We know that gasoline contains stored potential energy. When burned in the engine of a car, only about 20% of this energy is used in the actual process of moving the car. Where does the rest of the energy go? (The First Law of Thermodynamics says that it cannot just disappear.) In this particular case, about 75% is lost as heat energy—"lost" in the sense that this energy does not do any useful work. Figure 2.32 shows the ultimate uses of the rest of the energy released from burning a gallon of gasoline in a medium-sized car.

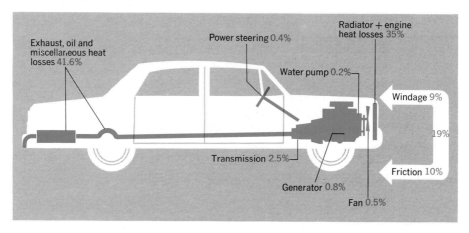

Figure 2.32 Only about 20% of the energy released in the burning of a gallon of gasoline goes toward moving the car. This diagram shows the ultimate uses of the rest of the energy released when a gallon of gasoline is burned in a medium-size car with a 120 horsepower engine cruising at 50 miles per hour. (From *Order and Chaos: Laws of Energy and Entropy* by Stanley W. Angrist and Loren G. Hepler. Drawings by Ed Fisher, Jr., © 1967 by Basic Books, Inc., Publishers, New York, Figure 4–6)

In the same way that gasoline represents a store of energy for a car, the food that we eat represents potential energy for our cells. But in this case too, the conversion of stored food energy into energy that our cells can use is not 100% efficient. For example, only about 38% of the potential energy contained within a molecule of sugar is converted into energy that can be used by our cells to do work. The rest ends up as heat energy that helps to maintain body temperature. Much to many people's displeasure, our bodies must obey the First Law of Thermodynamics; if we take in more food energy than our bodies need, this excess energy doesn't just disappear, but is stored within our tissues mostly in the form of unwanted fat. To get rid of this excess stored energy, we must take in less food energy than our bodies require so that our cells will start using the energy stored in the fatty tissue.

On a larger scale, the total energy of the earth must also be conserved.

The earth receives a tremendous amount of energy from the sun, estimated to be about 4×10^{13} kilocalories per second. This is comparable to the amount of energy that would be released from burning 6 million tons of coal each second. Some of this energy is reflected back into space by the atmosphere, but the rest is absorbed. This energy maintains the earth's temperature and provides energy for the winds, rains, ocean currents, and all forms of life. Living organisms have a very narrow range of temperatures in which they can survive. The adaptive mechanisms of living organisms react extremely slowly, and even very gradual increases or decreases in temperature can lead to the extinction of many species. If the earth's temperature is to remain constant, the earth must radiate back into space each night any unconsumed energy it absorbed from the sun during the day, plus the relatively small amount of energy generated in the core of the earth and the heat energy produced by man (Figure 2.33). Man's

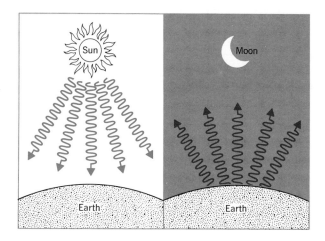

Figure 2.33 Each second, the earth receives from the sun energy equivalent to the burning of six million tons of coal. In order for the temperature of the earth to remain fairly constant, energy must be radiated back into space each night.

contribution to the total heat that the earth must radiate is increasing as his demands for energy increase, and many scientists are concerned about the possible long range effects of such heat production on the future climate of the earth.

2.27 Entropy*

You might be wondering why we need a constant replenishing supply of energy from the sun to maintain life on earth if energy is conserved in all

---

* Optional.

natural processes. Although the First Law of Thermodynamics tells us that the total quantity of energy remains unchanged, it tells us nothing about the quality of that energy. Only concentrated forms of energy can be used to do work, and work is essential to the maintenance of living organisms. The energy from the sun reaches us in the very concentrated form of light energy, energy that is very quickly dissipated as heat.

If you think about it for a moment, you will realize that all naturally occurring (or spontaneous) processes tend to go in one direction. Left alone, water always runs downhill, heat flows from a hot to a cold object, gases flow from regions of high pressure to regions of low pressure, and people grow old. We would be quite shaken if we were to observe water flowing uphill of its own accord, but we still might ask what it is that makes naturally occurring processes irreversible. The answer to this question is fairly obvious when the final state of the system is at a lower energy level than the initial state. Water flows downhill from higher to lower potential energy, and heat flows from a state of higher to lower kinetic energy. It is quite reasonable that there should be a general tendency for all substances to reach a state of lower energy.

But there are other spontaneous processes with which you are familiar that don't seem to depend upon the energy of the system. For example, if you were to put a drop of ink into a glass of water, the ink would immediately begin to disperse throughout the liquid, eventually giving the water a uniform tint (Figure 2.34). This process is certainly irreversible;

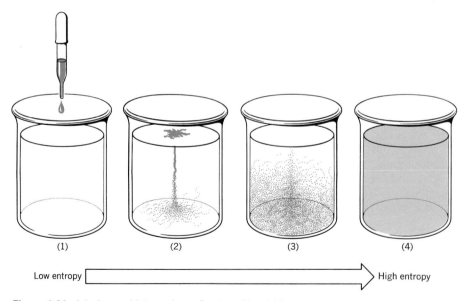

Figure 2.34 Ink dropped into a glass of water will quickly change from a state of low entropy (forming the droplet) to a state of high entropy (dispersed throughout the water).

you would be astonished if you ever saw a glass of tinted liquid suddenly change so that all of the color moved through the liquid to form a single concentrated spot of dye. Yet, it is impossible to identify any change of energy taking place in this process. But the situation can get even more complicated. Consider, for instance, an ice cube sitting on a plate at room temperature. Eventually the ice cube will melt, in a process that you will no doubt agree is irreversible as long as the plate remains at room temperature. But we know that the water molecules in the liquid state have more kinetic energy than the water molecules in the solid state, so it appears that this is an example of a naturally occurring process that moves from a state of lower energy to a state of higher energy. How could this be? The answer is that there is another factor besides energy levels that determines the direction of spontaneous processes; that factor is the amount of randomness or disorder of the system. The term used to describe the disorder of the system is **entropy.** The more random and disordered a situation is, the greater its entropy, and all spontaneous reactions go toward a condition of greater randomness, greater disorder, and greater entropy. We can now see that the drop of ink, which was originally in a small compact drop, will spontaneously disperse throughout the liquid into a random and disordered arrangement of ink molecules, thereby increasing the entropy of the system. In the same way, the water molecules forming ice exist in a highly ordered structure; they have low entropy. As the ice melts, the solid structure breaks down and the molecules take on a more random, disordered arrangement; they now have higher entropy (Figure 2.18). The driving force behind the dispersal of the drop of ink and the melting of the ice cube is the tendency for the entropy of the system to increase. Notice that in the case of the ice cube, the increase in entropy is sufficiently large to cause this process to occur even though it allows the water molecules to assume a state of higher energy.

The **Second Law of Thermodynamics** states that the entropy of the universe is increasing. This law is interwoven through all of science and each of our lives. Even the writers of nursery rhymes had an intuitive feeling for the Second Law when they wrote, "All the king's horses and all the king's men couldn't put Humpty together again." You now see that the breaking of an egg is an irreversible process that leads to increased disorder of the egg parts and therefore greater entropy (Figure 2.35).

By now it may have occurred to you that there are some naturally occurring processes that can be reversed. Water can be pumped uphill to a storage reservoir, heat can be pumped from cold to hot areas in a refrigerator, and air can be pumped from low pressure to high pressure in a tire. The key word in each of these examples is "pump." Each of these examples involves adding energy; that is, work must be done. This means that each of these processes must be coupled with, or related to, another reaction in which energy is produced. Electrical energy is needed to power the water pump or to run the refrigerator, and chemical energy is needed to give our muscles the strength to push a bicycle tire pump. But

Figure 2.35 The writer of this nursery rhyme had an intuitive understanding of the concept of entropy.

in the reactions that were required to produce this energy, entropy was increased. If we then look at both of the reactions required to pump the water uphill (that is, the process of actually pumping the water uphill and the process for generating the energy necessary to do this pumping), we would find that the total entropy of the combined processes has increased.

Entropy considerations are very important in understanding the functioning of living organisms. Such organisms are composed of cells, which are highly complex structures; there is, therefore, a natural tendency for these cells to break down, increasing the entropy of the system (Figure 2.36). In order to maintain the structure and functioning of the living system, energy must be added to counteract the natural drive toward increased entropy. We will see that this energy, which is extracted from the food that organisms eat, actually originates on the sun. If we want to consider the coupled reactions of this system, we see that the life processes that build up and maintain the intricate structure of the cells in living organisms depend upon energy-producing reactions on the sun. It is rather amazing to realize that each of our bodies maintains its highly complex structures and functions only at the expense of huge entropy increases on the far-off sun.

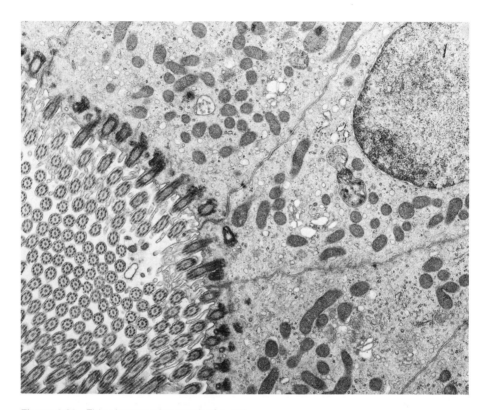

Figure 2.36 This electron micrograph of a cell from the retina of a rabbit illustrates the highly organized, complex structures found in living organisms. To function normally, the cell must constantly expend energy to counteract the natural tendency toward increased entropy and the breakdown of these complex structures. (Courtesy Perkin-Elmer Corporation)

Additional Reading

Books
1. Stanley W. Angrist and Loren G. Helper, *Order and Chaos,* Basic Books, New York, 1967.

2. Isaac Asimov, *Life and Energy,* Doubleday, Garden City, New York, 1962.

3. Harold F. Blum, *Time's Arrow and Evolution,* Princeton University Press, Princeton, 1951.

Articles

1. H. A. Bent, "Haste Makes Waste—Pollution and Entropy," *Chemistry*, October 1971, page 6.

2. R. Igor Gamow and John F. Harris, "The Infrared Receptors of Snakes," *Scientific American*, May 1973, page 94.

3. George Porter, "The Laws of Disorder," *Chemistry*, May 1968 to February 1969.

4. Richard J. Wurtman, "The Effects of Light on the Human Body," *Scientific American*, July 1975, page 69.

Questions and Problems

---

1. Define the following terms:

    (a) Matter
    (b) Mass
    (c) Weight
    (d) SI units
    (e) Force
    (f) Atom
    (g) Molecule
    (h) Element
    (i) Compound
    (j) Physical change
    (k) Chemical change
    (l) Energy
    (m) Kinetic energy
    (n) Potential energy
    (o) Kinetic-molecular theory
    (p) Melting point
    (q) Boiling point
    (r) Evaporation
    (s) Pressure
    (t) Atmosphere (atm)
    (u) Milliliters of mercury (mm Hg)
    (v) Standard atmospheric pressure
    (w) Boyle's law
    (x) Charles' law
    (y) General gas law
    (z) Graham's law
    (aa) Henry's law
    (bb) Dalton's law
    (cc) Heat
    (dd) Temperature
    (ee) calorie (cal)
    (ff) Electromagnetic energy
    (gg)* First law of Thermodynamics
    (hh)* Second Law of Thermodynamics

    (*See optional sections 2.26, 2.27.)

2. Is air matter? Why or why not?

3. What is the difference between mass and weight?

4. Make the following conversions (review Appendix 2 for a method of solving these problems).

    (a) 400 g to kg
    (b) 1500 m to km
    (e) 100 mm to m
    (f) 0.02 g to mg

(c) 10 cm to mm (g) 15 ml to l
(d) 0.2 l to ml (h) 1.5 kg to g

5. A prescribed injection of a drug is 1.5 ml. If the syringe is graduated in cubic centimeters, to what mark on the syringe would you draw the drug?

6. Wine in the United States is sold in bottles marked 4/5 of a quart. How many liters of wine are there in 4/5 of a quart? How many ml?

7. A sign on the freeway outside San Francisco reads "Los Angeles, 412 miles." How many kilometers would that be?

8. Give two examples of (a) an element; (b) a compound.

9. How would you distinguish between an atom and a molecule?

10. Give two examples of common foods in your kitchen that are

(a) Homogeneous (b) Heterogeneous

11. Identify each of the following as either a physical or chemical change:

(a) A log burning (d) Ski goggles fogging
(b) Bread becoming stale (e) Snow melting
(c) Salt dissolving in water (f) Brown smog forming

12. Discuss the potential and kinetic energy of a skier as he:

(a) Boards the ski lift.
(b) Travels to the top of the lift.
(c) Stops to adjust his goggles at the top of the hill.
(d) Skis down the hill to the bottom.

13. To treat acne, dermatologists wipe the patient's face with a sponge cooled to 77 K by liquid nitrogen. What is the temperature of the sponge in °C? °F?

14. A weatherman on the television news gives temperatures around the world. He says that it is 22°C in Paris and 32°C in Rome. What are the temperatures in Paris and Rome in °F?

15. When the barometer reads 720 mm Hg, a sample of oxygen occupies a volume of 250 ml. What volume will the sample occupy when the barometer reads 750 mm Hg?

16. A 1 liter sample of gas is collected under a pressure of 912 mm Hg. What will the volume of the gas be at standard atmospheric pressure?

17. A sample of nitrogen occupies 500 ml at −23°C. What volume will it occupy at 23°C?

18. A sample of ammonia occupies a volume of 1.2 liters at 45°C. To what temperature in °C must the gas be lowered to reduce its volume to 1.0 liter?

19. What volume will a gas occupy at 27°C and 760 mm Hg if it occupies 3.8 liters at 30°C and 909 mm Hg?

20. When collected under the conditions of 27°C and 800 mm Hg, a gas occupies 400 ml. To what temperature must the gas be cooled to reduce its volume to 320 ml when the pressure falls to 720 mm Hg?

21. Chlorine and ammonia have distinctly different odors, and chlorine gas molecules are about four times heavier than ammonia molecules. If equal quantities of these gases are released at the same time on the far side of the laboratory, which odor would you detect first?

22. You have two flasks containing one liter of water in which equal amounts of carbon dioxide have been introduced. The pressure in one container is 1 atm, and in the other container it is 2.5 atm. Which container would have more carbon dioxide dissolved in the water? Give the reason for your answer.

23. The pressure in a bottle containing a mixture of oxygen and nitrogen is one atmosphere. The partial pressure of the oxygen is 76 mm Hg. What is the partial pressure of the nitrogen?

24. How many calories are there in 3.5 kilocalories?

25. When four cashews are burned in a calorimeter, the temperature of the 1000 g of water changes from 25 to 69°C. How many kilocalories of energy are contained in one cashew? How many food Calories?

26. How many kilocalories of energy are contained in one gram of butter if burning 10 grams of butter raises the temperature of 1000 grams of water in a calorimeter from 20 to 90°C?

27. Which light contains more energy: orange light or yellow light?

28.* In a nuclear power plant only 40% of the energy released in the nuclear reaction is converted to electrical energy. Why is this not a violation of the First Law of Thermodynamics?

29.* Describe the changes in entropy of the molecules of a sugar cube as it dissolves in water.

Problems marked with an asterisk pertain to optional sections in the chapter.

chapter 3

Atomic Structure

Learning Objectives

By the time you have finished this chapter, you should be able to:

1. Describe the structure of an atom, listing three subatomic particles, their relative mass, their charge, and their location in the atom.

2. Define "atomic number."

3. Define "mass number."

4. Given the atomic number and mass number of any element, indicate the number of protons, neutrons, and electrons in the atom.

5. Describe how isotopes of an element differ.

6. Define "radioactivity."

7. Describe the three types of radiations given off by radioactive material.

8. Explain "half-life."

9. Define "nuclear transmutation."

10. Define atomic "fusion" and "fission."

11. Explain the difference between a breeder reactor and a burner reactor.

12. Describe three options open to scientists for handling radioactive wastes from nuclear power plants.

"Dawn, July 16, 1945, 5:29:35 Mountain War Time. The countdown had reached Zero minus ten seconds.

"In those milliseconds before the most awesome weapon devised by man created its first terrifying sunrise, the only sound on the desert wastes of southern New Mexico was the mating buzz of a colony of spadefoot toads. And, if one strained for it, the distant drone of a B-29 bomber. . . .

"A pinprick of a brilliant light punctured the darkness, spurted upward in a flaming jet, then spilled into a dazzling cloche of fire that bleached the desert to a ghastly white. It was precisely 5:29:45 A.M. . . .

"For a fraction of a second the light in that bell-shaped fire mass was greater than any ever produced before on earth. Its intensity was such that it could have been seen from another planet. The temperature at its center was four times that at the center of the sun and more than 10,000 times that at the sun's surface. The pressure, caving in the ground beneath, was over 100 billion atmospheres, the most ever to occur at the earth's surface. The radioactivity emitted was equal to one million times that of the world's total radium supply.

"No living thing touched by that raging furnace survived. Within a millisecond the fireball had struck the ground, flattening out at its base and acquiring a skirt of molten black dust that boiled and billowed in all directions. Within twenty-five milliseconds the fireball had expanded to a point where the Washington Monument would have been enveloped. At eight tenths of a second the ball's white-hot dome had topped the Empire State Building. The shock wave caromed across the roiling desert. . . .

"The sight beggared description.

"For a split second after the moment of detonation the fireball, looking like a monstrous convoluting brain, bristled with spikes where the shot tower and balloon cables had been vaporized. Then the dust skirt whipped up by the explosion mantled it in a motley brown. Thousands of tons of boiling sand and dirt swept into its maw only to be regurgitated seconds later in a swirling geyser of debris as the fireball detached itself from the ground and shot upward. As it lifted from the desert, the sphere darkened in places, then opened as fresh bursts of luminous gasses broke through its surface.

"At 2,000 feet, still hurtling through the atmosphere, the seething ball turned reddish yellow, then a dull blood-red. It churned and belched forth smoking flame in an elemental fury. Below, the countryside was bathed in golden and lavender hues that lit every mountain peak and crevasse, every arroyo and bush with a clarity no artist could capture. At 15,000 feet the fireball cleaved the overcast in a bubble of orange that shifted to a darkening pink. Now, with its flattened top, it resembled a giant mushroom trailed by a stalk of radioactive dust. Within another few seconds the fireball had reached 40,000 feet and pancaked out in a mile-wide ring of greying ash. The air had ionized around it and crowned it with a lustrous purple halo. As the cloud finally settled, its chimney-shaped column of dust drifted northward and a violet afterglow tinged the heavens above Trinity."*

Man's knowledge of nuclear energy has rapidly expanded since that awesome morning in July, and the technology being brought to bear on radioactive materials is now becoming increasingly important in upgrading standards of human life—from medical diagnosis and treatment, to prevention of food spoilage, to the production of useful energy for our society.

What makes a substance radioactive? How is the tremendous energy released in a nuclear reaction produced? Why is it that radioactivity can both cause and cure cancer? Are there significant dangers to living organisms from increased use of radioactive materials in power generation, industrial processes, and medical technology? It is important that we each know the answers to these questions, for in the next decade our local and national governments will be making policy decisions concerning nuclear energy that will have implications for many generations to come.

---

* Excerpted from Day of Trinity by Lansing Lamont. Copyright © 1965 by Lansing Lamont. Reprinted by permission of Atheneum Publishers.

Atomic Structure

3.1 The Parts of the Atom

To answer these questions we need to take a more detailed look at the structure of the atom. In the last chapter we stated that atoms are composed of many types of particles, and that we would be focusing our attention on three such particles: protons, neutrons, and electrons. **Protons** and **electrons** are electrically charged particles, while the **neutron** is neutral (that is, it has no charge). A proton is assigned the smallest unit of positive charge (+1) that will just nullify the negative charge on an electron (−1). Interestingly, the terms "positive" and "negative" that are used to describe the opposing effects of electrical charges were originated by Ben Franklin more than 50 years before the discovery of the electron and proton. A proton will repel other protons (like charges repel one another), and will attract electrons (unlike charges attract one another). See Table 3.1.

Table 3.1   Subatomic Particles

Name of Particle	Location in the Atom	Charge	Symbol	Relative Mass (amu)
Proton	Nucleus	+1	p, ^1_1H	1
Electron	Around the nucleus	−1	e, e^{-1}, $^0_{-1}e$	$\dfrac{1}{1837}$
Neutron	Nucleus	0	n, ^1_0n	1

An atom consists of a small dense **nucleus** that contains protons and neutrons, and a region surrounding the nucleus in which the electrons are found. For the most part, though, an atom is largely empty space. To give you an idea of the relative positioning of the subatomic particles in an atom, suppose you were in a large baseball stadium. If we were to let a flea standing on second base represent the nucleus of the atom, the nearest electron would be found somewhere in the top deck of the stands (Figure 3.1).

3.2 Atomic Number

The unique characteristic determining which element an atom represents is the number of protons in that atom. The **atomic number** of an element is the term used to denote the number of protons in the nucleus of any atom of that element. (The atomic numbers of each element are listed on the inside back cover of the book.) Since the electrical charge on a proton just cancels the electrical charge on an electron, we can see that the atomic number of an element also tells us how many electrons there are in a

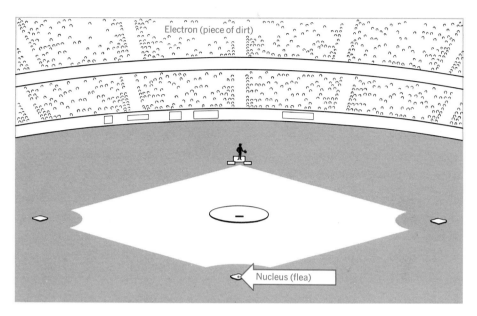

Figure 3.1   An atom is mostly empty space. If we
imagine a flea standing on second base to be the
nucleus of an atom, the nearest electron would be a
speck of dust somewhere in the top deck of the
stands. (From *Chemistry, An Experimental Science.*
Used with the permission of CHEM Study.)

neutral atom of that element. For example, the element sodium has an
atomic number of 11. Therefore, each neutral atom of sodium will have 11
protons in the nucleus and 11 electrons surrounding the nucleus.

3.3 Mass Number

Protons, electrons, and neutrons are extremely small particles. Protons and
neutrons have a mass of 1.7×10^{-24} grams; electrons are even lighter,
having a mass only $\frac{1}{1837}$ that of a proton. In fact, the mass of the electrons
in an atom is so small in comparison to the mass of the protons and
neutrons that it is ignored when calculating the mass of an atom. We define
the **mass number** of an atom to equal the number of protons plus the
number of neutrons in the nucleus of the atom.

$$\text{Mass number} = \text{Protons} + \text{Neutrons}$$

or

$$M \quad = \quad p \quad + \quad n$$

Chemists often use shorthand methods to express the mass number and
the atomic number of an atom. For example, an atom of carbon (C) has six
protons and six neutrons. The atomic number (Z) is 6, and the mass number

(*M*) is 12. Chemists may refer to atoms of carbon having the mass number of 12 as carbon-12, ^{12}C or $^{12}_{6}C$, where the upper number is the mass number and the lower number is the atomic number.

3.4 Isotopes

In the early nineteenth century, John Dalton developed an atomic theory based on the supposition that each atom of a given element was exactly alike. It was not until 100 years later that Fredrick Soddy disproved this theory by showing that the element neon consisted of not one, but two types of atoms, some with the mass number of 20 and some with the mass number of 22. (A third type of neon atom with the mass number of 21 also exists.) Neon has an atomic number of 10, so each atom of neon must have 10 protons in the nucleus. Therefore, the two different types of neon atoms must differ in the number of neutrons in their nuclei. Soddy invented the term **isotopes** to describe atoms of an element containing different numbers of neutrons in the nucleus (Figure 3.2). For neon, then, the isotopes are:

	Atomic Number	Mass Number	Number of *p*	Number of *n*
Neon-20	10	20	10	10
Neon-21	10	21	10	11
Neon-22	10	22	10	12

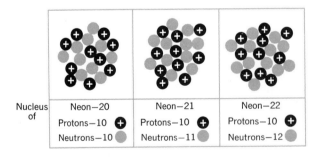

Figure 3.2   The element neon has three types of atoms that differ in the number of neutrons in the nucleus. These atoms are isotopes of neon.

Most elements have two or more naturally occurring isotopes; in fact, tin (Sn) has 10! Only 22 elements have but one type of atom, and therefore no naturally occurring isotopes.

3.5 Atomic Weight

Since atoms are so small and light, it is impossible to measure the mass of only a small number of atoms. However, it is possible to measure the relative masses of atoms of different elements. A scale based on these relative measurements has been devised, with the carbon-12 isotope being arbitrarily assigned a value of exactly 12 **atomic mass units (amu).** If an isotope of another element were $\frac{1}{2}$ as heavy as the carbon-12 isotope, then its mass in atomic mass units would be $\frac{1}{2} \times 12$, or 6; if an isotope were 2.84 times as heavy as the carbon-12 isotope, its mass would be 2.84×12, or 34.08 amu.

Since the vast majority of elements have at least two naturally occurring isotopes, a sample of any of these elements will contain a mixture of the different isotopes. Therefore, the exact mass of any sample will depend upon the percent abundance of each isotope in that particular sample. Table 3.2 lists the percent abundance of the naturally occurring isotopes

Table 3.2   The Relative Abundance of the Isotopes of Several Elements

Isotope	Percent Natural Abundance	Isotope	Percent Natural Abundance
Hydrogen-1	99.99%	Silicon-28	92.21%
Hydrogen-2	0.01%	Silicon-29	4.70%
		Silicon-30	3.09%
Carbon-12	98.89%		
Carbon-13	1.11%	Chlorine-35	75.53%
		Chlorine-37	24.47%
Nitrogen-14	99.63%		
Nitrogen-15	0.37%	Zinc-64	48.89%
		Zinc-66	27.81%
Oxygen-16	99.76%	Zinc-67	4.11%
Oxygen-17	0.04%	Zinc-68	18.57%
Oxygen-18	0.20%	Zinc-70	0.62%
Fluorine-19	100.00%	Bromine-79	50.54%
		Bromine-81	49.46%

of several elements. We would expect that the mass of a random sample of any element will be an average of the masses of its isotopes, weighted by the relative abundance of those isotopes. The name given to such a weighted average is **atomic weight.** That is, the atomic weight of an element is the weighted average of the masses of the naturally occurring isotopes of the element, expressed in atomic mass units. The atomic weight of each of the elements is listed on the inside back cover of this text.

The calculation of a weighted average is not a very difficult procedure. You are probably accustomed to having teachers use such a procedure in assigning final grades in a course. For example, suppose a teacher were to tell you that each of two mid-term exams would determine 25% of your final grade, and the final exam would constitute 50% of the final grade. If you received 76 on your first mid-term, 64 on your second mid-term, and 90 on your final exam, then your grade for the course would be 80. This final score is calculated by multiplying each grade by its fractional weight in the final grade, and summing these products.

$$76 \times 25\% = 76 \times 0.25 = 19$$
$$64 \times 25\% = 64 \times 0.25 = 16$$
$$90 \times 50\% = 90 \times 0.50 = \underline{45}$$
$$80$$

A similar procedure is followed in calculating the atomic weight of an element. The mass of each isotope is multiplied by its fractional abundance, and the products are then totaled to give the atomic weight. For example, boron has two isotopes, boron-10 and boron-11, whose percentage abundances are 19.6% and 80.4%, respectively. The atomic weight of boron is, then,

$$10.0 \text{ amu} \times 19.6\% = 10.0 \times 0.196 = 1.96$$
$$11.0 \text{ amu} \times 80.4\% = 11.0 \times 0.804 = \underline{8.844}$$
$$10.804 \text{ or } 10.8 \text{ amu}$$

Radioactivity

3.6 What Is Radioactivity?

Some atoms have nuclei that are unstable; this is especially true of elements that have high atomic numbers and, therefore, a high number of positive protons in the nucleus. (Remember that like charges repel one another.) Such an unstable nucleus will decay; that is, it will "spit out" nuclear particles, producing a new nucleus called a daughter nucleus. A daughter nucleus may or may not be stable. An unstable daughter nucleus will decay again, and this process will continue until a stable daughter nucleus is formed. Such a series of decays is called a **decay series** or **disintegration series** (Figure 3.3). The decay of a nucleus can give rise to several different forms of radiation (Table 3.3). **Radioactivity** is the term used to denote the emission of such radiation from samples of certain elements or their compounds.

3.7 Alpha Radiation (α Rays)

One way in which a nucleus can become more stable is by giving off alpha rays. **Alpha rays** are actually not rays at all, but rather are streams of alpha particles, which consist of two protons and two neutrons (the nucleus of a helium atom, $_2^4$He). By ejecting an alpha particle, the atomic number of the nucleus is reduced by two, and the mass number is reduced by four. A

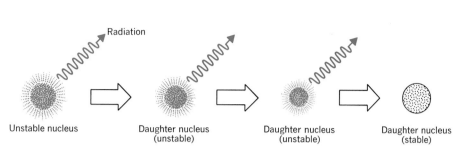

Figure 3.3 Radioactivity is the term used to describe the emission of radiation from an unstable nucleus.

Table 3.3   Natural Radioactive Radiation

Radiation	Composition	Charge	Symbol	Penetration (Rays Stopped by)
Alpha rays	Helium nucleus	$+2$	α, ^4_2He	Piece of paper
Beta rays	Electrons	-1	β, $^0_{-1}e$	Piece of wood
Gamma rays	High energy rays similar to X rays	0	γ	High density concrete

well-known source of alpha rays is the most abundant isotope of uranium, uranium-238, which decays by giving off an alpha particle to form an atom of thorium-234. A shorthand way of representing this decay is as follows:

$$^{238}_{92}\text{U} \longrightarrow \, ^{234}_{90}\text{Th} + \, ^4_2\text{He}$$

The original unstable nucleus is shown on the left-hand side of the arrow, and the products resulting from its radioactive decay are shown on the right-hand side. To be sure that this shorthand statement is correctly written, we must check that the number of protons and neutrons on one side of the arrow is equal to the number of protons and neutrons on the other side. In other words, the sum of the mass numbers on each side of the arrow must be equal, and the sum of the atomic numbers on each side of the arrow must be equal. For the radioactive decay above,

$$\text{Mass number:} \quad 238 = 234 + 4$$
$$\text{Atomic number:} \quad 92 = 90 + 2$$

Incidentally, the thorium atom produced by the decay of uranium-238 is itself unstable, and will decay to form a new nucleus.

Alpha particles are the largest particles emitted by radioactive substances, and have very little penetrating power. Even when traveling through air they lose energy very quickly through collisions with air molecules, and will stop within a few inches. They can be stopped by a piece of paper, and cannot penetrate even the dead layer of cells on the surface of your skin; however, an intense external dose of alpha rays would produce a burn on the skin. Nevertheless, alpha particles can do a great deal of damage if they are emitted inside the body, as a result of inhaling or swallowing an alpha emitter.

3.8 Beta Radiation (β Rays)

Beta rays, like alpha rays, are streams of particles rather than actual rays. In this case, the particle is an electron ($_{-1}^{0}e$) that is produced within the nucleus and then ejected. The net effect of beta decay is to change a neutron into a proton; the daughter nucleus will have the same mass number, but a different atomic number. For example, the thorium atom produced by the alpha decay of uranium-238 is a beta emitter.

$$^{234}_{90}\text{Th} \longrightarrow {}^{234}_{91}\text{Pa} + {}^{0}_{-1}e$$

Beta particles are seven thousand times smaller than alpha particles and, therefore, have much more penetrating power. (Think about the ability of a very fine needle to penetrate your skin as compared to a basketball.) Beta particles can pass through a piece of paper, but will be stopped by a piece of wood. Beta radiation can penetrate the dead outer layer of your skin; it will be stopped within the skin layer, causing damage to the skin tissue and making it appear burned (Figure 3.4). As with alpha particles,

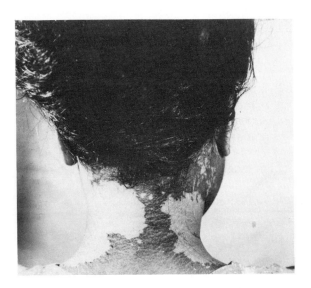

Figure 3.4 The burned area on the neck of a Rongelap native was caused by accidental exposure to over 2000 rads of beta radiation from radioactive fallout. (Courtesy Brookhaven National Laboratory)

beta particles hitting the skin from the outside cannot penetrate to internal organs, but the effect on internal organs can be severe if a beta emitter is taken internally.

3.9 Gamma Radiation (γ Rays)

Gamma rays are not particles, but are high-energy radiation similar to X rays. (You might refer again to the energy spectrum shown in Chapter 2, Figure 2.29.) Quite often the daughter nucleus produced by an alpha or beta emitter will be in a high energy, or excited, state; it can release this energy in the form of gamma radiation to become more stable. Therefore, the release of gamma rays often accompanies alpha or beta radiation. Radium-226 has a radioactive nucleus that releases alpha and gamma radiation when it decays.

$$\mathrm{^{226}_{88}Ra} \longrightarrow \mathrm{^{222}_{86}Rn} + \mathrm{^{4}_{2}He} + \gamma$$

Because of their high energy, gamma rays easily pass through paper and wood, but can be stopped by thick concrete walls. Gamma rays will completely penetrate the human body, causing cellular damage as they pass through (Figure 3.5).

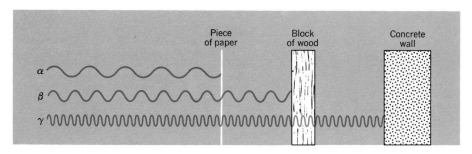

Figure 3.5 The penetration power of alpha (α), beta (β), and gamma (γ) radiation differs considerably.

3.10 Half-Life

The rate at which a radioactive substance decays (that is, the number of disintegrations that occur each minute) is totally independent of the normal things that will affect chemical change. Each isotope has its own characteristic rate of decay, which is designated as the **half-life ($t_{1/2}$)** of the isotope. The half-life of an element is the amount of time it takes for one half of the atoms in a given sample to undergo radioactive decay. This means that after one half-life, one half of a sample of the radioactive element will have decayed to form a new element, and one half will remain. After two half-lives, one half of one half (or one fourth) will remain. After three half-lives, one half of one fourth (or one eighth) will remain (Figure 3.6).

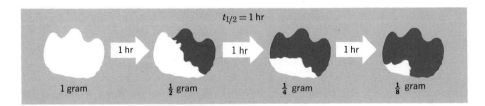

$t_{1/2} = 1$ hr

1 hr   1 hr   1 hr

1 gram $\frac{1}{2}$ gram $\frac{1}{4}$ gram $\frac{1}{8}$ gram

Figure 3.6 After one half-life, one half of a sample of the radioactive element will have decayed to form a new element and one half will remain. After two half-lives, one fourth of the original sample will remain.

For example, nitrogen-13 has a half-life of 10 minutes. If you have one gram of nitrogen-13, after one half-life 0.5 grams of nitrogen-13 will have decayed to form carbon-13 and 0.5 grams of nitrogen-13 will remain.

As a further example, an isotope of technetium is widely used in medical diagnosis. This isotope has a half-life of six hours, which is quite favorable for purposes of diagnosis but which requires that the laboratory constantly replenish its supply. If there were one gram of this isotope in the laboratory at 6:00 p.m. on Friday, only 0.25 grams would remain for use on Saturday morning (two half-lives later). Moreover, only one milligram would remain when the lab opened on Monday morning (Figure 3.7).

Half-lives of the different isotopes range from fractions of a second to billions of years. The half-life of an isotope is an indication of the stability of that isotope; most artificially produced radioisotopes are highly unstable and have very short half-lives. Table 3.4 gives some examples of isotopes and their half-lives.

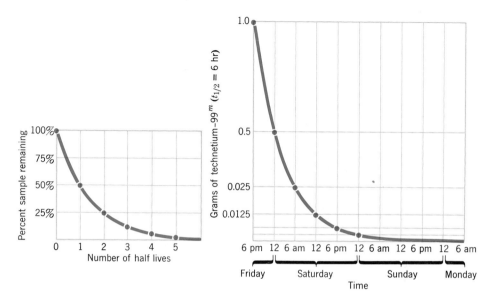

Figure 3.7 The half-life of an element is the amount of time it takes for one half of the atoms in a given sample to undergo radioactive decay. The half-life of technetium-99^m is 6 hours. If you had a one gram sample of technetium-99^m at 6 pm, Friday, you would have only one half of a gram of technetium-99^m at 12 midnight. By 6:00 am Monday morning, only one milligram (0.001 gm) would be left.

Table 3.4   Some Radioactive Isotopes and Their Half-Lives

Element	Isotope	Half-Life	Radiations Given Off
Hydrogen	$^3_1$H	12 years	Beta
Carbon	$^{14}_6$C	5730 years	Beta
Potassium	$^{40}_{19}$K	1.28×10^9 years	Beta and gamma
Cobalt	$^{60}_{27}$Co	5 years	Beta and gamma
Technetium	$^{99m}_{43}$Tc	6 hours	Gamma
Iodine	$^{131}_{53}$I	8 days	Beta and gamma
Polonium	$^{214}_{84}$Po	1.6×10^{-4} seconds	Alpha and gamma
Radium	$^{226}_{88}$Ra	1600 years	Alpha and gamma
Uranium	$^{235}_{92}$U	7.1×10^8 years	Alpha and gamma
	$^{238}_{92}$U	4.5×10^9 years	Alpha
Plutonium	$^{239}_{94}$Pu	24,400 years	Alpha and gamma

The half-lives of radioactive elements furnish a very useful tool for establishing the age of archeological objects. Carbon-14, which has a half-life of 5730 years, can be used to date organic material. This carbon isotope is created in the earth's upper atmosphere when nitrogen atoms are bombarded by cosmic rays (which are streams of particles pouring into the atmosphere from the sun and outer space). The procedure of carbon-14 dating assumes that the ratio of carbon-14 to the stable carbon-12 isotope is constant in living organisms. When such an organism dies, however, the total amount of carbon-12 it has accumulated during its life becomes fixed and unvarying, while half of the carbon-14 it has accumulated will be gone in 5730 years. Therefore, measuring the ratio of carbon-14 to carbon-12 in material that was once alive is a fairly accurate way to date objects that have died within the last 40,000 years.

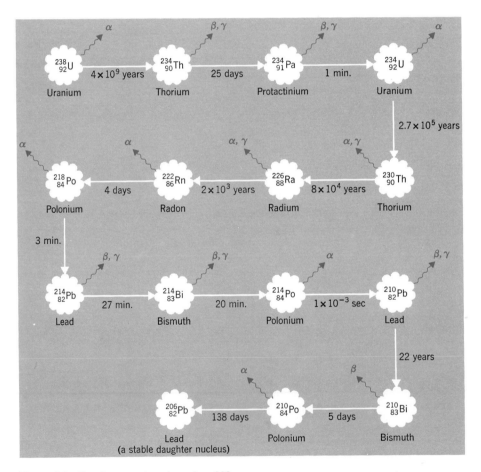

Figure 3.8 The decay series of uranium-238.

Although carbon-14 is very useful for dating human artifacts, isotopes with much longer half-lives must be used to date geologic periods in the earth's history. Uranium-238 decays through a decay series shown in Figure 3.8 to produce the stable lead-206 isotope. The half-life of uranium-238 is 4.5 billion years; thus if an original sample of rock contained one gram of uranium-238, after 4.5 billion years only 0.50 grams of ^{238}U will remain, and 0.43 grams of ^{206}Pb will have been produced. It is through uranium–lead radioactive dating of rocks that scientists have estimated the age of the planets in the solar system to be approximately 4.5 billion years.

Potassium-40 decays to form either the calcium-40 isotope or the argon-40 isotope. A measurement of the ratio of potassium-40 to argon-40 showed the jaw bone of a manlike creature found recently in Tanzania to be 5.5 million years old. Until this dating, the earliest man was thought to exist only 2.2 million years ago.

3.11 Nuclear Transmutation

A **nuclear transmutation** is a reaction in which a high speed nuclear particle collides with a nucleus to produce a different nucleus. Some transmutations occur in nature, and many are produced in laboratories. For example, carbon-14 is produced in the upper atmosphere, and can be produced in chemical laboratories, by bombarding nitrogen-14 with neutrons (Figure 3.9).

$$^{14}_{7}\text{N} \quad + \quad ^{1}_{0}n \quad \longrightarrow \quad ^{14}_{6}\text{C} \quad + \quad ^{1}_{1}\text{H}$$

Target　　　Projectile　　　　　　　　　　Proton

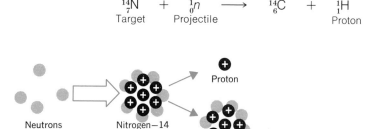

Neutrons　　　　Nitrogen—14

Proton

Carbon—14

Figure 3.9　Carbon-14 is produced in the upper atmosphere by the bombardment of nitrogen-14 with neutrons.

Carbon-14 is a beta emitter that can be incorporated into biological compounds and traced as it travels through a living system. By using carbon dioxide "labeled" with carbon-14, Melvin Calvin was able to obtain a detailed picture of the chemical pathways of photosynthesis, the process by which plants make sugar molecules from carbon dioxide and water.

A nuclear transmutation was produced in the laboratory for the first time in 1919 by Ernest Rutherford, who bombarded nitrogen gas with alpha particles.

$$^{14}_{7}N + ^{4}_{2}He \longrightarrow ^{18}_{9}F$$

The fluorine nucleus produced in this transmutation is very unstable and rapidly decays to form oxygen-17 and a proton.

$$^{18}_{9}F \longrightarrow ^{17}_{8}O + \underset{\text{Proton}}{^{1}_{1}H}$$

This experiment led to the discovery of the proton. The existence of the neutron also was shown by means of a nuclear transmutation. In 1932 James Chadwick bombarded beryllium-9 with alpha particles, causing the following transmutation:

$$^{9}_{4}Be + ^{4}_{2}He \longrightarrow ^{12}_{6}C + \underset{\text{Neutron}}{^{1}_{0}n}$$

The neutrons that were produced had enough energy to cause additional nuclear reactions in nuclei with which they collided.

Prior to 1940, the uranium atom was the heaviest atom known. However, the invention of the cyclotron and other nuclear accelerators allowed scientists to create very high energy projectiles, and in 1940 E. M. McMillan and P. H. Abelson produced neptunium ($_{93}$Np) by bombarding uranium with a stream of high energy deuterons (the nucleus of hydrogen-2, $^{2}_{1}H$).

$$^{238}_{92}U + ^{2}_{1}H \longrightarrow ^{239}_{92}U + ^{1}_{1}H$$

$$^{239}_{92}U \xrightarrow[t_{1/2} = 23.5 \text{ min}]{} ^{239}_{93}Np + ^{0}_{-1}e$$

The neptunium produced is a radioactive beta emitter, and decays to form plutonium-239. Plutonium is an alpha emitter with a half-life of 24,400 years. It is an important fuel for use in atomic reactors and bombs, and is one of the most toxic substances known to man.

$$^{239}_{93}Np \xrightarrow[t_{1/2} = 2.33 \text{ days}]{} ^{239}_{94}Pu + ^{0}_{-1}e$$

The Nucleus and Energy

Nuclear power is being acclaimed as the answer to the world's energy crisis. There is no doubt that as we deplete our supply of fossil fuels, something must take their place to supply the energy needs of our technological society. Although nuclear fuel can be the answer, great care and control must be exercised in expanding the use of nuclear power to minimize its impact on the environment and on future generations. In order to participate in decisions regarding the use of this nuclear technology, it is important that you understand how such a vast amount of energy can be produced from the nuclei of atoms.

3.12 Nuclear Fission

When bombarded with neutrons, the nuclei of several isotopes (^{235}U, ^{233}U, and ^{239}Pu) are capable of breaking apart or undergoing **fission** to form smaller, more stable nuclei (Figure 3.10). This fission process yields

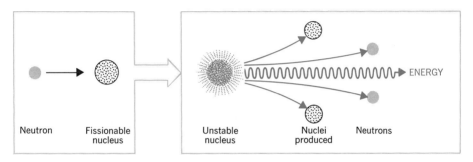

Figure 3.10 Nuclear fission occurs when a fissionable nucleus is struck by a neutron and breaks apart into fragments.

great amounts of energy; when one kilogram of ^{235}U or ^{239}Pu is fissioned, energy equivalent to 20,000 tons of TNT is released. Bombs dropped on Japan in World War II contained about one kilogram of fissionable material.

In nuclear fission, an atom of fissionable fuel such as uranium-235 is struck by a neutron and breaks into two small fragments that fly apart at high speeds. The kinetic energy of these fast-moving particles is converted into heat as they collide with surrounding molecules. In addition to these two fragments, two or three neutrons are released that can react with other ^{235}U nuclei. If this process continues, a **chain reaction** occurs which, if uncontrolled, could result in an explosion (Figure 3.11). However, in order for this chain reaction to occur, at least a minimum number of fissionable nuclei (called the **critical mass**) must be present.

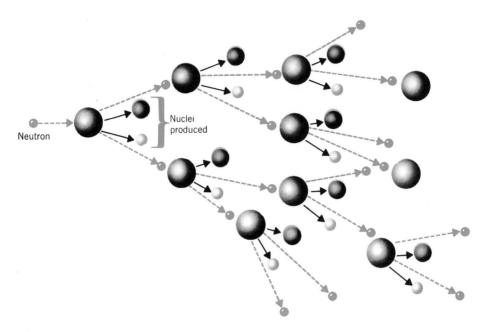

Figure 3.11 A nuclear chain reaction. When a nucleus undergoes fission, two or three neutrons are produced which are capable of reacting with other nuclei. If a critical mass of fissionable nuclei is present, a chain reaction will occur.

3.13 Nuclear Reactors: The Burner Reactor

In 1942 Enrico Fermi used a converted squash court at the University of Chicago as a place to build the first atomic reactor. Today, several hundred reactors are being built in various sizes and designs; however, the basic components of all nuclear reactors are the same. These components include:

1. The fuel, which contains significant amounts of one of the fissionable isotopes (^{235}U, ^{233}U, or ^{239}Pu), of which ^{235}U is the only naturally occurring isotope.

2. A moderator, such as graphite or water, which slows down the neutrons produced in the fission process.

3. Control rods made of cadmium or boron steel, which will absorb neutrons and which are moved in and out of the reactor to control the rate of the reactions.

4. A heat transfer fluid such as water, liquid sodium, or pressurized gas, which removes the heat from the reactor core and transfers it to a steam-generating system.

5. Shielding, which consists of an internal thermal shield to protect the walls of the reactor from radiation damage, and an external biological shield of high-density concrete to protect the personnel from radiation.

The exact reactor design will depend upon its use, which could be electric power generation, the production of plutonium-239, the propulsion of ships or rockets, or the production of heat for desalinization, drying, evaporation, or other industrial uses (Figures 3.12 and 3.13).

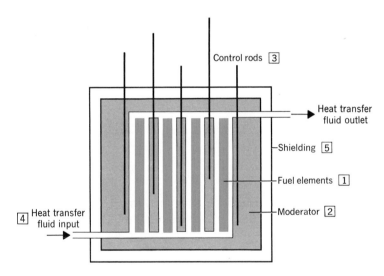

Figure 3.12 The core of a nuclear reactor.

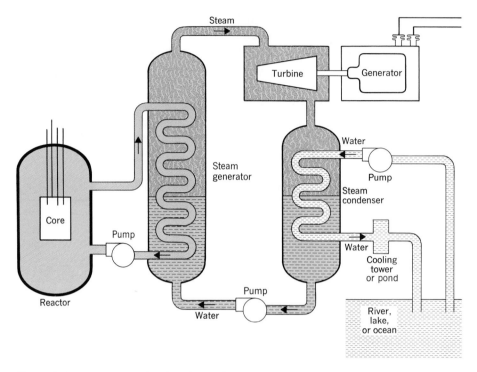

Figure 3.13   A schematic drawing of a nuclear power plant.

3.14 Breeder Reactors

In burner reactors, the ^{235}U fuel is consumed as energy is produced. This uranium isotope is the only naturally occurring fissionable fuel, and is found only in trace amounts (0.71%) in uranium ore. Since the world's supply of this fuel may be relatively quickly exhausted, the true promise of nuclear power hinges on the development of **breeder reactors.** A breeder reactor is one in which more fissionable fuel is produced than is used up in the heat generation process.

A breeder reactor uses uranium-238 and thorium-232, both of which are relatively abundant isotopes; by positioning them so that they are bombarded by some of the neutrons produced in the fission of uranium-235, new fissionable isotopes are formed.

$$^{238}_{92}U + ^{1}_{0}n \longrightarrow ^{239}_{92}U \xrightarrow[t_{1/2} = 24 \text{ min}]{} ^{239}_{93}Np + ^{0}_{-1}e \xrightarrow[t_{1/2} = 2.3 \text{ days}]{} ^{239}_{94}Pu + ^{0}_{-1}e$$

Fertile isotope

Fissionable isotope

$$^{232}_{90}Th + ^{1}_{0}n \longrightarrow ^{233}_{90}Th \xrightarrow[t_{1/2} = 22 \text{ min}]{} ^{233}_{91}Pa + ^{0}_{-1}e \xrightarrow[t_{1/2} = 27 \text{ days}]{} ^{233}_{92}U + ^{0}_{-1}e$$

Fertile isotope

Fissionable isotope

If the technology of breeder reactors can be sufficiently developed, the supplies of fertile uranium-238 and thorium-232 will assure a production of energy equal to hundreds of times the world's fossil fuel reserves.

One of the critical problems in the development of breeder reactors is the safety factor. The fissionable plutonium and uranium that are produced must be periodically extracted from the reactor products. Extreme care must be exercised in the transport and extraction of plutonium since it is extremely toxic to man. Plutonium is an alpha emitter with a half-life of 24,400 years. If this material gets into the body, it accumulates in the bones and liver, and is only very slowly excreted; 80 percent of the amount taken in will still remain in the body after 50 years. In addition to plutonium, other long-lived radioactive wastes are produced in the fission process, and must be adequately stored for long periods of time. For example, strontium-90, a common and biologically dangerous waste (with a half-life of 28.1 years) should be kept in safe storage for 600 years! Various options for waste disposal are now being carefully studied. Technology is being developed to trap radioactive gases and store them until they have decayed, to concentrate liquid wastes, and to dispose of highly radioactive wastes in solid form (Figure 3.14).

Figure 3.14 These huge million-gallon tanks shown under construction will store radioactive waste materials. Before storage, the highly radioactive liquid wastes are reduced to solid salt cakes to reduce the volume. The double-walled carbon steel tanks pictured here will be encased with three feet of concrete, and then will be completely surrounded with up to 40 feet of dirt. (Courtesy Energy Research and Development Administration)

3.15 Nuclear Fusion—A Captured Sun

Enormous amounts of energy are given off when small nuclei, such as those of hydrogen, helium, or lithium combine to form heavier, more stable nuclei; this process is called **nuclear fusion.** Nuclear fusion reactions are the source of the energy released by the sun and by the hydrogen bomb.

$$4^1_1H \longrightarrow \, ^4_2He \, + \, 2\,^0_{+1}e \, + \, Energy$$
$$\text{Positron}$$

The reaction shown above is thought to be the overall reaction occurring on the sun; it releases four times as much energy per gram of fuel as does a fission reaction (Figure 3.15).

$$^2_1H \, + \, ^3_1H \longrightarrow \, ^4_2He \, + \, ^1_0n \, + \, Energy$$
$$\text{Deuterium} \quad \text{Tritium}$$

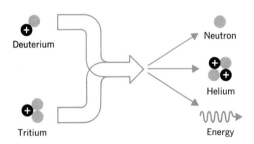

Figure 3.15 This fusion reaction between two isotopes of hydrogen is the reaction most likely to be harnessed by man.

This fusion reaction is the one that is most likely to be harnessed for use by man since it occurs at the lowest temperature and because deuterium can be obtained from the sea at a very low cost.

Making use of nuclear fusion is proving to be many times more difficult than harnessing nuclear fission because temperatures of one hundred million degrees Celsius or more are required for the fusion reaction to occur. However, if the technology can be developed to control a fusion reaction, this process has significant advantages over fission as the source of the world's energy supply: it would be more efficient; the fuel would be cheap and almost inexhaustible; there is no possibility of runaway accidents from having the reactor core melt as could occur in fission reactors; and there is very little radioactive waste. Tritium, which is one of the least toxic radioactive substances, may be produced, but it can be returned to the system as a fuel.

Additional Reading

Books
1. C. M. Lederer, J. M. Hollander, and I. Perlman, *Table of Isotopes,* 6th Edition, Wiley, New York, 1967.

2. K. Z. Morgan and J. E. Turner, ed., *Principles of Radiation Protection,* Wiley, New York, 1967.

3. Glenn T. Seaborg and William R. Corliss, *Man and Atom,* E. P. Dutton, New York, 1971.

4. United States Atomic Energy Commission, *Understanding the Atom Series.* These are short interesting books on nuclear chemistry.

Articles
1. W. J. Blair and R. C. Thompson, "Plutonium: Biomedical Research," *Science,* February 22, 1974, page 715.

2. Gregory R. Choppin, "Nuclear Fission," *Chemistry,* July–August 1967, page 25.

3. Herbert M. Clark, "The Origin of Nuclear Science," *Chemistry,* July–August 1967, page 8.

4. "Great Moments in Chemistry: Discovery of Radioactivity," *Chemistry,* May 1974, page 19.

5. Russell H. Johnsen, "Radiation Chemistry," *Chemistry,* July–August 1967, page 31.

6. Yelena Knorre, "Expanding Mendeleev's Table," *Chemistry,* July–August 1973, page 27.

7. John W. Landis, "Fusion Power," *Journal of Chemical Education,* October 1973, page 659.

8. *Science,* April 19, 1974. The entire issue is devoted to the energy crisis.

9. *Scientific American,* September 1971. This issue is on energy and power.

10. Joan Zimmerman, "Answering the Question When?" *Chemistry,* July–August 1970, page 22.

Questions and Problems

1. Define the following terms:

   (a) Proton
   (b) Neutron
   (c) Electron
   (d) Atomic number
   (e) Mass number
   (f) Isotope
   (g) Atomic weight
   (h) Amu
   (i) Decay series
   (j) Radioactivity

   (k) Alpha radiation
   (l) Beta radiation
   (m) Gamma radiation
   (n) Half-life
   (o) Nuclear fission
   (p) Nuclear fusion
   (q) Critical mass
   (r) Breeder reactor
   (s) Fertile isotope
   (t) Fissionable isotope

2. Complete the following chart:

Subatomic Particle	Location	Charge	Relative Mass
(a)_____	_____	+1	_____
(b)_____	_____	_____	_____
(c)_____	_____	_____	$\frac{1}{1837}$

3. State the number of protons and the number of electrons for a neutral atom of the elements whose atomic numbers are 3, 20, 53, and 92.

4. What is the atomic number and mass number of the following atoms?

   (a) Atom A with 6 protons and 8 neutrons

   (b) Atom B with 90 protons and 142 neutrons

   (c) Atom C with 40 protons and 51 neutrons

5. Calculate the atomic number and the mass number, and use two different shorthand methods to denote an atom of each of the following elements.

Name	Number of p	Number of n
(a) Chlorine (Cl)	17	18
(b) Cobalt (Co)	27	33
(c) Hydrogen (H)	1	2

6. State the number of protons, neutrons, and electrons in a neutral atom of each of the five isotopes of zinc (Table 3.2).

7. Using the information with which you answered question 6, describe how isotopes differ.

8. Calculate the atomic weight of chlorine to three digits (see Table 3.2).

9. What makes a substance radioactive?

10. How do the three radiations given off by radioactive material differ?

11. Radon-222 is an alpha particle emitter that decays to polonium-218. Write the shorthand representation of this radioactive decay.

12. Bismuth-210 is a beta emitter that decays to form an isotope of polonium. Write the shorthand representation of this radioactive decay.

13. When the thorium-230 isotope decays, it gives off an alpha particle and gamma radiation. Write the shorthand representation of this radioactive decay.

14. If you had $10 to gamble in a slot machine in Las Vegas, but could only bet half of it the first hour, and half of the remainder in each succeeding hour, at the end of what hour would you have $1.25 left? (Neglect any earnings.) How is this problem analogous to the half-life of a radioactive substance?

15. What is the meaning of the term "nuclear transmutation"? Give an example.

16. Insert the correct bombarding particle for the following nuclear transmutations.

(a) $^{27}_{13}Al + \underline{\hspace{1cm}} \longrightarrow ^{30}_{15}P + ^{1}_{0}n$

(b) $^{239}_{94}Pu + \underline{\hspace{1cm}} \longrightarrow ^{240}_{95}Am + ^{0}_{-1}e$

17. What is the difference between atomic fission and atomic fusion? Which process is presently being used in power plants? Why?

18. Why is the breeder reactor preferable to the burner reactor for meeting the world's long-range energy requirements?

chapter 4

Radioactivity and the Living Organism

Learning Objectives

By the time you have finished this chapter you should be able to:

1. Describe two ways in which ionizing radiation can be harmful to living tissue.

2. Define the following terms: curie, rad, rem, LET.

3. Describe two means by which radioactivity can be detected.

4. List three sources of background radiation.

5. Describe two ways in which radioactive substances can help in medical diagnosis.

6. Describe three different techniques for using radioactive materials in medical therapy.

7. Explain two ways in which radioactive materials have aided agricultural research.

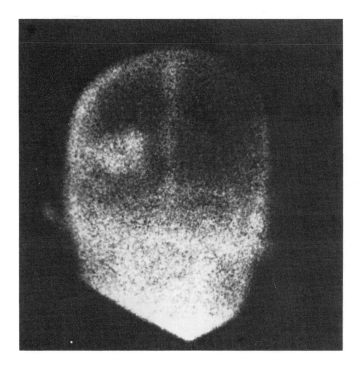

George Downing, a 34-year-old advertising executive, tried to ignore the familiar throbbing in his head. For weeks he had been suffering persistent headaches that made his work impossible and his life miserable. His doctor had performed a long series of routine examinations and tests, but nothing unusual had been found. At his doctor's suggestion, George was now on his way to a New York hospital to have a brain scan done by the Department of Radiology.

Thirty minutes before the brain scan was to begin, George was given an intravenous injection of a small amount of radioactive material. He was positioned under an instrument called a scanner, which repeatedly passed back and forth over his head to detect the high-energy rays emitted by the radioactive element. The scanner recorded these rays as black dots on a sensitized photographic film. In less than an hour the doctor was able to repeat the scanning process three times, once on the front of his head and once on each side. Upon being developed, the photographic film revealed that the radioactive material had concentrated in one spot in George's brain, indicating to the doctor that a tumor was present. This diagnosis was based on the fact that cells of a brain tumor will absorb that particular radioactive material at a higher rate than normal cells in the brain; hence a scan will detect a localized concentration of the radioactive substance in the region of the tumor. Several days later George underwent surgery to remove the tumor. Using all three scans, the surgeon was able to pinpoint the tumor exactly and to remove it successfully.

The Effects of Radiation on Living Cells

4.1 Ionizing Radiation

The use of radioactive materials is increasingly becoming standard procedure in many areas of medical diagnosis. In addition, such materials are also playing important roles in the treatment of disease. It may seem just a bit contradictory, then, to realize that these same materials can also critically damage and kill living organisms.

Why are radioactive substances potentially so harmful to man? When they hit living tissue, alpha, beta, gamma, and cosmic radiation are each capable of producing unstable and highly reactive charged particles called **ions;** hence these radiations are called **ionizing radiation.** In addition, such radiations can transfer so much energy to the molecules in living tissue that these molecules literally vibrate apart to form high-energy, uncharged fragments called **free radicals.** Free radicals are even more reactive than ions; they are capable of pulling other molecules apart, and can create havoc in living cells. Ions and free radicals may recombine with one another, causing little damage, or may combine with other molecules to form new substances foreign to the cell and, therefore, potentially dangerous (Figure 4.1).

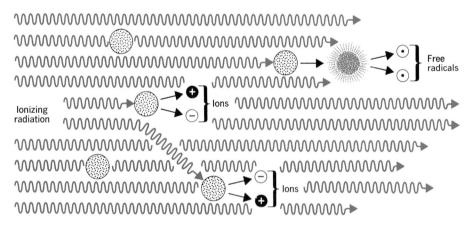

Figure 4.1 When ionizing radiation hits living tissue, it is capable of producing unstable and highly reactive charged particles called ions, and even more reactive uncharged particles called free radicals.

To be more specific, ionizing radiation can cause damage within the cell in two ways. First, the damage may be caused by direct action—a direct hit on a biologically important molecule, causing the molecule to split into biologically useless fragments. The most vital molecule in a living cell is the DNA molecule, which carries all the "blueprints" necessary for the cell to divide and reproduce. If the DNA is destroyed, the cell cannot divide, and will die. When such cells die without replacement, the entire irradiated tissue will eventually die. Furthermore, if this tissue is essential to the organism, the entire organism may die prematurely.

   Even if the DNA molecule is not destroyed completely, this damaged molecule may cause the cell to divide abnormally into new cells with altered DNA. Such cells are known as **mutant** cells. A mutant cell may have its DNA so altered that it is no longer under the body's control, and may begin to grow and divide in an uncontrolled fashion, destroying the normal cells around it. Cells that behave in this manner are called **cancerous** or **malignant** (Figure 4.2).

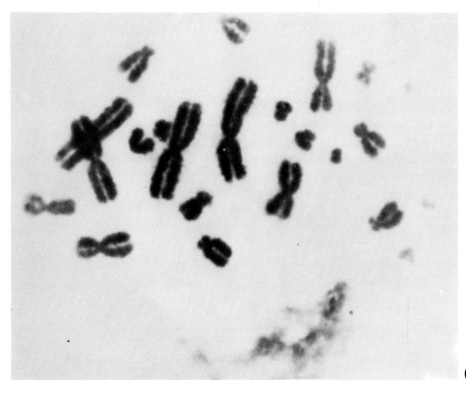

(a)

Figure 4.2   The effect of radiation on living cells. (a) A normal cell nucleus containing 23 pairs of chromosomes, composed of DNA molecules and protein. Photos (b) and (c) are on page 86. (Courtesy Argonne National Laboratory.) Photos (b) and (c) are continued on page 86.

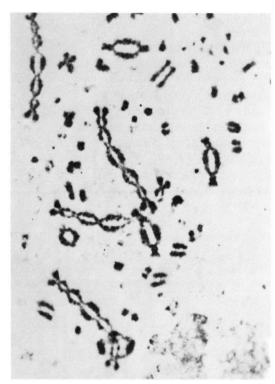

Figure 4.2 (cont'd.) The effect of radiation on living cells. (b) Ionizing radiation has caused the chromosomes in this cell nucleus to double and triple abnormally. The cell could easily become cancerous. (c) Exposure to radiation resulted in this cell's inability to divide. The cell contains more than 700 chromosomes and has grown to ten times the size of a normal cell. (Courtesy Argonne National Laboratory)

(b)

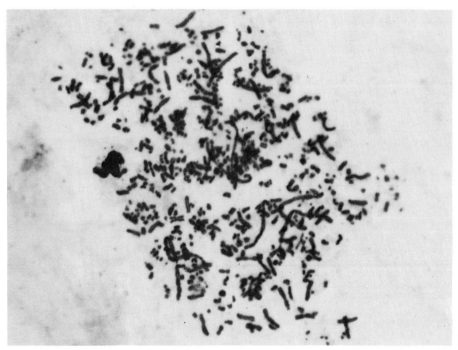

(c)

Direct hits of ionizing radiation on biologically important molecules are not the only way that radiation can damage cells. Ionizing radiation can also cause damage through indirect action. Animal cells are about 80% water, and ionizing radiation can cause water molecules in living tissue to form positive and negative ions, or to form highly reactive free radicals. Such free radicals can recombine to form water, can combine to form hydrogen, which can be tolerated in small amounts by living cells, or can combine to form hydrogen peroxide, which is a highly toxic substance. This may be the reason that radiation sickness resembles hydrogen peroxide poisoning in many respects. Free radical fragments can also react with oxygen in the cell to produce another free radical that is even more undesirable than hydrogen peroxide (Figure 4.3).

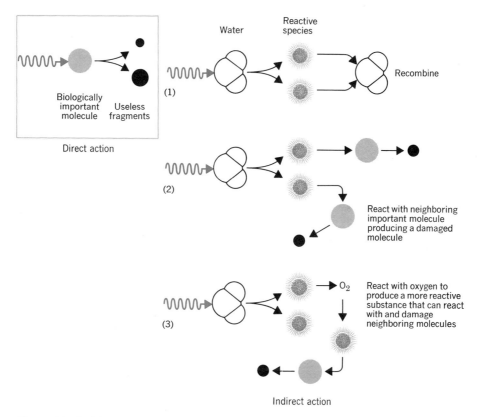

Figure 4.3 Ionizing radiation can damage living tissue by direct action, such as a direct hit on a biologically important molecule, or by indirect action, such as reacting with water to produce highly reactive substances which can then react with biologically important molecules.

4.2 Radiation Dosage

There are several scientific units used to measure various aspects of ionizing radiation, units that help answer specific questions about the nature of the radioactive source and its effects on living tissue. The following pages will describe a few of these terms: the curie, the rad, the rem, and the LET.

The first aspect of ionizing radiation that one might like to describe is the activity level of the source of that radiation. That is, how often do disintegrations occur per unit of time? The unit used to describe the activity of a radioactive source is the **curie (Ci).** One curie is equal to 3.7×10^{10} disintegrations per second, the rate of disintegration exhibited by one gram of radium. It is important to note that this unit describes neither the type of radiation produced nor the effect this radiation will have on tissue or other matter. It identifies only the frequency of radiation. In radiation therapy, hospitals often use a sample of cobalt-60 as their source of gamma radiation. If a source is rated at 1.4 curies, it will deliver $1.4 \times 3.7 \times 10^{10}$, or 5.18×10^{10} disintegrations per second; a 2 Ci cobalt-60 source will deliver to a patient twice as many gamma rays per unit of time as a 1 Ci cobalt-60 source (Table 4.1).

Table 4.1   Units to Measure Ionizing Radiation

Unit	Measures	
Curie (Ci)	Frequency of radiation	1 Ci $= 3.7 \times 10^{10}$ disintegrations per second
rad (D)	Absorbed dose of radiation	1 rad $=$ absorbed radiation that liberates 100 ergs of energy/ gram of tissue
rem	Absorbed dose of radiation	1 rem $=$ absorbed dose that will produce same biological effect as 1 rad of therapeutic X ray
LET	Energy transferred to tissue by absorbed dose	LET $=$ amount of energy transferred to tissue per unit of path length

To study the effect of ionizing radiation on living tissue we must try to identify the amount of energy that has been absorbed by the tissue. The dose actually absorbed by a tissue will be influenced by many factors, including the nature of the radioactive source, the energy of the radiation, the distance of the tissue from the source, the nature of the tissue itself, and the duration of exposure. The **rad (radiation absorbed dose)** is the unit used to measure the amount of energy that is absorbed by irradiated tissue.

Regardless of the type of radiation, one rad is an absorbed dose of radiation that results in the liberation of 100 ergs of energy per gram of irradiated tissue. Most mammals will not survive a dose of 1000 rads of radiation to the entire body; this means the delivery of 1000 × 100 ergs, or 100,000 ergs of energy per gram of the mammal's tissue. Now, the erg is a very small unit of energy — 100,000 ergs equals 0.0024 calories. If this is all the energy needed for a killing dose of radiation, why is it so lethal? Again, it is the ability of ionizing radiation to change important biological molecules through direct or indirect action that results in the death of cells.

Scientists have discovered that the absorption of one rad of different types of ionizing radiation will produce different effects on identical biological systems. For example, it takes twice the dose of cobalt-60 gamma radiation as it does a specific level of X rays to produce the same disruption in a certain type of pollen grain. Therefore, in order to study the effect of different types of radiation on man, a unit called the **rem** (**r**oentgen **e**quivalent **m**an) was devised. For each type of radiation, the rem is defined as the absorbed dose of radiation that will produce the same biological effect as one rad of therapy X rays. When a dose is stated in rems, there is no need to specify the type of radiation since the biological effects will be the same for each type. For example, 700 rems of any type of radiation is the dose sufficient to kill 50% of a population of rats within 30 days. One rad of X rays, gamma rays, or beta rays has a rem of one, but one rad of alpha rays has a rem of 10 to 20.

Why does the same dose in rads of different types of radiation have widely different effects on identical biological systems? One factor influencing the effect of radiation on living tissue is the way in which the energy of the radiation is deposited in the tissues. The **linear energy transfer (LET)** of radiation is the amount of energy transferred per unit of path length traveled by the radiation. High LET radiation deposits its energy in a tight dense path (Figure 4.4). Alpha particles and neutrons are high LET radiation; the molecules they ionize are so close together that they quickly recombine, allowing very few to react with other molecules such as oxygen. So high LET radiation does most of its damage by direct

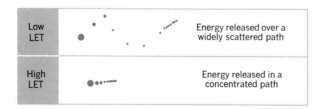

Figure 4.4 High LET radiation transfers its energy to tissues in a highly concentrated path. Low LET radiation, on the other hand, transfers its energy to tissues in a widely scattered path.

action, and when biologically important molecules are hit, the damage is so great that it usually can't be repaired.

X rays, beta rays, and gamma rays are low LET radiation; they distribute their energy over wide and less dense paths. Such radiations do most of their damage by indirect action. The effect of this radiation is greatly increased in the presence of oxygen because of the production of the highly reactive free radical. However, damage done to biologically critical molecules by sublethal doses of low LET radiation is much more likely to be repaired by the cell than is damage done by high LET radiation.

4.3 Detecting Radiation: Radiation Dosimetry

You cannot hear, feel, taste, or smell low levels of radiation. (Do you feel anything when you have a dental X ray?) Because of the potential dangers of such radiation, various instruments and detection systems have been designed to measure radiation dose and exposure. Depending on the particular instrument, radiation dosage may be measured by recording the amount of ionization in a gas, the amount of chemical reaction produced, the heat produced when the radiation gives off energy, or the light produced in a luminescent material.

A **Geiger-Müller** counter is a widely used radiation-detecting instrument. The detector consists of a gas-filled ionization chamber in the shape of a tube. Radiation that enters the tube produces ions in the gas, allowing an electrical current to flow. This current will either generate a click in a speaker or will turn a numerical counter to indicate the number of radiations that have entered the tube (Figure 4.5). The Geiger-Müller counter is sensitive to beta radiation; most gamma rays will pass through the gas without interacting with it, and all but the most energetic alpha rays will be stopped before entering the tube. Although the counter will indicate the number of rays entering the tube, it reveals nothing about the energy of those rays.

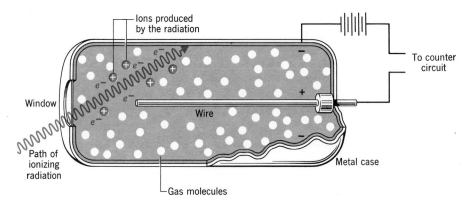

Figure 4.5 The detection tube of a Geiger-Müller counter.

The **scintillation counter** is the most widely used instrument for simultaneously determining the number of radiations and the dose rate. Incoming radiations strike a surface coated with special chemicals, producing tiny flashes of light. The number of these flashes and their intensity (which is proportional to the energy of the radiation) is electronically detected and converted to electric pulses that can be counted, measured, and recorded.

It is important to carefully monitor the exposure of researchers, X-ray technicians, nurses, and others who work with ionizing radiation. Small portable devices have been designed to accurately measure radiation dose, the most common being a **film badge** worn by the individual. The film is periodically developed, and the degree of darkening in the resulting negative indicates the total amount of radiation to which the individual has been exposed (Figure 4.6).

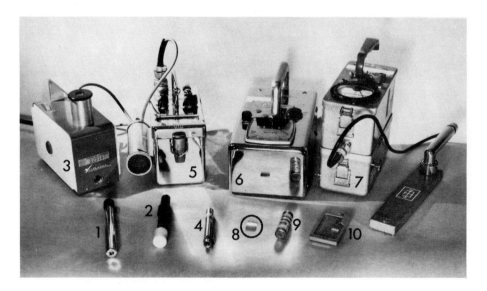

Figure 4.6 Various dosimeters for detecting and measuring radiation. (a) Numbers 4 and 10 are designed to be worn by a person. Number 4 is a small tissue equivalent chamber with a tiny readable meter inside, and number 10 is a film badge. (b) Number 5 is a portable Geiger-Müller counter. (c) Number 6 is a portable air ionization chamber with a meter, and number 7 is a portable proportional counter. (d) Number 3 is a meter that can read the radiation detected by number 1, an air ionization chamber, and number 2, a tissue equivalent ionization chamber. (e) Number 8 is a lithium fluoride luminescent dosimeter, and number 9 is a chlorinated hydrocarbon liquid dosimeter. (Courtesy U.S. Energy Research and Development Administration, Technical Information Center, Oak Ridge, Tennessee)

Photographic film is also used to pinpoint the position of radioactive material in living tissue. This technique, known as **autoradiography,** involves introducing radioactive elements or compounds into material, which is then exposed to a photographic emulsion. The emulsion is developed after an exposure time that varies with the different materials being tested. This technique has been used for such purposes as determining the location of calcium-45 in a leaf, or of strontium-90 in the body of a mouse (Figures 4.7 and 4.8). On a microscopic level this technique can help pinpoint the specific location of a radioactive chemical in a living cell.

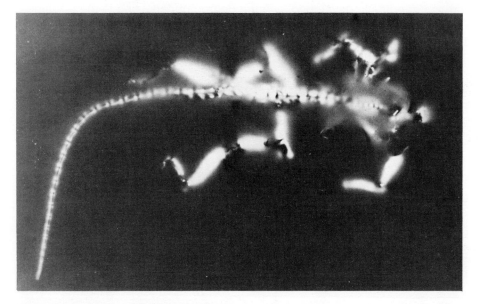

Figure 4.7 A radioautogram of the skeleton of a mouse that had been injected with a solution of strontium-90. (Courtesy Victor Arena. From V. Arena *Ionizing Radiation and Life,* The C. V. Mosby Co., St. Louis, 1971.)

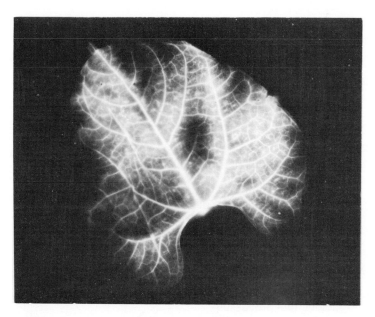

Figure 4.8 A radioautogram of the leaf of a bean plant that had absorbed calcium-45 through its root system. (Courtesy Victor Arena. From V. Arena, *Ionizing Radiation and Life,* The C. V. Mosby Co., St. Louis, 1971.)

4.4 Background Radiation

There is no way that we can totally escape exposure to ionizing radiation. In fact, the average human body undergoes several hundred thousand radioactive disintegrations per minute as a result of the natural radioisotopes (mostly potassium-40) found in the body. Seventy of the 350 naturally occurring isotopes are radioactive, and small amounts of radioactive materials are found in the soil we stand on, the food we eat, the water we drink, and the air we breathe.

On the average, we each receive a very small dose (0.211 rems, or about 200 millirems) of radiation per year, of which 63% is from outer space and natural materials in the soil, water, and air; 35% from medical radiation such as medical and dental X rays; and 2% from radioactive fallout and pollution from nuclear power plants (Table 4.2).

Table 4.2   Summary of the Annual Per Capita Radiation Dose in the United States

Radiation Source	Dose (Millirem/Year, 1970)	
Environmental		
Natural		130
Cosmic rays	(45)	
Terrestrial radiation		
External	(60)	
(Soil, potassium-40, decay products of uranium and thorium)		
Internal	(25)	
(mainly potassium-40)		
Global fallout		4
All other		0.06
(Nuclear reactors, fuel processing, nuclear tests, etc. By the year 2000 it will equal about 0.5 mrem/year)		
Medical		
Diagnostic		72
Radiography	(58)	
Fluoroscopy	(14)	
Radiopharmaceutical		2
(By the year 2000 it will equal about 16 mrem/year)		
Occupational		0.8
Miscellaneous		2.7
(TV, consumer products, air transport)		
Total		211

Data from "Estimates of Ionizing Radiation Doses in the United States, 1960-2000," Environmental Protection Agency, Rockville, Maryland, 1972.

We cannot control the natural background radiation, but we can control radiation exposure from man-made sources. Very few people would advocate the stoppage of all uses of radioactive materials, for these materials are playing a large role in improving living conditions around the world. But when making a decision that might increase the level of background radiation, one must weigh the benefits to mankind against the risk. Such decisions are made even harder because of the lack of knowledge about the long-term effects on man of very low levels of radiation.

Radioisotopes and Medicine

One area in which the benefit to mankind appears to far outweigh the risk is the use of radioisotopes in medical diagnosis, treatment, and research.

4.5 Medical Diagnosis

Major advances in medical diagnosis have been made possible by the ability to trace the paths taken by specific chemical compounds in living systems. **Radioactive tracers** are chemicals that contain radioactive atoms and that have the same chemical nature and behavior as the compounds they trace. A living system treats the tracer just as it would the normal compound, but the radioactive atoms make it possible to follow the compound's path through the system.

In 1923 Georg von Hevesy was the first to use a radioactive tracer in his study of the uptake of lead-212 in plants. Artificially produced radioisotopes were first used in 1936, when Joseph G. Hamilton and Robert S. Stone of the University of California at Berkeley used radioactive sodium produced in a cyclotron to study the uptake and excretion of sodium. Since then, the use of natural and artificially produced radioisotopes has greatly increased; in 1970 about eight million "atomic cocktails" of some 30 different radioisotopes were administered to patients in diagnostic and therapeutic treatments. Some of these radioisotopes and their medical uses are listed in Table 4.3, but a few of these isotopes deserve discussion in greater detail.

Table 4.3   Some Radioisotopes and Their Uses in Medical Diagnosis

Radioisotope	Compound Given	Diagnostic Procedure
Iodine-131	Sodium iodide	Thyroid uptake and scanning
	Labeled albumen	Blood volume determination
		Cardiac output
		Scanning of lungs, placenta, brain, heart, and liver
	Sodium iodohippurate	Kidney function and scanning
	Labeled fats	Fat metabolism
	Rose Bengal	Liver function and scanning
Cobalt-60	Labeled vitamin B_{12}	Vitamin B_{12} absorption
Chromium-51	Sodium chromate	Red blood cell survival, blood volume
	Heat-treated red blood cells	Spleen scanning
Iron-59	Iron chloride or citrate	Iron turnover
Technetium-99^m	Pertechnetate	Brain and thyroid scanning
	Technetium sulfur colloid	Liver scanning
	Labeled albumen	Placenta and lung scanning
Gold-198	Colloid gold	Liver scanning
Mercury-203	Labeled mercurials	Brain and kidney scanning
Mercury-197	Mercurihydroxypropane	Spleen scanning
Strontium-85	Strontium nitrate	Bone scanning
Selenium-75	Selenomethionine	Pancreas scanning

4.6 Iodine-131

Early studies using iodine as a tracer relied on iodine-128, which has a half-life of 25 minutes. In 1938, searching for a tracer with a longer half-life, Hamilton was able to produce iodine-131 with the use of a cyclotron. This isotope is a beta and gamma emitter with a half-life of eight days; it is a very versatile tracer and accounts for more than half of all diagnostic tests using radioactivity.

In early tracer diagnosis procedures, the movement of the radioiodine was followed with a geiger counter. The 1950s saw the development of instruments called scanners, which use a moving measuring head to detect the radioactivity in specific locations, and which then produce a picture showing, for instance, the position of a brain tumor or the size and shape of an organ that cannot be seen well by means of X rays. The 1970s have seen further advancement in scanner technology. Computer

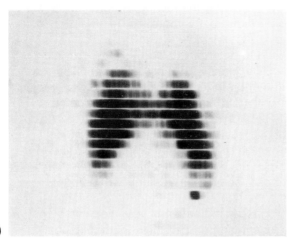

(a)

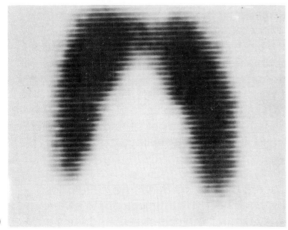

(b)

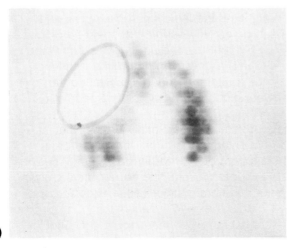

(c)

Figure 4.9 These pictures of (a) a normal thyroid, (b) an enlarged thyroid, and (c) a cancerous thyroid were taken by a linear photo scanner using iodine-131. (Courtesy U.S. Energy Research and Development Administration, Technical Information Center, Oak Ridge, Tennessee)

programs have been developed to analyze the raw data produced by the scanner so as to display pictures with greatly improved detail and resolution.

Iodine-131 is commonly used to determine blood volume, cardiac output, plasma volume, liver activity, kidney function, fat metabolism, thyroid function, and the location of brain tumors. Of the above, its greatest value lies in the measurement of thyroid function. Thyroxin, an iodine compound manufactured in the thyroid, is released in the blood to help control the overall utilization of nutrients by the body. To test the functioning of the thyroid gland, a patient can be given a small amount of iodine-131 in the form of the iodide ion. Since any iodine in the body is concentrated in the thyroid gland, the doctor can monitor the amount of radioiodine in the patient's blood to determine the rate at which the iodine is taken up by the thyroid. This will give the doctor an indication of how well this gland is functioning. If the thyroid is functioning properly, it should take up 12% of the radioiodine within a few hours. Too rapid an uptake means the patient is suffering from an overactive or hyperthyroid; too slow an uptake means the patient could be suffering from an underactive or hypothyroid. Scans taken of the thyroid will yield pictures such as those in Figure 4.9.

Iodine-131 is a valuable tool in determining if thyroid cancer has spread throughout the body, or **metastasized.** Several hours after a carefully measured sample of ^{131}I is administered, the patient is placed under a whole-body scanner, which produces a picture of the distribution of the radioisotope in the body. If the cancer has not spread, the radioiodine will be concentrated in the thyroid. If it has metastasized, the scanner picture will show "hot spots" or areas of radioactivity in other parts of the body (Figure 4.10).

Liver function can be measured using a chemical dye called Rose Bengal that has been tagged with iodine-131 and injected into a vein. The liver normally removes the dye from the blood stream and transfers it to the intestines for excretion. The speed with which the dye is removed can be followed by detectors monitoring the liver, small intestines, and blood stream. This test, which can be quickly performed, could help save the life of an auto accident victim brought into the emergency room with suspected liver damage, or could help a physician determine without additional surgery whether, say, the artificial bile duct connecting the liver to the intestines of a three-year-old is still functioning properly (Figure 4.11).

The normal brain contains a protective **blood-brain barrier** that prevents albumin in the blood from entering the brain; this blood-brain barrier is unique since the albumin from blood readily passes into muscles and other tissues. When a brain tumor is present, this barrier breaks down and allows the albumin to penetrate the brain tumor tissue. Blood albumin tagged with iodine-131 can be given to a patient with a suspected brain tumor. If the tumor is present, a brain scan will show a "hot spot" in the area of the brain where the tumor is located.

Figure 4.10 Several scans made by a whole body counter were put together to provide this picture of a patient with thyroid cancer that had spread to the lung. The patient was given one millicurie of iodine-131 72 hours before the scans were taken. (Courtesy Information Division, Lawrence Radiation Laboratory)

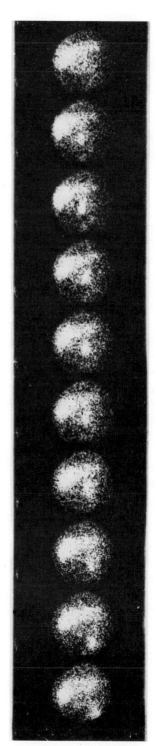

Figure 4.11 A three-year-old child who had been born without a bile duct and who had had an artificial duct surgically created was displaying symptoms indicating that this duct may have become blocked. These pictures are time lapse photographs made with a scintillation camera one hour after the child was injected with Rose Bengal dye tagged with iodine-131. The small light area, seen moving downwards and to the right in these pictures, showed that the artificial duct was still open. (Courtesy U.S. Energy Research and Development Administration)

4.7 Technetium-99^m

Used in at least 2000 diagnoses a day in the United States, technetium-99^m is very useful in the study of brain tumors, the thyroid, and the liver. Technetium is one of the four elements lighter than uranium that do not occur naturally, and so must be man-made. One advantage in its use is its very short half-life of six hours. The "m" after the mass number means that the isotope is **metastable,** or in an energy state higher than normal. When it decays it releases this excess energy as gamma rays, and forms the lower-energy isotope technetium-99.

$$^{99m}_{43}Tc \xrightarrow[t_{1/2} = 6 \text{ hours}]{} \, ^{99}_{43}Tc + \gamma$$

4.8 Chromium-51

Chromium, in the form of sodium chromate, attaches to red blood cells, making it useful in several types of tests. Red blood cells tagged with chromium-51 can be used in measuring blood volume to determine whether transfusions are needed in cases involving bleeding, burns, or surgical shock. They can be used to measure the lifetimes of red blood cells, aiding in the detection of certain types of anemia. A monitor placed over the heart can detect radioactive red blood cells and determine the blood flow rate through the heart. Obstetricians use red blood cells tagged with chromium-51 to find the exact location of the placenta in a pregnant woman; in a condition known as placenta previa, the placenta may be placed in such a position that fatal bleeding may occur. (Radioiodine was originally used for this procedure, but red blood cells tagged with chromium-51 have the advantage of not being transferred to the fetus.)

4.9 Phosphorus-32

In many types of tumors, phosphorus is found in abnormally high concentrations; this fact often enables doctors to distinguish between normal and cancerous cells. Radioactive phosphorus can be used to help locate such tumors, but since phosphorus-32 gives off only beta radiation, a counter must be very close to the tissue in which the phosphorus is present. Phosphorus-32 finds its greatest use in the detection of skin cancers and in brain surgery when the cancerous tissue is very hard to distinguish from normal tissue. If the patient is given ^{32}P, the surgeon can measure the radioactivity of the brain cells during surgery to determine which tissues to remove.

4.10 Radiation Therapy

Cancer cells, by their very nature, divide much more rapidly than do normal cells and are, therefore, more sensitive to radiation. This is the key fact that enables doctors to use radiation in killing cancerous tissue while leaving normal tissue relatively unharmed (Table 4.4).

Table 4.4   Some Radioisotopes Used in Medical Therapy

Teletherapy		
Treatment		Source
Beam therapy		Cobalt-60
		Cesium-137

Brachytherapy		
Treatment	Radioisotope	Form
Eye treatment	Strontium-90	Eye applicator
	Radium-226	
Skin treatment	Radium-226	Plaques and molds
	Strontium-90	Plaques
Cancer within an organ	Radium-226	Needles
	Radon-222	Seeds
	Iridium-192	Seeds and nylon ribbon
	Gold-198	Seeds
Cancer within a cavity	Cobalt-60	Needles and seeds
	Radium-226	Nasopharyngeal applicator
	Cesium-137	Sealed sources
Noncancerous conditions	Radium-226	Nasopharyngeal applicator
	Strontium-90	

4.11 Teletherapy

Teletherapy is the term used to denote the use of high intensity radiation to destroy cancerous tissue that can't be removed by surgery. In such therapy, radiation from an X-ray machine, a cobalt-60 source, or a particle accelerator is forced into a thin beam and directed at the cancerous tissue. In the treatment of internal cancers, the patient is carefully positioned so that the tumor is at the center of a large circular framework. The source of radiation is rotated about the patient so that the radiation will produce minimal skin tissue damage, but yet will be constantly focused on the tumor cells (Figure 4.12).

4.12 Brachytherapy

Brachytherapy is the medical procedure of inserting a radioisotope by means of a needle, or in the form of a seed, into the area to be treated. Radium-226 and strontium-90 are two isotopes often used in this way for the treatment of skin and eye cancers (Figure 4.13). One frequent target of brachytherapy is the pituitary gland. This small gland at the base of the brain produces secretions that stimulate cell reproduction throughout the body; if part of the pituitary gland can be destroyed, then runaway growth of cancer cells elsewhere in the body can usually be slowed down, although

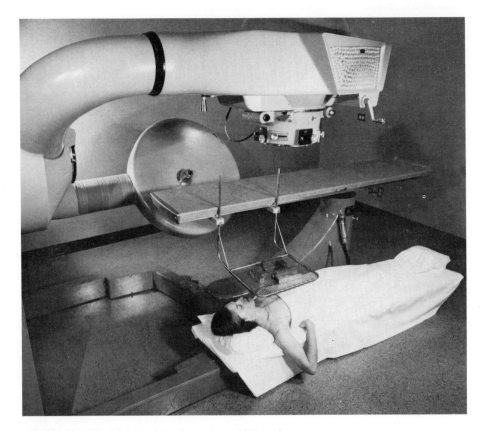

Figure 4.12 This cobalt-60 unit containing a 3000-curie source can be rotated around the stationary patient, allowing for many possible treatment positions. This patient is undergoing irradiation of the portions of the neck and chest outlined by the lines painted on her. The tray above the patient contains blocks of lead that absorb gamma rays and control the field of treatment. (Courtesy the University of Texas at Houston, M.D. Anderson Hospital & Tumor Institute)

not cured. Trying to destroy part of the pituitary gland surgically is very risky since the gland is so close to the brain. However, it is possible to implant small glass-like beads of yttrium-90 oxide into the gland. Yttrium-90 is a beta emitter; its radiation has very little penetrating power and will destroy the surrounding pituitary cells without affecting the brain tissue (Figure 4.14).

4.13 Radiopharmaceutical Therapy

Radiopharmaceutical therapy involves administering radioisotopes in chemical forms designed so that they will be concentrated in specific areas of the body. For example, since radioiodine is concentrated in the thyroid, it can be used to destroy thyroid cells in the treatment of an

Figure 4.13 This applicator contains the beta emitter strontium-90 in its tip, and is used in the treatment of eye lesions. The plastic shield protects the therapist's hand from backscatter radiation. (Courtesy Radiation Therapy Division, Nuclear Associates, Inc.)

overactive or hyperthyroid. Phosphorus-32 is used to treat the disease polycythemia vera, which causes an abnormal increase in the number of red blood cells. There is no known cure for this disease, but the radiophosphorus destroys the cells that produce red blood cells, thereby slowing the formation of red blood cells and temporarily alleviating the symptoms of the disease.

Certain types of cancerous tumors cause malignant effusions, which are large accumulations of excess fluids in the chest and abdominal cavities. To treat this condition, gold-198 in a suspension of very finely divided particles is introduced into the abdominal cavity. Gold-198 is chemically inert and nontoxic, and cannot pass through the abdominal wall; it has a half-life of 2.7 days, and emits beta and gamma radiation that is absorbed by tissues in the immediate neighborhood. This treatment destroys the cancerous cells and stops the oozing of fluid.

4.14 Radioisotopes and Agriculture

Agricultural researchers have made extensive use of radioisotopes in studying plant and animal nutrition and metabolism, plant diseases, and weed control. Radiation, rather than pesticides, is gaining increasing popularity in the control of insects. One success story is the almost total eradication of the screwworm fly in the United States. The female screwworm fly lays her eggs in the open wounds of livestock, and the burrowing maggots that emerge almost always kill their host. The eradication was accomplished by flooding the countryside with sterile male and female flies; the sterilization was accomplished by irradiating the flies with gamma rays from a cobalt-60 source.

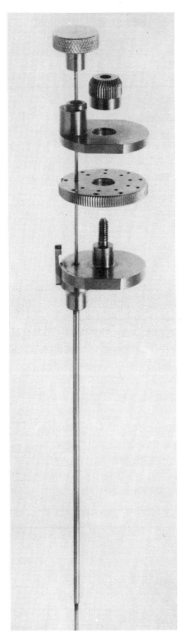

(a)

Figure 4.14 Irradiation of the pituitary with yttrium-90 pellets. (*a*) The needle used to implant the pellets. (*b*) An X ray showing the needle in place and the pellets that have just been passed through the needle into the bone area surrounding the pituitary gland. (*c*) An X ray showing the needle removed and the pellets in place around the pituitary. (Courtesy The Franklin McLean Memorial Research Institute, operated by the University of Chicago for ERDA)

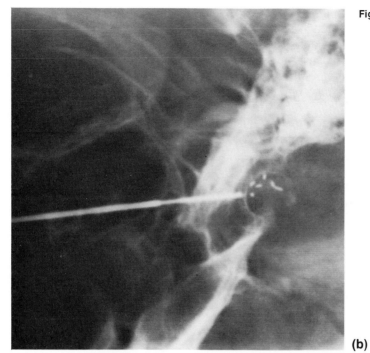

Figure 4.14 (cont'd.)

(b)

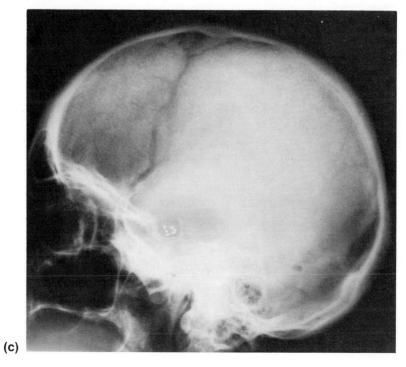

(c)

Significant advances in developing desirable traits in plants have been made through mutations in existing strains caused by exposure to radiation. The mutant plants have characteristics making them disease resistant, high yielding, machine harvestable, and more appealing in color to the consumer. To help improve the world's food supply, gamma radiation has been used to produce new varieties of wheat and rice that have a much higher yield and are more disease resistant.

Food spoilage is a critical problem in underdeveloped countries; the radiation treatment of food can help control insects and prevent spoilage. Very low doses of radiation will prevent sprouting in potatoes and onions; moderate doses will delay ripening in such perishable foods as bananas, strawberries, and papayas; low to moderate doses will control insect infestations in stored foods such as wheat, rice, and oats; and moderate to high dosages can be used to destroy the microorganisms in food (Figure 4.15). A wide variety of irradiated food is available to the consumer in Russia, but the Food and Drug Administration has still not approved the use of such foods in the United States.

Figure 4.15 Potatoes 16 months after exposure to gamma radiation. The potato on the top left received no radiation and the dose was progressively increased on each potato. The potato at the bottom left received 20,000 R, and the potato at the bottom right 106,250 R (which was too much). In some countries radiation is used commercially to retard sprouting in potatoes. (Courtesy Brookhaven National Laboratory)

Additional Reading

Books

1. Victor Arena, *Ionizing Radiation and Life,* C. V. Mosby, St. Louis, 1971.

2. United States Atomic Energy Commission, *Understanding the Atom Series.* These are short, interesting books on nuclear chemistry.

Article

1. "Radiation for Preventing Food Spoilage," *Chemistry,* July–August 1967, page 41.

Questions and Problems

---

1. Define the following terms:

    (a) Ionizing radiation
    (b) Curie
    (c) Rad
    (d) Rem
    (e) LET
    (f) Geiger-Müller counter
    (g) Film badge
    (h) Autoradiography
    (i) Background radiation
    (j) Radioactive tracer
    (k) Teletherapy
    (l) Brachytherapy
    (m) Radiopharmaceutical therapy

2. Explain how ionizing radiation can both be the cause of cancer and the means of arresting it.

3. Beta radiation is low LET radiation. Explain the way in which beta radiation can damage living cells. What are the chances of a cell recovering from a dose that is sublethal to the organism?

4. Alpha radiation is high LET radiation. Explain how its effects on living cells differ from beta radiation. Is a cell more likely to recover from alpha or beta radiation?

5. You are working in a laboratory that you fear has been contaminated with a radioactive beta emitter. How might you try to detect this contamination?

6. Would you be exposed to more background radiation living in Denver or San Francisco? Give reasons to support your answer.

7. After giving his patient a complete examination, a doctor feels that the patient may be suffering from an abnormal thyroid. How might radioisotopes be used to confirm the doctor's diagnosis?

8. What are three ways in which radioisotopes can be used in medical therapy?

9. You, as a leading agricultural researcher in a small country, are faced with the problem of an insect infestation that is destroying the nation's major crop, rice. Indicate two ways in which you might use radioactive substances to help you eliminate the problem.

chapter 5

Electron Configuration and the Periodic Table

Learning Objectives

By the time you have finished this chapter, you should be able to:

1. Describe Bohr's model of the atom.

2. Indicate the maximum number of electrons possible in energy levels 1 through 7.

3. Explain the difference between an atom in the ground state and an excited atom.

4. Explain, on an atomic level, what is happening in a luminous material.

5. Indicate the electron configuration by energy level for each of the elements with atomic number less than 18.

6. Define "periodicity."

7. Use the periodic table to state the symbol, atomic number, atomic weight, and electron configuration of any element.

8. Given one element in a period or chemical family, pick out another member of that same period or family.

9. State the number of valence electrons in an atom of any representative element.

10. Identify on a periodic table the elements that are metals, nonmetals, representative elements, transition metals, and metalloids.

11. Draw the electron dot diagram for any representative element.

12. Predict how atomic size changes across a period and down a group.

13. Define "ionization energy," and predict how ionization energy will change across a period or down a group.

14. Define "electron affinity," and predict how it will change across a period or down a group.

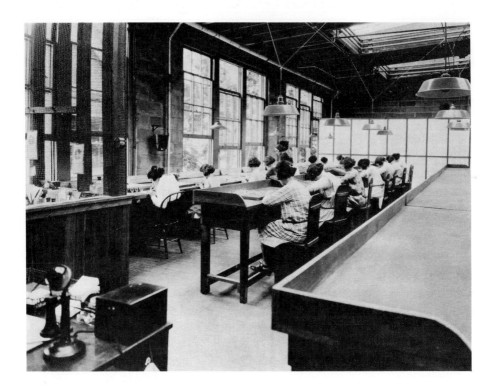

In 1918 Ruth Adams, then 16 years old, was elated to have finally found a job. She was hired by the Radium Luminous Materials Corporation of Orange, New Jersey, to apply luminous paint to watch dials and instrument dials. As you can imagine, this was painstaking work requiring great precision. To help her get started, the other women showed Ruth how to keep a fine point on her brush by turning the bristles on her tongue and lips. The paint she used was invented by the President of the company, and contained phosphorescent zinc sulfide with a small amount of radium and adhesive added.

Neither Ruth nor anyone else at that time realized the dangers that radioactive material posed to human tissues. However, in 1925 the potential harm of internal exposure to radium was forceably brought to the public's attention when the *New York Times* reported that five watch dial painters had died, and ten others had been stricken with "radium necrosis," a general breakdown of the bone tissues. In fact, so little was known about the nature of radium and its effects on human tissues that the company defended itself in subsequent court actions with the incredible assertion that small quantities of radium were actually beneficial to health!

The watch dial painters who died in the 1920s suffered symptoms including severe anemia, tumors of the mouth and nose, and inflammation of the bone marrow in the jaw or other bone structures. Although many of the women who worked in this industry during that time are still living and in good health, many others died after developing bone cancer 20 to 30 years later. Ruth Adams worked in the watch dial factory for six years before leaving to get married. She showed no adverse effects until some 25 years later, when she died of bone cancer that had spread throughout her body.

The disabilities or deaths of the women who were exposed to radium resulted from destruction of their bone tissue, especially of the bone marrow, which produces the blood cells for the body. You might wonder what special characteristic of radium makes its adverse effects so apparent in the bones of the body as opposed to other tissues. We will see that radium has chemical properties that are very similar to those of calcium, which is the chief component of bones and teeth. Unfortunately, the body is unable to distinguish the toxic radium from the important and essential calcium, and so deposits both in the bones, where the ionizing radiation produced by the radium does its damage.

We have stated that radium and calcium exhibit similar chemical properties; these properties are determined for each element by the arrangement of electrons surrounding the nucleus of the atom. Therefore, a good place to start this chapter is to take a closer look at the arrangement of electrons in the various elements.

Electron Configuration

In a previous chapter we described an atom of any element as containing a small, dense, positive nucleus surrounded by negative electrons. You will remember that a neutral atom contains the same number of electrons as there are protons in the nucleus and, therefore, the number of electrons in a neutral atom of an element will be equal to the atomic number of that element. Since protons and electrons are too small to be seen even with the most powerful microscopes, we must rely on various theoretical models to describe the arrangement of the electrons around the nucleus.

5.1 The Bohr Model

One of the first models of the atom, called the planetary model, was developed in the early 1900s by Niels Bohr, who based his theory on the experimental data of Ernest Rutherford (Figure 5.1). Over the past 75 years this theory has been modified considerably but, for the purposes of our discussions, we will find it convenient to use this less sophisticated model. Bohr pictured the atom as consisting of a small, dense nucleus surrounded by electrons traveling in orbits, similar to planets traveling around the sun. In his model the electrons could occupy only certain positions around the nucleus, positions that corresponded to certain energy values. Such allowed energy positions were called **energy levels** or **energy shells.** Electrons closest to the nucleus are in the lowest possible energy states, and electrons have correspondingly higher energy values if they are found farther from the nucleus.

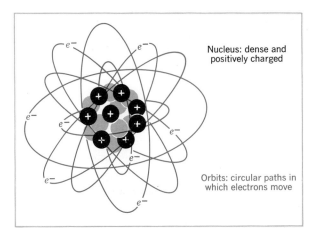

Figure 5.1 Bohr's planetary model of the atom has a dense, positively charged nucleus, with electrons moving in circular paths, called orbits, around the nucleus.

We might compare this concept of electron energy levels to the potential energy you possess as you climb the stairs to the balcony of a theater. At the bottom of the stairs you are at the lowest possible energy position. You must exert a specific amount of energy to climb up one stair and, once there, you then possess greater potential energy relative to the ground floor. Also, there is no way that you can stand between two steps; you must exert a specific amount of energy to get from one step to another step, and you are not able to stop part way between them (Figure 5.2). Similarly, an electron is not permitted to exist in between any of the specified energy levels of the atom. To move from one energy level to another, an electron must gain (or lose, depending upon whether it is moving away from or closer to the nucleus) an amount of energy exactly equal to the energy difference between the two levels.

The energy levels within an atom are assigned both number (1, 2, 3, 4, . . .) and letter (K, L, M, N, . . .) designations. The first, or K, energy level is the one closest to the nucleus. As the number or letter of the energy level increases, it is found farther from the nucleus and contains electrons with increasingly greater energy. Each energy level can contain only a certain maximum number of electrons, but in nature this maximum number is, in most cases, never realized (Table 5.1).

Figure 5.2 We can compare the potential energy one possesses as one climbs the stairs to a seat in the balcony to that possessed by the electrons in an atom. The higher one goes the more potential energy one possesses; one must exert a specific amount of energy to climb each stair, and there is no way to stand between steps.

Table 5.1   Energy Levels in the Atom

Energy level number (n)	1	2	3	4	5	6	7
Energy level letter	K	L	M	N	O	P	Q
Maximum number of electrons allowed in theory ($2n^2$)	2	8	18	32	50	72	98
Maximum number of electrons actually found in nature	2	8	18	32	32	18	8

It is easy to see that the energy levels close to the nucleus can accommodate only a few electrons, while those farther from the nucleus have more room available and, therefore, can accommodate more negative electrons. (Remember that like charges repel one another.)

In a normal atom, all of the electrons will be found in the lowest allowable energy levels; such an atom is said to be in the **ground state.** If the atom absorbs energy (such as heat or light) from an external source, some electrons might jump to a higher energy level, with each such electron absorbing an amount of energy exactly equal to the energy difference between the two levels. An atom having one or more electrons in a higher energy state than normal is called an **excited atom.** Excited atoms are unstable, and in such atoms electrons will drop back to lower energy levels by giving off energy, usually in the form of radiant energy (or light). The light given off by an atom as an electron drops from one energy level to another is of a specific energy or wavelength corresponding to the energy difference between the two levels (Figure 5.3). If enough

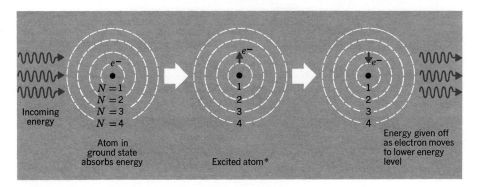

Figure 5.3   Electrons can move to energy levels farther away from the nucleus by absorbing energy, forming an excited atom. Electrons return to lower energy levels by giving off energy, usually in the form of light.

energy is supplied to an atom, one of its electrons may jump completely away from the atom, leaving the atom with one less electron than it has protons. This new particle is then no longer a neutral atom, but is rather a charged **ion.** Ionizing radiation, such as gamma radiation, can give an atom enough energy that some of its electrons break completely away. This, then, explains why one of the effects of ionizing radiation on living cells is to produce positive ions.

5.2 Luminescence

When an element or compound becomes excited by absorbing energy from an external source, and then gives off that energy as visible light, it is said to be **luminescent.** There are two kinds of luminescence: fluorescence, in which the element or compound stops giving off light as soon as the energy being supplied to the compound is stopped; and phosphorescence, in which the element or compound continues to give off light for a short time after the incoming energy has stopped. There are many examples of luminescence with which you are familiar. Laundry whiteners and brighteners make use of fluorescent dyes that become attached to clothing in the laundry process. These compounds absorb energy from the ultraviolet light in sunlight, and then give off light in the visible region. Through reflection and fluorescence, then, the clothing gives

off more visible light than falls on it and, therefore, appears "brighter."
Red reflective tape, which is often attached to vehicles to increase their
visibility, contains compounds that absorb nonred light and give off the
red–orange light we see (Figure 5.4). The luminous dial you may have on
your wristwatch contains tiny amounts of radioactive material mixed with a
powdered phosphor. The radioactive material releases a small but steady
amount of energy, which is absorbed by the atoms in the phosphor. When
the electrons in these atoms return to the ground state, they release the
visible light that you perceive as a faint glow on the watch dial.

Living organisms make use of luminescence for sexual attraction,
protection, and predation. The firefly uses bioluminescence to attract a
mate; chemical energy is used to excite electrons in a special compound
that produces the blink of the firefly as the electrons return to the ground
state. Green plants use the reverse of this process in the production of food:

Figure 5.4 The reflective trim on these bicycle tires
contains a fluorescent material, greatly increasing the
visibility of these bicycles at night. (Courtesy 3M
Company)

energy from sunlight excites electrons in molecules of the green pigment chlorophyll. As these electrons return to the ground state the energy is not released in the form of light, but rather is ultimately trapped as chemical energy in a molecule of sugar. This process of using the sun's energy to produce a sugar molecule from carbon dioxide and water is called **photosynthesis,** and will be studied in detail in Chapter 13.

5.3 The Quantum Mechanical Model*

The simple planetary model of the atom developed by Bohr did not explain some experimental data later collected, and in the late 1920s Erwin Schrödinger, P. A. M. Dirac, Werner Heisenberg, and others developed a new model of the atom called the **quantum mechanical model.** This model is based on fairly complicated mathematical concepts, and we will look only at some of the ways this theory modified the concept of the atom developed by Bohr.

Although the Bohr atom described electrons orbiting around the nucleus, it was found that one could not accurately pinpoint the position and speed of an electron. Therefore, the electron's position could be described only in terms of a probability distribution. (That is, rather than specifying exact locations, one speaks of regions in which there is a great probability of finding an electron.) Furthermore, it was found that each energy level above the first is divided into two or more energy **sublevels,** and each sublevel consists of one or more energy states or probability

---

* This section is optional and may be skipped without loss of continuity.

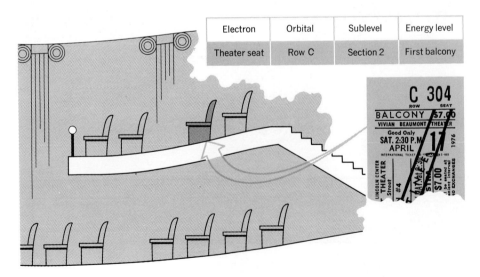

Electron	Orbital	Sublevel	Energy level
Theater seat	Row C	Section 2	First balcony

Figure 5.5 The position of an electron in an atom can be described in much the same way as a seat in a theater.

regions called **orbitals.** An orbital can contain zero, one, or two electrons. Therefore, just as a theater seat is found in a specific row in a specific section in a specific area of the theater, so will each electron in an atom be associated with an orbital that is in a particular sublevel of a particular energy level of that atom (Figure 5.5).

Each atom of a given element has a specific number of electrons characteristic of that element, and those electrons will occupy positions in energy levels according to the following rules:

Rule 1. The number of electrons in a neutral atom is equal to its atomic number.

	Atomic Number	Number of Electrons
Hydrogen	1	1
Oxygen	8	8
Sodium	11	11

Rule 2. The electrons will fill in the lowest possible energy positions available to them; when one level is filled, electrons go on to fill the next level. For the first 18 elements this pattern is closely followed. The electron configurations for these elements are shown in Table 5.2.

For example, lithium has three electrons, two of which will be on the first energy level; since these two electrons completely fill the first level, the third electron will be found on the second energy level.

Rule 3. After the element argon (atomic number 18), the assignment of electrons becomes a bit more complicated because of the complex overlapping of the higher energy levels. In this book we will not be concerned with these more complicated rules, and will simply list the electron configurations of the rest of the naturally occurring elements in Figure 5.7.

Table 5.2 Electron Configuration of Elements with Atomic Numbers under 19

Element	Atomic Number	Electron Configuration						
		1 K	2 L	3 M	4 N	5 O	6 P	7 Q
Hydrogen, H	1	1						
Helium, He	2	2						
Lithium, Li	3	2	1					
Beryllium, Be	4	2	2					
Boron, B	5	2	3					
Carbon, C	6	2	4					
Nitrogen, N	7	2	5					
Oxygen, O	8	2	6					
Fluorine, F	9	2	7					
Neon, Ne	10	2	8					
Sodium, Na	11	2	8	1				
Magnesium, Mg	12	2	8	2				
Aluminum, Al	13	2	8	3				
Silicon, Si	14	2	8	4				
Phosphorus, P	15	2	8	5				
Sulfur, S	16	2	8	6				
Chlorine, Cl	17	2	8	7				
Argon, Ar	18	2	8	8				

The Periodic Table of Elements

5.4 Mendeleev's Periodic Table

When talking about radium we mentioned that the chemical properties of each element depend upon that element's electron configuration. It would certainly be quite a chore to memorize the electron configuration of all 106 chemical elements. But take heart; there is a way of displaying the

Figure 5.6(a) Dmitri Mendeleev (1834–1907), a prolific writer, popular teacher, inventor and chemist, is best known for establishing the framework for the periodic table of elements. (Culver Pictures)

chemical elements so that their similarities are easily seen. This method
makes use of the **periodicity,** or repeating nature, of the chemical properties
of the elements, which was first noticed by a Russian scientist named
Dmitri Mendeleev in 1869 (Figure 5.6). He had devised a table showing
each of the then-known elements arranged according to their atomic weight,
and noticed that the chemical properties of the elements seemed to
exhibit periodic relationships; that is, similarities in chemical properties

	Group I	Group II	Group III	Group IV	Group V	Group VI	Group VII	Group VIII
1	H 1							
2	Li 7	Be 9.4	B 11	C 12	N 14	O 16	F 19	
3	Na 23	Mg 24	Al 27.3	Si 28	P 31	S 32	Cl 35.5	
4	K 39	Ca 40	— 44	Tc 48	V 51	Cr 52	Mn 55	Fe 56, Co 59 Ni 59, Cu 63
5	(Cu 63)	Zn 65	— 68	— 72	As 75	Se 78	Br 80	
6	Rb 85	Sr 87	?Yt 88	Zr 90	Nb 94	Mo 96	— 100	Ru 104, Rh 104 Pd 105, Ag 100
7	(Ag 108)	Cd 112	In 113	Sn 118	Sb 122	Te 125	I 127	
8	Cs 133	Ba 137	?Di 138	?Ce 140	—	—	—	— — — —
9	—	—	—	—	—	—	—	
10	—	—	?Er 178	?La 180	Ta 182	W 184	—	Os 195, Ir 517 Pt 198, Au 199
11	(Au 199)	Hg 200	Tl 204	Pb 207	Bi 208	—		
12	—	—	—	Th 231	—	U 240	—	— — — —

Figure 5.6(b) Mendeleev's periodic table.

repeated themselves among the elements as he read down the table. However, Mendeleev also realized that there seemed to be gaps in this table that would have to be filled in if the periodic relationships were to hold exactly. He wisely left these spaces blank convinced that these gaps represented elements that had not yet been discovered. About 45 years later it was found that if Mendeleev's table were slightly rearranged so that the elements were listed by atomic number rather than by atomic weight, some of the discrepancies left in the table would be eliminated. The resulting arrangement of elements is known as the **periodic table of the elements,** and modern-day periodic tables can be drawn so as to yield a vast amount of information about each element. A periodic table is shown in Figure 5.7 and on the inside front cover of this textbook.

Let's examine the periodic table closely. You see that each element appears in a box containing specific information about that element. In this case the box lists the name and symbol of the element, the atomic number, the atomic weight, and the electron configuration for the principle energy levels.

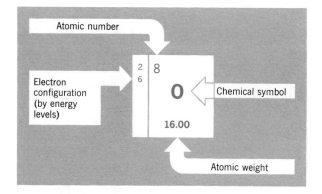

Explanation Example

Figure 5.7 The periodic table of the elements.

Group → / Period ↓	I	II	TRANSITION ELEMENTS							

H (1): config 1 — 1.008

TRANSITION ELEMENTS

Period	I	II							
2	3 Li (2,1) 6.94	4 Be (2,2) 9.01							
3	11 Na (2,8,1) 22.99	12 Mg (2,8,2) 24.31							
4	19 K (2,8,8,1) 39.10	20 Ca (2,8,8,2) 40.08	21 Sc (2,8,9,2) 44.96	22 Ti (2,8,10,2) 47.90	23 V (2,8,11,2) 50.94	24 Cr (2,8,13,1) 52.00	25 Mn (2,8,13,2) 54.94	26 Fe (2,8,14,2) 55.85	
5	37 Rb (2,8,18,8,1) 85.47	38 Sr (2,8,18,8,2) 87.62	39 Y (2,8,18,9,2) 88.91	40 Zr (2,8,18,10,2) 91.22	41 Nb (2,8,18,12,1) 92.91	42 Mo (2,8,18,13,1) 95.94	43 Tc (2,8,18,13,2) (97)	44 Ru (2,8,18,15,1) 101.07	
6	55 Cs (2,8,18,18,8,1) 132.91	56 Ba (2,8,18,18,8,2) 137.34	57-71 *	72 Hf (2,8,18,32,10,2) 178.49	73 Ta (2,8,18,32,11,2) 180.95	74 W (2,8,18,32,12,2) 183.85	75 Re (2,8,18,32,13,2) 186.2	76 Os (2,8,18,32,14,2) 190.2	
7	87 Fr (2,8,18,32,18,8,1) (223)	88 Ra (2,8,18,32,18,8,2) (226)	89-103 **	104 Kut (2,8,18,32,32,10,2) (261)	105 Hat (2,8,18,32,32,11,2) (260)	106			

H (1) — 1.008

***Lanthanide Series**

57 La (2,8,18,18,9,2) 138.91	58 Ce (2,8,18,20,8,2) 140.12	59 Pr (2,8,18,21,8,2) 140.91	60 Nd (2,8,18,22,8,2) 144.24	61 Pm (2,8,18,23,8,2) (147)

****Actinide Series**

89 Ac (2,8,18,32,18,9,2) (227)	90 Th (2,8,18,32,18,10,2) 232.04	91 Pa (2,8,18,32,20,9,2) (231)	92 U (2,8,18,32,21,9,2) 238.03	93 Np (2,8,18,32,22,9,2) (237)

III	IV	V	VI	VII	VIII
					2 2 **He** 4.00
2,3 5 **B** 10.81	2,4 6 **C** 12.01	2,5 7 **N** 14.01	2,6 8 **O** 16.00	2,7 9 **F** 19.00	2,8 10 **Ne** 20.18
2,8,3 13 **Al** 26.98	2,8,4 14 **Si** 28.09	2,8,5 15 **P** 30.97	2,8,6 16 **S** 32.06	2,8,7 17 **Cl** 35.45	2,8,8 18 **Ar** 39.95
2,8,18,3 31 **Ga** 69.72	2,8,18,4 32 **Ge** 72.59	2,8,18,5 33 **As** 74.92	2,8,18,6 34 **Se** 78.96	2,8,18,7 35 **Br** 79.90	2,8,18,8 36 **Kr** 83.80
2,8,18,18,3 49 **In** 114.82	2,8,18,18,4 50 **Sn** 118.69	2,8,18,18,5 51 **Sb** 121.75	2,8,18,18,6 52 **Te** 127.60	2,8,18,18,7 53 **I** 126.90	2,8,18,18,8 54 **Xe** 131.30
2,8,18,32,18,3 81 **Tl** 204.37	2,8,18,32,18,4 82 **Pb** 207.19	2,8,18,32,18,5 83 **Bi** 208.98	2,8,18,32,18,6 84 **Po** (210)	2,8,18,32,18,7 85 **At** (210)	2,8,18,32,18,8 86 **Rn** (222)

Group VIII transition elements (left of III):

2,8,15,2 27 **Co** 58.93	2,8,16,2 28 **Ni** 58.71	2,8,18,1 29 **Cu** 63.55 · 2,8,18,2 30 **Zn** 65.37
2,8,18,16,1 45 **Rh** 102.91	2,8,18,18 46 **Pd** 106.4	2,8,18,18,1 47 **Ag** 107.87 · 2,8,18,18,2 48 **Cd** 112.40
2,8,18,32,15,2 77 **Ir** 192.2	2,8,18,32,17,1 78 **Pt** 195.09	2,8,18,32,18,1 79 **Au** 196.97 · 2,8,18,32,18,2 80 **Hg** 200.59

Atomic weights are based on carbon-12;
values in parentheses are for the most stable or the most familiar isotope.
† Symbol is unofficial

2,8,18,24,8,2 62 **Sm** 150.35	2,8,18,25,8,2 63 **Eu** 151.96	2,8,18,25,9,2 64 **Gd** 157.25	2,8,18,27,8,2 65 **Tb** 158.92	2,8,18,28,8,2 66 **Dy** 162.50	2,8,18,29,8,2 67 **Ho** 164.93	2,8,18,30,8,2 68 **Er** 167.26	2,8,18,31,8,2 69 **Tm** 168.93	2,8,18,32,8,2 70 **Yb** 173.04	2,8,18,32,9,2 71 **Lu** 174.97

2,8,18,32,23,9,2 94 **Pu** (242)	2,8,18,32,24,9,2 95 **Am** (243)	2,8,18,32,25,9,2 96 **Cm** (247)	2,8,18,32,26,9,2 97 **Bk** (247)	2,8,18,32,27,9,2 98 **Cf** (249)	2,8,18,32,28,9,2 99 **Es** (254)	2,8,18,32,29,9,2 100 **Fm** (253)	2,8,18,32,30,9,2 101 **Md** (256)	2,8,18,32,31,9,2 102 **No** (254?)	2,8,18,32,32,9,2 103 **Lr** (257)

5.5 Periods and Families

Each horizontal row of the periodic table is called a **period.** The periods are numbered from one to seven corresponding to the seven energy levels of an atom that can contain electrons. This means that any element in, say, period four will have its outermost electrons located in the fourth energy level of the electron shell. For example, find sodium (symbol Na, atomic number 11) on the table. Sodium has only one electron in its outermost energy level and, since sodium appears on period three of the periodic table, this indicates that sodium contains that one electron on the third energy level.

Each column of the periodic table is called a **group** or **chemical family.** Mendeleev noticed that the various members of a family exhibit similar chemical behavior. We will see that it is the number of electrons in the outermost energy level of an atom that determines its chemical behavior. These outermost energy level electrons are called **valence electrons;** each member of a chemical family has the same number of valence electrons (Table 5.3).

Table 5.3   Electron Configurations of Three Chemical Families

Group Number	Chemical Family	Element	Atomic Number	Electron Configuration						
				1	2	3	4	5	6	7
I	Alkali metals	Lithium	3	2	1					
		Sodium	11	2	8	1				
		Potassium	19	2	8	8	1			
		Rubidium	37	2	8	18	8	1		
		Cesium	55	2	8	18	18	8	1	
		Francium	87	2	8	18	32	18	8	1
II	Alkaline earth metals	Beryllium	4	2	2					
		Magnesium	12	2	8	2				
		Calcium	20	2	8	8	2			
		Strontium	38	2	8	18	8	2		
		Barium	56	2	8	18	18	8	2	
		Radium	88	2	8	18	32	18	8	2
VIII	Noble gases	Helium	2	2						
		Neon	10	2	8					
		Argon	18	2	8	8				
		Krypton	36	2	8	18	8			
		Xenon	54	2	8	18	18	8		
		Radon	86	2	8	18	32	18	8	

Notice the columns on the periodic table labeled with the Roman numerals
I to VIII (on some periodic tables these are identified as the A group
families); these eight families are called the **representative elements**
(Figure 5.8). The Roman numeral above these columns indicates the

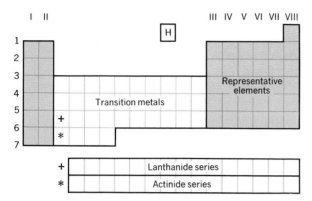

Figure 5.8 The location of the representative elements,
the transition metals, and the metals of the lanthanide and
actinide series on the periodic table.

number of valence electrons for each member of the family. For example,
the members of group II are beryllium, magnesium, calcium, strontium,
barium, and radium. Each neutral atom of these elements will contain two
valence electrons. These two electrons will be located on the fourth energy
level in an atom of calcium, and on the seventh energy level in an atom
of radium.

The elements of group VIII are called the **noble gases.** These gases
can be forced to form compounds with oxygen or with members of the
halogen family (the elements in group VII), but otherwise are extremely
unreactive. Each gas, with the exception of helium, has an electron
configuration of eight electrons in the outermost energy level, and the
chemical stability of these elements stems from this stable arrangement
of electrons. Although helium contains only two electrons, these valence
electrons completely fill the first energy level of this atom, giving helium its
stability. At the present time, chemists are still uncertain of the complete
explanation for the great stability imparted to the noble gases by their
particular arrangement of valence electrons.

We stated that members of the same chemical family will exhibit
similar chemical behavior. For example, chlorine and iodine, both members
of the halogen family, are highly effective in killing bacteria, and are used
as disinfectants and antiseptics, respectively. Germanium and silicon,
members of group IV, are both widely used as semiconductors in transistors.
Hard water problems can be caused by either calcium or magnesium ions

because both of these group II elements undergo similar chemical reactions with soap. Two radioactive wastes of special concern are strontium-90 and cesium-137. Strontium, like radium, is in group II and will therefore exhibit chemical properties similar to calcium. Strontium-90 in the atmosphere falls onto the grass, is eaten by cows, and is ingested by humans when they drink milk. As with radium, the human body will mistake strontium-90 for calcium, and will deposit it in bones and teeth. Cesium is in group I, and will behave in the body in much the same way as sodium, an element that is essential for many body functions.

You have probably noticed that hydrogen is separated slightly from the other elements on the periodic table, indicating that hydrogen actually belongs in a chemical family of its own. It exhibits chemical properties similar to elements of group I and to elements of group VII.

5.6 Metals and Nonmetals

Various properties of the elements allow them to be classified as **metals** or **nonmetals,** and the periodic table can be divided into two large sections based on this classification (Figure 5.9). Metals are generally shiny, dense,

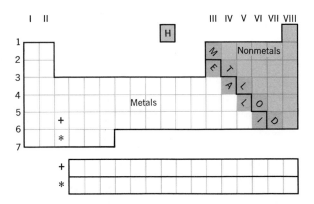

Figure 5.9 The location of the metals, nonmetals, and metalloids on the periodic table.

and malleable. They have high melting points, and are good conductors of electricity. Nonmetals, on the other hand, tend to be of low density and brittle. Most nonmetals have low melting points, and are poor conductors of electricity. The heavy, jagged line on the periodic table divides those elements that are metals from the nonmetals, with the metals appearing to the left of the line and clearly making up the majority of elements. However, there is no sharp distinction between the elements having metallic properties and those having nonmetallic properties; the elements next to the jagged line may exhibit properties of both metals and nonmetals. Such elements are known as **metalloids** or **semimetals.** For example, arsenic (symbol As, atomic number 33) is a brittle gray nonmetal. However, when freshly cut, it has a bright metallic luster. Arsenic acts as a metal in forming compounds with oxygen and chlorine, but acts as a nonmetal in other chemical reactions. Aluminum (symbol Al, atomic number 13) is a silvery metal, but it also exhibits some nonmetallic properties.

The metals of the periodic table can be further classified. The elements of group I are soft metals that can be cut with a knife. They react very rapidly with water, and are called **alkali metals.** Because of their high reactivity, they are found only in a combined form in nature. The elements of group II are harder metals that react less rapidly with water, and are called the **alkaline earths.** The three rows of 10 elements in the middle of the table are called **transition metals,** and they often show similar chemical characteristics not only within groups, but also along periods (Figure 5.8). This similarity in chemical behavior results from the fact that as we move along a period across the transition metals, electrons are not being added to the outermost energy level of the atoms, but rather are being added to the next-to-the-outermost level (Figure 5.7). The outermost energy level of each of the transition metals contains only one or two electrons. For example, iron, cobalt, and nickel, located on the fourth row of the periodic table, all have two electrons on the same outermost energy level. Although they contain different total numbers of electrons, they are very similar in chemical behavior, and are often grouped together for this reason.

We have seen that the same number of electrons in the outermost energy level of the transition metals masks the difference in the number of electrons in the next-to-outermost level. This is doubly true for the **lanthanide** and **actinide series,** the two rows of 14 elements at the bottom of the table, whose differences in electron makeup are hidden beneath two identical outer shell electron configurations. The naturally occurring elements in these two rows were once known as the rare earths because they appeared to exist only in minute quantities. These very similar elements were finally separated and identified using a technique called chromatography, and it is now known that many of these elements are in fact more abundant than either gold or silver (Figure 5.8).

5.7 Electron Dot Diagrams

We have stated that the chemical behavior of an atom is related to its electron configuration, especially to the arrangement of its valence electrons. A simple way of representing the valence electron configuration of an atom is by the use of an electron dot diagram. As you may guess from the name, in these diagrams a dot representing each valence electron is placed around the symbol of the element. On each side of the element's symbol there is room for two dots; beginning on any side, the dots are first placed singly, then paired for elements having more than four valence electrons.

$$\begin{array}{c} \square\square \\ \boxminus \; X \; \boxminus \\ \square\square \end{array}$$

Let's look at the electron dot diagrams for the elements in period 2. Remember that the group number of a representative element tells you the number of valence electrons for each member of that group or family.

Element	Symbol	Atomic Number	Group	Valence e	Electron Dot Diagram
Lithium	Li	3	I	1	Li·
Beryllium	Be	4	II	2	Be·
Boron	B	5	III	3	·B·
Carbon	C	6	IV	4	·Ċ·
Nitrogen	N	7	V	5	·N:
Oxygen	O	8	VI	6	·Ö:
Fluorine	F	9	VII	7	:F:
Neon	Ne	10	VIII	8	:Ne:

Periodic Properties

Various physical characteristics of atoms vary periodically with their atomic number; that is, a specific pattern of properties is repeated as atomic number increases. An examination of three of these properties will help us better understand the chemical behavior of the elements, and will explain the difference between the chemical behavior of metals and nonmetals.

5.8 Atomic Size

Figure 5.10a on page 130, shows a periodic table listing the atomic radius of each element. Figure 5.10b is a graph on which the atomic radii have been plotted; the atomic number of the element is listed on the bottom of the graph, and the height of the line corresponds to the radius of an atom of that element. How does the atomic radius change as the atomic number increases across any period?

Examine period 3. Sodium has the largest radius; then the radii of the atoms decrease across the period until we get to chlorine. Argon's radius increases again. Can we explain this trend? You will remember that as you read across a row, electrons are being added to the same outermost energy level. Also, as we move from element to element, an additional proton is added to the nucleus. Thus, as the atomic number increases, the charge of the nucleus also increases; therefore, so does the attraction of the nucleus for the electrons. Since the valence electrons added as we move along the period are all contained in the same outermost energy level, the electrons are drawn closer to the nucleus by this added attraction of the nucleus. As a result, the atomic radius grows smaller. But what about the noble gases? We have stated that this family is unique in its chemical stability, which comes from its arrangement of valence electrons. It is this stable arrangement of electrons in the outermost energy level that leads to the larger atomic radius. This explains the graph shown in Figure 5.10b. You can see that with only a few irregularities, the atomic radius decreases as we move across a period of the periodic table.

We might then ask how the atomic radius changes as the atomic numbers increase down a chemical family. Examine group II. As the atomic number increases from beryllium to radium, the atomic radius increases. Can we explain this trend in the same manner as before? As the atomic number increases down a group, we know that the number of protons in the nucleus increases. Therefore the charge of the nucleus will increase, intensifying the attraction of the nucleus for its electrons. The number of electrons also increases; but, as we move down a family, each element contains an additional outermost energy level, so that the valence electrons are found farther and farther from the nucleus. As a result, although the attraction of the nucleus for its electrons increases, this increased attraction is not enough to overcome the increased distance of the outermost energy level from the nucleus. Therefore, the radius of the atom increases as we move down a family.

We can summarize the above discussion by stating that the atomic radius of the elements decreases as the atomic number increases across a period, and the radius increases as the atomic number increases down a family. One result of this pattern is that the metallic elements on the left-hand side of the table will have large radii, and the nonmetals on the upper right-hand portion of the table will have small radii.

5.9 Ionization Energy

A second important property of an element is its ionization energy; we will see that this property is closely related to the atomic radius of the element. **Ionization energy** is the amount of energy that must be added to an atom to remove one electron from its outermost energy level. Put another way, the ionization energy indicates how strong the attraction of the nucleus is for the valence electrons; the stronger the attraction for these valence electrons, the greater is the ionization energy. The size of an atom determines how strong the attraction of the nucleus is for the valence

FIGURE 5.10a
Periodic Table of Atomic Radii

	I	II											III	IV	V	VI	VII	VIII
1	1 H 0.30																	2 He 0.93
2	3 Li 1.23	4 Be 0.89											5 B 0.80	6 C 0.77	7 N 0.70	8 O 0.66	9 F 0.64	10 Ne 1.12
3	11 Na 1.57	12 Mg 1.36											13 Al 1.25	14 Si 1.17	15 P 1.10	16 S 1.04	17 Cl 0.99	18 Ar 1.54
4	19 K 2.02	20 Ca 1.74	21 Sc 1.44	22 Ti 1.32	23 V 1.22	24 Cr 1.19	25 Mn 1.18	26 Fe 1.17	27 Co 1.16	28 Ni 1.15	29 Cu 1.18	30 Zn 1.21	31 Ga 1.25	32 Ge 1.24	33 As 1.21	34 Se 1.17	35 Br 1.14	36 Kr 1.69
5	37 Rb 2.16	38 Sr 1.91	39 Y 1.62	40 Zr 1.45	41 Nb 1.34	42 Mo 1.30	43 Tc 1.27	44 Ru 1.25	45 Rh 1.25	46 Pd 1.28	47 Ag 1.34	48 Cd 1.38	49 In 1.42	50 Sn 1.42	51 Sb 1.39	52 Te 1.37	53 I 1.33	54 Xe 1.90
6	55 Cs 2.35	56 Ba 1.98	57-71 *	72 Hf 1.44	73 Ta 1.34	74 W 1.30	75 Re 1.28	76 Os 1.26	77 Ir 1.26	78 Pt 1.30	79 Au 1.34	80 Hg 1.39	81 Tl 1.44	82 Pb 1.50	83 Bi 1.51	84 Po 1.65	85 At	86 Rn 2.20
7	87 Fr	88 Ra 2.20	89-103 **	104 Ku	105 Ha	106												

*	57 La 1.56	58 Ce 1.65	59 Pr 1.64	60 Nd 1.64	61 Pm 1.63	62 Sm 1.62	63 Eu 1.85	64 Gd 1.62	65 Tb 1.61	66 Dy 1.60	67 Ho 1.58	68 Er 1.58	69 Tm 1.58	70 Yb 1.70	71 Lu 1.56
**	89 Ac 2.0	90 Th 1.65	91 Pa 1.65	92 U 1.43	93 Np	94 Pu	95 Am	96 Cm	97 Bk	98 Cf	99 Es	100 Fm	101 Md	102 No	103 Lr

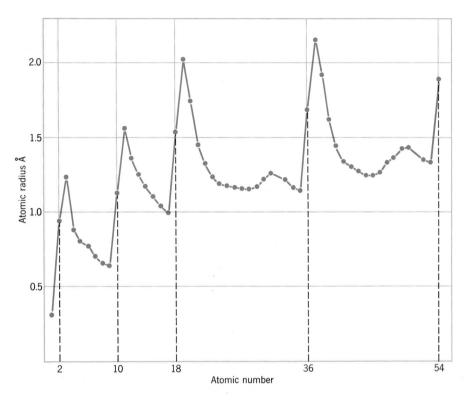

Figure 5.10 (a) Periodic table of atomic radii (in Angstroms). (b) Graph of the periodic trends of atomic radii. With a few irregularities, atomic radii decrease across a row and increase as the atomic numbers increase down a chemical family.

electrons. In small atoms, the nucleus and the valence electrons are close together and, therefore, the electrical attraction is strong. In larger atoms, the outermost energy level is farther from the nucleus and, therefore, the attraction for the valence electrons is weaker. If this is the case, how would we expect the ionization energy to change as the atomic number increases across a period? The periodic table in Figure 5.11a shows the ionization energies of each of the elements. We can see that the ionization energy increases across a period as the radius decreases. For example, the ionization energy of sodium is 5.1, and the ionization energy generally increases as we read across the period to chlorine, with an ionization energy of 13.0. The "sawtooth" variations in Figure 5.11b result from the fact that certain arrangements of electrons are more stable than others. Notice that although argon has a larger radius than does chlorine, it also has a larger ionization energy. All of the noble gases, with their very stable configuration of electrons, have very high ionization energies since this stability makes it hard to remove an electron from the outermost energy level of the atom.

FIGURE 5.11a
Periodic Table of Ionization Energies

Period	I	II											III	IV	V	VI	VII	VIII
1	1 H 13.6																	2 He 24.6
2	3 Li 5.4	4 Be 9.3											5 B 8.3	6 C 11.3	7 N 14.5	8 O 13.6	9 F 17.4	10 Ne 21.6
3	11 Na 5.1	12 Mg 7.6											13 Al 6.0	14 Si 8.1	15 P 11.0	16 S 10.4	17 Cl 13.0	18 Ar 15.8
4	19 K 4.4	20 Ca 6.1	21 Sc 6.6	22 Ti 6.8	23 V 6.7	24 Cr 6.8	25 Mn 7.4	26 Fe 7.9	27 Co 7.9	28 Ni 7.6	29 Cu 7.7	30 Zn 9.4	31 Ga 6.0	32 Ge 8.1	33 As 10.5	34 Se 9.7	35 Br 11.8	36 Kr 14.0
5	37 Rb 4.2	38 Sr 5.7	39 Y 6.6	40 Zr 7.0	41 Nb 6.8	42 Mo 7.2	43 Tc 7.5	44 Ru 7.5	45 Rh 7.7	46 Pd 8.3	47 Ag 7.6	48 Cd 9.0	49 In 5.8	50 Sn 7.3	51 Sb 8.6	52 Te 9.0	53 I 10.4	54 Xe 12.1
6	55 Cs 3.9	56 Ba 5.2	57–71 *	72 Hf 5.5	73 Ta 6.0	74 W 8.0	75 Re 7.9	76 Os 8.7	77 Ir 9.2	78 Pt 9.0	79 Au 9.2	80 Hg 10.4	81 Tl 6.1	82 Pb 7.4	83 Bi 8.0	84 Po	85 At	86 Rn 10.7
7	87 Fr	88 Ra	89–103 **	104 Ku	105 Ha	106												

* 57 La 5.6 | 58 Ce 6.9 | 59 Pr 5.8 | 60 Nd 6.3 | 61 Pm | 62 Sm 5.6 | 63 Eu 5.7 | 64 Gd 6.2 | 65 Tb 6.7 | 66 Dy 6.8 | 67 Ho | 68 Er | 69 Tm | 70 Yb 6.2 | 71 Lu 5.0

** 89 Ac | 90 Th | 91 Pa | 92 U 4.0 | 93 Np | 94 Pu | 95 Am | 96 Cm | 97 Bk | 98 Cf | 99 Es | 100 Fm | 101 Md | 102 No | 103 Lr

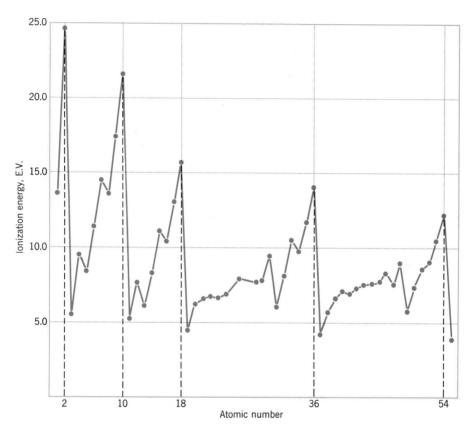

Figure 5.11 (a) Periodic table of ionization energies (in electron volts). (b) Graph of the periodic trends of ionization energies. Ionization energies tend to increase across a row and to decrease down a chemical family.

As we move down any family, the ionization energy will decrease. This happens because the atomic radii are increasing and, therefore, the attraction of the nucleus is becoming weaker. For example, beryllium has an ionization energy of 9.3, and this energy decreases as we read down to barium, with an ionization energy of 5.2. This indicates that it is easier to remove a valence electron from an atom of barium than it is from an atom of beryllium. Notice that the metals tend to have lower ionization energies than do the nonmetals, so that it is easier to remove a valence electron from a metal than it is from a nonmetal.

5.10 Electron Affinity

A third property that exhibits a periodic relationship among the elements is electron affinity. The **electron affinity** of an atom is the amount of energy that is released when an electron is added to a neutral atom of that element. This property indicates the ability of an atom to attract additional electrons. The larger the electron affinity, the larger the attraction of the atom for additional electrons. Determination of electron affinities is a complex process, and values are not yet available for all of the elements; however, enough values have been determined to establish trends in electron affinity.

From the data in Figure 5.12 we can see that metals have low electron affinities and nonmetals have high electron affinities. As we would expect from their stable electron configurations, the noble gases have low electron affinities.

Now that we have examined the relationship between the atomic radius, ionization energy, and electron affinity of the elements, we can begin to understand what makes metals different from nonmetals. The main difference between these two classifications of elements is the way in which they enter into chemical reactions, and we will closely examine this difference in the next chapter.

I	II			III	IV	V	VI	VII	VIII
H 0.75									**He** (-0.22)
Li 0.62	**Be** (-2.5)			**B** 0.86	**C** 1.27	**N** 0.0	**O** 1.46	**F** 3.34	**Ne** (-0.30)
Na 0.55	**Mg** (-2.4)			**Al** (0.52)	**Si** 1.24	**P** 0.77	**S** 2.08	**Cl** 3.61	**Ar** (-0.36)
K 0.50	**Ca** (-1.62)			**Ga** (0.37)	**Ge** 1.20	**As** 0.80	**Se** 2.02	**Br** 3.36	**Kr** (-0.40)
Rb 0.49	**Sr** (-1.74)			**In** 0.35	**Sn** 1.25	**Sb** 1.05	**Te** 1.90	**I** 3.06	**Xe** (-0.42)
Cs 0.47	**Ba** (-0.54)			**Tl** 0.5	**Pb** 1.05	**Bi** 1.05	**Po** (1.8)	**At** (2.8)	**Rn** (-0.42)
Fr (0.46)	**Ra**								

Figure 5.12 Periodic table of electron affinities (in electron volts; values in parentheses are not determined experimentally). (From "Experimental Values of Atomic Electron Affinities," by E. C. M. Chen and W. E. Wentworth, JOURNAL OF CHEMICAL EDUCATION, *52*, 488 (1975))

Additional Reading

Books
1. Ronald Rich, *Periodic Correlations,* W. A. Benjamin, New York, 1965.

2. Harry H. Sisler, *Electronic Structure, Properties, and the Periodic Law,* Reinhold, New York, 1963.

3. J. W. van Spronsen, *The Periodic System of Chemical Elements, a History of the First Hundred Years,* Elsevier, New York, 1969.

Articles
1. W. M. Allen, "The Diagonal Periodic Relationship," *Chemistry,* April 1970, page 23.

2. G. B. Kauffman, "American Forerunners of the Periodic Law," *Journal of Chemical Education,* March 1969, p. 128.

3. Eugene Keller, "Early Days of Radioactivity in Industry, Part II," *Chemistry,* May 1969, page 16.

4. O. L. Keller, Jr., "Predicted Properties of Elements 113 and 114," *Chemistry,* November 1972, page 8.

5. R. S. Mulliken, "Electrons, What They Are and What They Do," *Chemistry,* April 1967, page 13.

6. Glenn Seaborg, "From Mendeleev to Mendelevium and Beyond," *Chemistry,* January 1972, page 7.

7. Paul Seybold, "Luminescence," *Chemistry,* February 1973, page 6.

8. J. W. van Spronsen, "The Priority Conflict Between Mendeleev and Meyer," *Journal of Chemical Education,* March 1969, page 136.

9. Florence Wall, "Early Days of Radioactivity in Industry, Part I," *Chemistry,* April 1969, page 17.

10. H. G. Wallace, "The Atomic Theory, a Conceptual Model," *Chemistry,* November 1967, page 8.

11. Joan Zimmerman, "Mendeleev—His Own Man," *Chemistry,* December 1969, page 33.

Questions and Problems

1. Define the following terms:

   (a) Energy level
   (b) Ground state
   (c) Periodicity
   (d) Period
   (e) Group
   (f) Valence electron
   (g) Representative element
   (h) Noble gas
   (i) Metal
   (j) Nonmetal
   (k) Metalloid
   (l) Transition metals
   (m) Electron dot diagram
   (n) Ionization energy
   (o) Electron affinity
   (p) Family

2. Describe the Bohr model of the atom.

3. What are the maximum number of electrons found on the fourth energy level? The sixth?

4. Look at the following electron configuration for a magnesium atom, atomic number 12.

Energy level, n	1	2	3	4	5
Electron configuration	2	8	1	1	

   Is this atom in the ground state? Why or why not?

5. Why does a watch dial glow in the dark?

6. What is the electron configuration of a ground state atom containing 16 electrons?

7. Using the periodic table in Figure 5.7, fill in the following table.

Element	Symbol	Atomic Number	Atomic Weight	Electron Configuration
,a. Germanium				
,b. Xenon				
,c. Barium				

8. State the number of valence electrons in a neutral atom of each of the following elements.

   (a) Rubidium
   (b) Indium
   (c) Phosphorus
   (d) Krypton

9. Identify the following as (a) either a metal, a metalloid, or nonmetal and (b) either a representative element or transition metal.

   (a) Strontium
   (b) Selenium
   (c) Iron
   (d) Germanium

10. Arrange the following elements in order from the most metallic to the least metallic: sulfur, chlorine, silicon, and phosphorus.

11. Draw the electron dot diagrams for each of the following atoms.

    (a) Cesium   (c) Selenium
    (b) Silicon   (d) Bromine

12. For each of the following pairs predict which element has (a) the larger radius and (b) the larger ionization energy.

    (a) Na and P   (c) Ca and Ba
    (b) C and O   (d) Cl and I

13. Which has a higher ionization energy?

    (a) Na or Na^+ (b) Mg^+ or Mg^{2+}

    Give a reason for your answer.

14. Which has the higher electron affinity, Cl or Cl^-? Give a reason for your choice.

15. You are running a carefully controlled experiment monitoring the use of sodium ions by a single cell organism. You discover that your culture medium has been contaminated by trace amounts of the following ions: K^+, Ba^{2+}, and I^-. Do you think any of these ions would interfere with your data? Why or why not? Design an experiment that would support your conclusion.

chapter 6

Combinations of Atoms

Learning Objectives

By the time you have finished this chapter, you should be able to:

1. State the octet rule.

2. Describe an ionic bond and an ionic compound.

3. Given its name, write the formula for an ionic compound.

4. Given its formula, write the name of an ionic compound.

5. Describe a covalent bond and a covalent compound.

6. Describe a single, double, and triple covalent bond.

7. Draw the electron dot diagram for molecules of covalent compounds formed from the representative elements.

8. Write the formula for an ionic compound containing complex ions.

9. Define "electronegativity" and show how the electronegativity of the elements changes on the periodic table.

10. Describe the difference between a polar and a nonpolar covalent bond.

11. Predict whether a simple molecule will be polar or nonpolar.

12. Describe the hydrogen bond.

13. Define a "mole."

14. Calculate the formula weight of a substance given its formula.

15. Perform simple calculations using the mole.

16. Write a balanced chemical equation given the reactants and the products.

A light wind was blowing as the afternoon freight train rumbled through a small town in southeastern Louisiana. The train was just starting to gather speed about two miles outside of town when a freight car suddenly jumped the tracks, pulling 18 cars down an embankment with it. One of those 18, a tank car containing 30 tons of liquid chlorine, lay on its side with a huge gash ripped open under the weight of the other cars. The chlorine, which had immediately vaporized to a greenish-yellow gas, now poured out of the tank car and was carried by the breeze back toward the town.

The Harrison family lived in a nearby farmhouse, unaware of the approaching danger. Within minutes of the derailment the irritating odor of chlorine began to fill the house. Mrs. Harrison suddenly found she was having difficulty breathing, and her two small children began retching and vomiting. Her husband came running from the barn with his eyes streaming and quickly loaded his family into their car. As they rushed off, Mrs. Harrison noticed that her 11-month-old son Randy was having distinct trouble breathing. Their desperate rush to the hospital was accompanied by the sound of the volunteer fire department's siren screaming out the emergency signal, and the sight of the sheriff's car helping to evacuate nearby farms.

The deadly cloud of chlorine gas forced nearly 1000 people to flee their homes, offices, and schools. Scores of farm animals died from exposure to the gas, and several square miles of countryside were uninhabitable for several days. Fifty people were treated at the hospital for severe irritation caused by the chlorine, and 10 people, including the Harrison family, were hospitalized with critical poisoning. Unfortunately,

Combinations of Atoms

Randy died from the effects of the chlorine gas before his family could even reach the hospital.

Chlorine, at room temperature, is a greenish-yellow gas that has a characteristic irritating and suffocating odor. In low concentrations it will irritate mucous membranes and the respiratory system, and in high concentration will cause difficulty breathing — leading in extreme cases to death from suffocation.

Another dangerous substance is the element sodium. This element is an alkali metal that is so highly reactive that it is never found in a pure state in nature. (However, when isolated in pure form, it is a soft, silvery metal that can easily be cut by a knife.) Great care must be taken in handling sodium to be sure it is kept away from water, for when it contacts water sodium reacts extremely vigorously, releasing hydrogen gas which can be ignited by the heat accompanying this reaction.

Chlorine and sodium, then, are both highly reactive elements that are potentially dangerous to living tissue. Now, if you drop a piece of sodium in a container of chlorine gas, and warm the container, a white powder will form — a chemical reaction will have occurred. The product of this reaction is sodium chloride, the substance more commonly known as table salt (Figure 6.1). But table salt has none of the properties of the reactants, sodium and chlorine; in fact, sodium chloride is an essential nutrient in our diets. It plays a critical role in maintaining the proper amount of water in our cells and tissues, and is essential to the contraction of muscles and the transport of nerve impulses.

Figure 6.1 When it is heated, sodium reacts readily with chlorine gas, forming sodium chloride — table salt. (Courtesy D. O. Johnston. From Johnston, et al., *Chemistry and the Environment,* W. B. Saunders, 1973.)

Sodium chloride is a chemical compound, a homogeneous substance produced by the reaction of different elements. This chemical reaction, like many other spontaneous processes in nature, results in the formation of a more stable substance. In this case the elements sodium and chlorine both relatively unstable, will react spontaneously (the heat is applied only to speed up the reaction) to form sodium chloride, which is very stable.

About 1920, W. Kossel and G. N. Lewis observed that the representative elements (elements in groups I to VII) enter into chemical combinations that involve the loss, gain, or sharing of electrons in such a way as to attain a total of eight valence, or outer energy level, electrons. (You will remember that this is the electron configuration of the noble gases, which are extremely stable.) This tendency of the representative elements to attain an outer octet, or eight valence electrons, is called the **octet rule.** Although there are many exceptions to this rule among the heavier elements, it is quite useful in predicting the composition of chemical compounds produced from the reactions between lighter elements (atomic numbers 1 to 20).

Ionic Bonds and Ionic Compounds

6.1 Transfer of Electrons

As we have just indicated, chemical elements can attain a stable octet of electrons in two ways: by gaining or losing electrons, or by sharing electrons. The first of these ways, the transfer of electrons from one atom to another, results in the formation of ions — and the force of attraction between these ions is called an **ionic bond.** Such a transfer occurs when an element such as chlorine, which has a very strong attraction for additional electrons (that is, high electron affinity), reacts with an element such as sodium, which has a weak attraction for its valence electron (that is, low ionization energy). If an atom of sodium loses its single valence electron, it will form a positive sodium ion, Na^+; a sodium ion has 10 electrons — the same number as an atom of neon, the noble gas nearest to sodium on the periodic table. By gaining an electron, such as might be lost from a sodium atom, an atom of chlorine can form a negative chloride ion, Cl^-; the chloride ion will have 18 electrons, the same as the nearest noble gas, argon (Table 6.1 and Figure 6.2).

$$Na \longrightarrow Na^+ + e^-$$
$$e^- + Cl \longrightarrow Cl^-$$

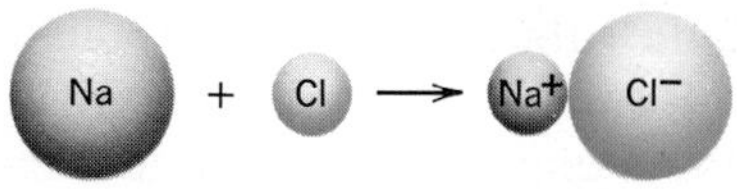

Figure 6.2 In the formation of sodium chloride, a sodium atom will lose one electron to a chlorine atom, producing the positive sodium ion Na^+ and the negative chloride ion Cl^-.

Table 6.1 Electron Configurations of Atoms and Ions

	Atomic Number	Electron Configuration				Atomic Number	Electron Configuration		
		K	L	M			K	L	M
Sodium, Na	11	2	8	1	Chlorine, Cl	17	2	8	7
Sodium ion, Na$^+$	11	2	8		Chloride ion, Cl$^-$	17	2	8	8
Neon, Ne	10	2	8		Argon, Ar	18	2	8	8

In general, we can expect that elements with few valence electrons (the metals in groups I, II, III) will lose electrons when reacting with elements that have close to eight valence electrons (the nonmetals in groups VI and VII). The ions that are thus created are attracted to each other — remember that unlike charged particles attract — and it is this attraction between ions that forms the ionic bond. When, for example, a chunk of sodium is placed in a container of chlorine gas, many billions of atoms will be involved in the ensuing reaction (that is, in the transfer of electrons). The resulting attraction between the positive and negative ions causes the grouping of these ions into an orderly three-dimensional pattern called a **crystal lattice.** This entire aggregation of ions is then called an **ionic compound** (Figure 6.3). In a crystal of sodium chloride each sodium ion is surrounded by six chloride ions, and each chloride ion is surrounded by six sodium ions. Since there is one sodium ion, Na$^+$, for every chloride ion, Cl$^-$, a crystal of sodium chloride is electrically neutral. And, in general, the ratio of ions in the crystal lattice of an ionic compound will always result in an electrically neutral compound.

An ionic compound contains no unique molecule; no particular ion is attracted exclusively to another ion but, rather, is attracted to all the ions surrounding it. You can see that the ionic bond is not a thing or a substance, but is simply the force of attraction between oppositely charged ions.

6.2 Electrovalence

The electrical charge carried by an ion is a measure of the ion's valence, or combining capacity. The valence that results from an electron transfer is called **electrovalence** or **ionic valence.** The electrovalence number indicates the kind of charge (positive or negative), and the amount of charge (the number of electrons gained or lost), on the ion.

To obtain a stable octet of electrons, elements in group I of the periodic table will lose one electron, forming ions with a charge of +1; therefore, the electrovalence of the elements in group I is +1. Elements in group II and

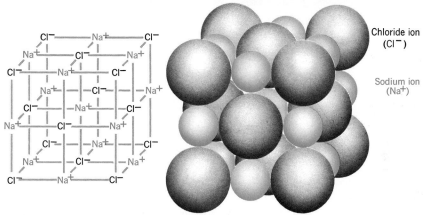

Figure 6.3 A sodium chloride crystal consists of sodium and chloride ions in a closely packed arrangement. (Courtesy American Museum of Natural History)

group III have electrovalences of +2 and +3, respectively. Group VI elements have an electrovalence of −2, and group VII elements have −1. Elements in groups IV and V rarely form ions, so we will not concern ourselves with their electrovalence numbers. Some elements can have more than one electrovalence number; that is, they may appear in different stable compounds as ions having different charges. The electrovalence numbers of some common ions are listed in Table 6.2. In our study of living organisms we will be concerned with compounds containing these elements, so it is important that you become familiar with their electrovalence numbers.

Table 6.2 Electrovalence Numbers of Some Common Ions

Name of Ion	Electro-valence Number of Its Common Ion	Symbol of Ion	Name of Ion	Electro-valence Number of Its Common Ion	Symbol of Ion
Lithium ion	+1	Li^+	Bromide ion	−1	Br^-
Potassium ion	+1	K^+	Chloride ion	−1	Cl^-
Silver ion	+1	Ag^+	Fluoride ion	−1	F^-
Sodium ion	+1	Na^+	Iodide ion	−1	I^-
Barium ion	+2	Ba^{2+}	Oxide ion	−2	O^{2-}
Calcium ion	+2	Ca^{2+}	Sulfide ion	−2	S^{2-}
Magnesium ion	+2	Mg^{2+}			
Zinc ion	+2	Zn^{2+}			
Aluminum ion	+3	Al^{3+}			
Copper(I) ion (Cuprous)	+1	Cu^+	Iron(II) ion (Ferrous)	+2	Fe^{2+}
Copper(II) ion (Cupric)	+2	Cu^{2+}	Iron(III) ion (Ferric)	+3	Fe^{3+}

6.3 Names and Symbols

Positive ions carry the name of their parent element; Na^+ is called the sodium ion and Ca^{2+} is the calcium ion. If an element can form more than one type of positive ion, the different electrovalence numbers are indicated by Roman numerals after the name of the element. For example, Fe^{3+} is the iron(III) ion and Cu^+ is the copper(I) ion. As shown in Table 6.2, an older practice of nomenclature identifies these two ions as the ferric ion and the cuprous ion, respectively; laboratory instruction manuals and labels on bottles of chemicals often make use of this practice of naming ions and compounds. The negative ion is named by using a prefix derived from the name of the parent element, and adding the suffix -ide. For example, Cl^- is the chloride ion and O^{2-} is the oxide ion.

Ionic compounds are named for the ions from which they are formed, with the name of the positive ion always appearing first. For example, NaCl is sodium chloride and FeO is iron(II) oxide.

Chemical formulas are a shorthand way of representing chemical compounds. The formula of a compound yields two pieces of information:

the types of atoms* present, and the ratio of these atoms in the compound. The type of atom is indicated by the symbol of the element, and the ratio of atoms is shown by the use of subscripts following the symbol.

Symbols show the type of atoms present:
calcium and chlorine

$CaCl_2$

Subscript shows the ratio of atoms:
2 chlorines to 1 calcium

As we mentioned before, all ionic compounds are electrically neutral, and the ratio of atoms used in the chemical formula is the lowest whole number ratio that will assure this electrical neutrality. This means that the formula for calcium chloride would not be expressed as, say, $Ca_{\frac{1}{2}}Cl$ or Ca_6Cl_{12}. Let's look at some other examples:

1. *Sodium bromide.* From Table 6.2 we see that the sodium ion is Na^+ and the bromide ion is Br^-. Therefore, to maintain neutrality we need one sodium ion for every bromide ion; the ratio will be 1:1. The correct formula is NaBr. (Notice that the subscript "1" is not written, and the charges on the ions are not included.)

2. *Magnesium chloride.* From Table 6.2 we see that the magnesium ion is Mg^{2+} and the chloride ion is Cl^-. Therefore, to maintain neutrality we need two chloride ions for each magnesium ion; the ratio is 1:2. The correct formula is $MgCl_2$. (The subscript 2 refers only to the chloride ion.)

3. *Iron(II) oxide.* From Table 6.2 we see that the iron(II) ion is Fe^{2+} and the oxide ion is O^{2-}. Therefore, to maintain neutrality we need one iron(II) ion for every oxide ion. The ratio will be 1:1. The correct formula is FeO.

4. *Iron(III) oxide.* From Table 6.2 we see that the iron(III) ion is Fe^{3+} and the oxide ion is O^{2-}. Therefore, to maintain neutrality we need two iron(III) ions (a total charge of +6) for every three

* We have just explained that in an ionic compound you will not find intact atoms, but rather ions that may be thought of as pieces of atoms with parts rearranged. However, it is easier to talk about chemical formulas in this way, and we will continue to do so now that we understand what is actually meant.

oxide ions (a total charge of -6); the ratio is 2:3. The formula is Fe_2O_3. (*Hint:* an easy rule to help you write formulas is to "crisscross" valence numbers.)

$$Fe^{3+} \quad O^{2-} \longrightarrow Fe_2O_3$$

Covalent Bonds and Covalent Compounds

6.4 Sharing Electrons

Elements that have similar attracting power for their electrons cannot reach stability by pulling electrons away from one another to form an ionic bond. Rather, these elements must reach stability by sharing electrons, and forming a **covalent bond.** The covalent bond results when two positive nuclei attract the same electrons, thus holding the nuclei close together. When two or more atoms share electrons through covalent bonds, a single (electrically neutral) unit called a **molecule** is formed. Covalent compounds are composed of molecules, which in turn are composed of different atoms held together by covalent bonds. We will see that covalent bonds are normally formed between nonmetallic elements. (Remember that nonmetals are elements having high ionization energies and, therefore, strong attraction for their valence electrons.)

Some nonmetallic elements exist in nature not as atoms, but as **diatomic molecules** — two atoms of the element covalently bonded together (Table 6.3).

Table 6.3   Elements That Exist as Diatomic Molecules

Hydrogen	H_2	Fluorine	F_2
Nitrogen	N_2	Chlorine	Cl_2
Oxygen	O_2	Bromine	Br_2
		Iodine	I_2

Chlorine, for example, exists as the diatomic molecule Cl_2. Each chlorine atom has seven valence electrons, and needs one more electron to reach the stable electron configuration of argon. Two chlorine atoms can share a pair of electrons (thus forming a covalent bond) to reach this stable octet. A shorthand way of representing a covalent bond is to draw a dash between the chemical symbols of the elements involved (Figure 6.4).

Hydrogen is another element found as a diatomic molecule, H_2. Each

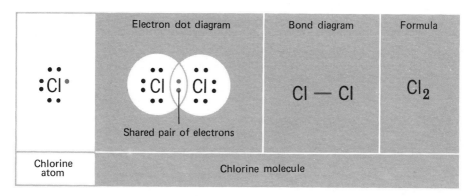

Figure 6.4 The bond which is formed between the chlorine atoms in a diatomic molecule of chlorine is covalent.

hydrogen atom has one valence electron, and needs one more electron to reach the stable electron configuration of the nearest noble gas, helium. Perhaps you can now visualize how two hydrogen atoms are able to share a pair of electrons to become stable (Figure 6.5).

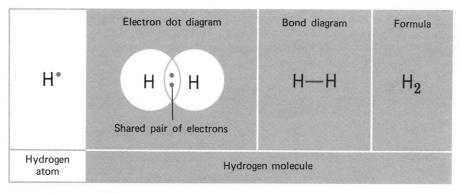

Figure 6.5 The bonding in the diatomic molecule of hydrogen.

6.5 Multiple Bonds

Often a stable octet of electrons can be attained only if more than one pair of electrons is shared between two nuclei. A single shared pair of electrons results in a **single covalent bond,** which we have represented by a dash between the symbols of the elements. A **double bond** is formed when two pairs of electrons are shared between two nuclei, and is represented by two dashes, $=$. Carbon dioxide is an example of a compound containing double covalent bonds; the carbon atom must share two pairs of electrons with each oxygen atom to reach a stable octet of electrons (Figure 6.6).

The element nitrogen is found in nature as a diatomic nitrogen molecule, N_2. Each nitrogen atom has five valence electrons, and needs to share three electrons to reach a stable octet. This is accomplished by having each of

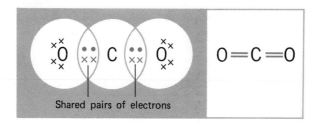

Shared pairs of electrons

Figure 6.6 The bonding in carbon dioxide, CO_2. The carbon atom has four valence electrons and needs to share four electrons to reach a stable octet. Each oxygen atom has six valence electrons and needs to share two electrons. There are two double bonds in a molecule of CO_2, resulting in an octet of electrons around each atom.

two nitrogen atoms share three electrons with the other atom. These three shared pairs of electrons form a **triple bond,** represented by three dashes (Figure 6.7). Quadruple bonds, which would be four pairs of electrons shared between two atoms, are never found since they are structurally impossible.

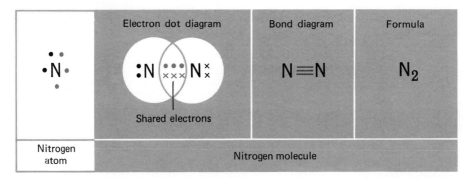

Figure 6.7 The bonding in the diatomic molecule of nitrogen. By forming a triple covalent bond, each atom of nitrogen attains an octet of electrons.

6.6 Covalence Numbers

The **covalence number** of an element indicates the number of electrons an atom of that element will share in a covalent bond. Since ions are not formed in covalent compounds, we will not specify signs for covalence numbers in our discussion. Some covalence numbers are shown in Table 6.4.

Table 6.4 Covalence Numbers of Some Common Elements

Group	Element	Symbol	Electron Dot Diagram	Common Covalence* Number
	Hydrogen	H	H·	1
IV	Carbon	C	·Ċ·	4
V	Nitrogen	N	·N̈·	3
VI	Oxygen	O	·Ö:	2
	Sulfur	S	·S̈:	2
VII	Chlorine	Cl	·C̈l:	1
	Bromine	Br	·B̈r:	1
	Iodine	I	·Ï:	1

* Some of these elements have more than one covalence number.

For example, carbon's covalence number is 4; in each compound it forms, carbon must share four pairs of electrons. Several carbon compounds are shown in Figure 6.8, and you will notice that there are four covalent bond lines attached to each carbon atom.

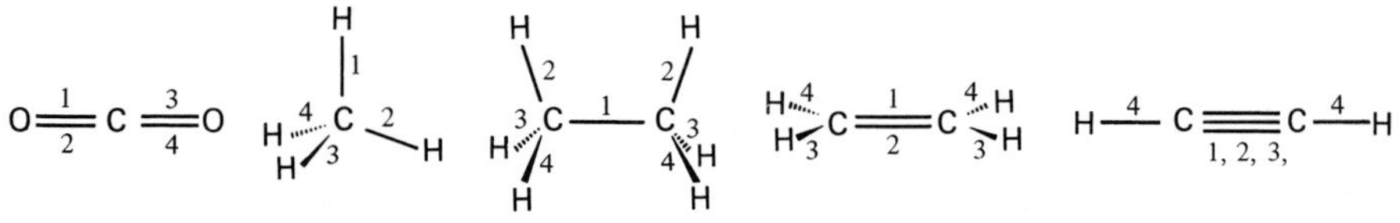

Figure 6.8 Some carbon compounds. In each compound the carbon atom has four pairs of electrons associated with it.

6.7 Complex Ions

A **complex ion** (or polyatomic ion) is a group of covalently bonded atoms that, as a group, carries an electrical charge, but that is so stable that it will go through most chemical reactions as a unit and won't come apart. Some complex ions and their electrovalence numbers are listed in Table 6.5.

Table 6.5   Some Complex Ions and Their Electrovalence Numbers

Name of Ion	Formula	Electro-valence	Name of Ion	Formula	Electro-valence
Ammonium Ion	NH_4^+	+1			
Acetate ion	$C_2H_3O_2^-$	−1	Hydrogen sulfite (bisulfite ion)	HSO_3^-	−1
Chromate ion	CrO_4^{2-}	−2	Sulfite ion	SO_3^{2-}	−2
Dichromate ion	$Cr_2O_7^{2-}$	−2	Hydroxide ion	OH^-	−1
Cyanide ion	CN^-	−1	Nitrate ion	NO_3^-	−1
Hydrogen carbonate (bicarbonate ion)	HCO_3^-	−1	Nitrite ion	NO_2^-	−1
Carbonate ion	CO_3^{2-}	−2	Permanganate ion	MnO_4^-	−1
Hydrogen sulfate (bisulfate ion)	HSO_4^-	−1	Phosphate ion	PO_4^{3-}	−3
Sulfate ion	SO_4^{2-}	−2	Monohydrogen phosphate ion	HPO_4^{2-}	−2
			Dihydrogen phosphate ion	$H_2PO_4^-$	−1

Many ionic compounds consist of a metal ion and a complex ion. The naming of these compounds follows the same rules as for ionic compounds; Na_2CO_3 is sodium carbonate, and $MgSO_4$ is magnesium sulfate. One additional rule is quite helpful when writing the chemical formula of a compound containing complex ions. If more than one complex ion appears in the formula, the symbol of the complex ion is put in parentheses, with the subscript following the parentheses.

For example:

1. *Magnesium hydroxide.* This compound has one Mg^{2+} ion and two OH^- ions, and is written $Mg(OH)_2$. (The 2 indicates two of the ion OH^-.)

2. *Calcium phosphate.* This compound has three Ca^{2+}

ions and two PO_3^{3-} ions, and is written $Ca_3(PO_4)_2$. (The 4 within the parentheses refers only to the oxygen, but the subscript 2 that appears outside the parentheses indicates two of the complex ion PO_4^{3-}.)

Ionic compounds containing complex ions are so common and play such a central role in our daily lives that it is important for you to become familiar with the names and formulas of the complex ions shown in Table 6.5. Examples of such familiar ionic compounds are sodium hydrogen carbonate (sodium bicarbonate) $NaHCO_3$, which is baking soda, and magnesium hydroxide $Mg(OH)_2$, which is a common antacid.

6.8 Polar Covalent Bonds

Imagine two identical atoms sharing a pair of electrons in a covalent bond. We would expect the electrons to be shared absolutely equally between the two nuclei, resulting in the center of positive charge and the center of negative charge occurring at the same place along the bond; this is called a **pure covalent bond** (Figure 6.10a).

An atom of each element has a specific tendency to attract a shared pair of electrons, and the strength of this attraction is known as **electronegativity.** In each row of the periodic table, the element with the smallest radius (and needing the fewest number of electrons to become stable) is the most electronegative; among all the elements, fluorine is the most electronegative. The electronegativity of the elements increases to the right along a row, and decreases from top to bottom down each group, as shown in Figure 6.9.

Now imagine that two different atoms are sharing electrons. We would expect that one atom might attract the electrons to a greater extent than

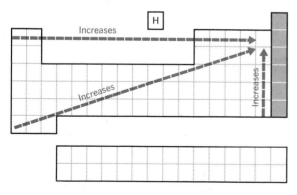

Figure 6.9 Trends in electronegativity on the periodic table.

the other, so the electrons will be found closer to the more electronegative (or electron-attracting) atom. Therefore, in such a bond the center of positive charge does not coincide with the center of negative charge, resulting in an unequal distribution of charge (Figure 6.10b). Such a bond is called a **polar covalent bond.**

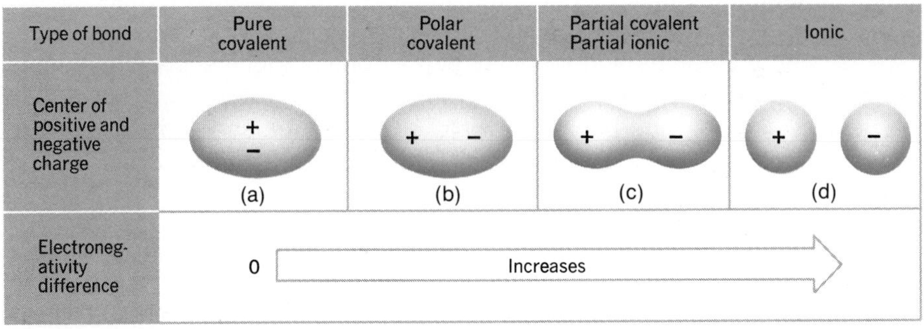

Type of bond	Pure covalent	Polar covalent	Partial covalent Partial ionic	Ionic
Center of positive and negative charge	(a)	(b)	(c)	(d)
Electroneg- ativity difference	0	Increases		

Figure 6.10   The bonding continuum. There is no absolute break between ionic and covalent bonding, but rather a continuous range from pure covalent bonding to complete ionic bonding.

For example, consider hydrogen chloride, HCl. Chlorine has a greater electron affinity than hydrogen and is the more electronegative element, so the shared electrons will be pulled closer to the chlorine nucleus in the hydrogen-chlorine bond. This will give the chlorine end of the bond a slightly negative charge, and the hydrogen end of the bond a slightly positive charge (Figure 6.11). In general then, we would expect the negative

$\delta+$ $\delta-$

H :Cl:

Figure 6.11   The polar covalent bond in hydrogen chloride. The Greek letter delta (δ) is used to indicate a small amount of positive or negative charge.

portion of a polar covalent bond to be in the region of the more electronegative atom, and the positive portion in the region of the less electronegative atom.

Atoms of the same element will form pure covalent bonds, with centers of positive and negative charges coinciding. Atoms of unlike elements that have differences in electronegativity will result in the centers of electrical charges moving apart. When the difference in electronegativity becomes very large, the electrical charges will be completely separated to the two opposite ends of the bond; but this is exactly the description of the ionic bond that we discussed earlier. That is, we can imagine a continuous range from pure covalent bonding to complete ionic bonding, with different compounds appearing at specific places all along this continuum (Figure 6.10). Even though we can identify no absolute break between ionic bonding and polar covalent bonding, it is still convenient to try to classify compounds as either ionic or covalent. In general, bonds formed between metals and either nonmetals or complex ions will be ionic, and bonds formed between nonmetals will be covalent — they will be pure covalent when the electronegativity of the atoms is the same, and polar covalent when one atom has greater electronegativity than the other.

6.9 Shapes of Molecules

The properties and behavior of a molecular compound depend not only on the combination of atoms of which it is composed, but also the shape or spatial arrangement of these atoms in the molecule. The properties affected by the shape of the molecule range from the odor of the compound to the role that the molecule plays in regulating the chemical reactions in living organisms. For example, it has been found that all psychedelic drugs seem to affect one particular site in the brain. One current theory is that this site is sensitive to, or keyed to, a specific three-dimensional shape. According to the theory, the molecules of the psychedelic drugs each have particular

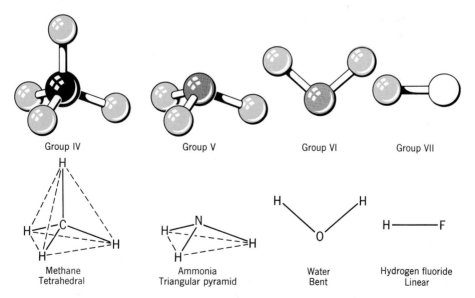

| Group IV | Group V | Group VI | Group VII |

| Methane | Ammonia | Water | Hydrogen fluoride |
| Tetrahedral | Triangular pyramid | Bent | Linear |

Figure 6.12 The shapes of some covalently bonded molecules. The dotted line indicates an outline of the shape formed by each molecule.

regions that closely resemble this three-dimensional shape and, therefore, will each trigger hallucinogenic symptoms.

We have previously seen that elements within the same group or family will possess similar chemical properties. Such similarities also hold with respect to the arrangement of atoms covalently bonded to nonmetals in groups IV through VIII. Assuming that all bonds to the central atom are single bonds, the arrangement around atoms of elements in group IV will be tetrahedral in shape, group V will be triangular pyramidal, group VI bent, and group VII linear (Figure 6.12).

6.10 Polar and Nonpolar Molecules

A molecule containing many different polar bonds may itself be polar or nonpolar, depending upon the shape of the molecule. As was the case in our discussion of molecular shape, the polarity or nonpolarity of a molecule also plays a large part in its behavior and role in living organisms. We will see that polar molecules will be found with other polar molecules, and nonpolar molecules will be found with other nonpolar molecules in living systems.

As with covalent bonds, the centers of positive and negative charges will coincide in a nonpolar molecule, and will not coincide in a polar molecule. It is important to realize that a polar molecule is still electrically neutral, even though there will be a separation of charge in the molecule resulting in regions of positive and negative charge.

Several examples will illustrate how the shape of a molecule can determine whether the molecule will be polar or nonpolar. Note that in each of our examples, however, the bonds that are formed are polar (Table 6.6).

Table 6.6   Molecular Shape and the Polarity of Molecules

Compound	Formula	Shape	Bond Diagram	Center of Charge	Type of Molecule
Hydrogen chloride	HCl	Linear	$\overset{\delta+}{H}\!\!-\!\!\overset{\delta-}{Cl}$	+ −	Polar
Carbon dioxide	CO_2	Linear	$\overset{\delta-}{O}\!\!=\!\!\overset{\delta+}{C}\!\!=\!\!\overset{\delta-}{O}$	±	Nonpolar
Water	H_2O	Bent	$\overset{\delta-}{O}$ $\overset{\delta+}{H}$ $\overset{\delta+}{H}$	− +	Polar

Hydrogen chloride, as we have already seen, is a linear molecule containing one polar bond; the molecule is polar with the hydrogen at the positive end and the chlorine at the negative end. Carbon dioxide is also a linear molecule containing, in this case, two polar double bonds. However, since the arrangement of atoms in the molecule is such that the centers of positive and negative charge coincide, the entire carbon dioxide molecule is nonpolar. A molecule of water contains three atoms and two polar bonds just as we found in carbon dioxide. But the water molecule has a bent (rather than linear) shape, and the centers of positive and negative charge do not coincide; thus, a water molecule is polar.

6.11 Hydrogen Bonding

Molecules that contain hydrogen attached to a highly electronegative atom such as fluorine, oxygen, or nitrogen will display an intermolecular (between molecules) type of bonding called **hydrogen bonding.** The slightly positive hydrogen region on one molecule will be attracted to the slightly negative fluorine, oxygen, or nitrogen region on another molecule. Such a hydrogen bond has a strength about $\frac{1}{10}$ that of an ordinary covalent bond. In some very large biological molecules, hydrogen bonding may occur between different parts of the same molecule, causing the molecule to bend back on itself.

Hydrogen bonding plays an important role in nature. Hydrogen bonds are responsible for the unusual properties of water (such as high melting and boiling points) that make this fluid so important to all living organisms — these properties will be discussed in detail in Chapter 8 (Figure 6.13).

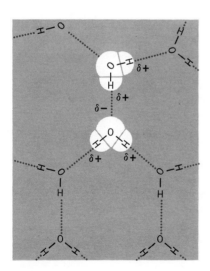

Figure 6.13   The hydrogen bonding in ice. The hydrogen bond is indicated by the colored dotted line.

Hydrogen bonds also determine the shapes of large biological molecules in living substances and, on a more familiar level, account for the hand-softening property of the lanolin in skin creams, for hard candy getting sticky, and for cotton fabrics taking longer to dry than synthetic fabrics.

Chemical Reactions

6.12 The Mole

We have seen that when sodium is placed in a container of chlorine, sodium atoms and chlorine atoms will undergo a chemical reaction to produce the compound sodium chloride. One sodium atom will react with each chlorine atom—or, more precisely, two sodium atoms will react with each chlorine molecule, Cl_2, to produce two sodium chloride ion groups. (Remember that ionic compounds do not exist as discrete molecules.) If a chemist wanted to produce a certain exact amount of sodium chloride, it would be impossible for him to measure out the sodium and chlorine atoms individually; atoms are too small to count out one at a time, even millions at a time. Therefore, a unit of measurement had to be devised that would allow us to measure out equal numbers of sodium and chlorine atoms. This unit is the **mole.** The mole is defined as the number of atoms in 12.0000 grams of carbon-12. Since atoms are so small, this number of atoms is very large—6.0238×10^{23} to be exact. This number is called **Avogadro's number,** after the nineteenth-century scientist Amadeo Avogadro who contributed a great deal to our knowledge of atomic weights. A mole, then, contains 6×10^{23} units; these units can be atoms, molecules, ions, electrons, or anything else. A mole of sodium atoms contains 6×10^{23} sodium atoms; a mole of chlorine molecules contains 6×10^{23} chlorine molecules; a mole of garden peas contains 6×10^{23} garden peas (enough to cover our planet and almost ten others like it with a layer of garden peas four feet deep*).

How much does one mole weigh? That depends upon the nature of the particles, just as a dozen lemons, grapefruit, or pumpkins each have a different total weight. **The atomic weight in grams of an element will contain one mole of atoms of that element.** For example, the atomic weight of sodium is 23; 23 grams of sodium will contain one mole of sodium atoms. Likewise, aluminum has an atomic weight of 27; therefore, 27 grams of aluminum will contain one mole, or 6×10^{23} atoms, of aluminum (Figure 6.14).

It has become accepted to use the term mole to refer not only to Avogadro's number of particles, but also to the number of grams of a substance that will contain 6×10^{23} particles. Therefore, we will refer to 23 grams of sodium as one mole of sodium, and will think of one mole of aluminum as 27 grams of aluminum.

6.13 Formula Weight

When atoms react to form compounds, their nuclei do not change; therefore, there is no net gain or loss of weight. The particle that forms, whether it be a molecule or ion group, will have a **formula weight** equal

---

* This calculation was made by Professors D. H. Andrews and R. J. Kokes of Johns Hopkins University.

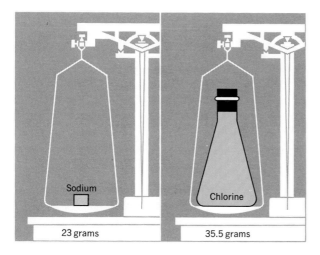

Figure 6.14 Although they do not weigh the same, the number of atoms in this sample of sodium equals the number of atoms in this sample of chlorine.

to the sum of the atomic weights of each atom in the formula of the compound formed.

For example, the formula weight for sodium chloride, NaCl, is 58.5 amu—the atomic weight of one sodium atom, 23 amu, plus the atomic weight of one chlorine atom, 35.5 amu. The formula weight of carbon tetrachloride, CCl_4, equals the sum of the atomic weight of one carbon atom plus the atomic weight of four chlorine atoms.

$$C + (4 \times Cl) \quad = CCl_4$$
$$12 + (4 \times 35.5) = 154$$

One mole of any substance will have a mass equal to the formula weight of that substance expressed in grams. For example, one mole of chlorine molecules, Cl_2, will weigh 2×35.5 or 71 grams. One mole of the compound sodium chloride, NaCl, will weigh $23 + 35.5$ or 58.5 grams. If a chemist weighs out 58.5 grams of NaCl, which is one mole of table salt, he knows that his table salt contains 6×10^{23} NaCl ion groups.

6.14 Solving Problems with the Mole

Problems involving mole calculations can be solved using the same techniques shown in Appendix 2 for the solution of problems involving the metric system. Before continuing with this chapter, it might be helpful for you to stop and reread Appendix 2.

Now let's do some examples of mole calculations:

Example 1:
You have a flask containing 9 grams of water. How many moles of water are in the flask?

Step 1: First you will need to establish the relationship between grams and moles of water.
Formula weight of $H_2O = (2 \times 1) + 16 = 18$. Therefore,

$$1 \text{ mole of water} = 18 \text{ grams}$$

Step 2: From the equality in Step 1, we are able to write two unit factors:

$$\frac{1 \text{ mole of water}}{18 \text{ grams}} \quad \text{or} \quad \frac{18 \text{ grams}}{1 \text{ mole of water}}$$

Step 3: Our problem asks:

$$9 \text{ grams} = (?) \text{ moles of water}$$

Therefore, for our answer to appear in the correct units we must use the first of the two unit factors shown in Step 2.

$$9 \text{ grams} \times \frac{1 \text{ mole of water}}{18 \text{ grams}} = \frac{9}{18} \text{ moles of water}$$

$$= \frac{1}{2} \text{ moles of water}$$

Example 2:
How many molecules of chlorine gas will there be in a tank containing 7.1 grams of chlorine?

Step 1: To solve this problem we need to establish a relationship between grams of chlorine and molecules of chlorine. We know that

$$1 \text{ mole of } Cl_2 = 71 \text{ grams}$$

and

$$1 \text{ mole of } Cl_2 = 6 \times 10^{23} \text{ molecules}$$

Therefore,

$$71 \text{ grams of } Cl_2 = 6 \times 10^{23} \text{ molecules}$$

Step 2: From the equality in Step 1, we can write two unit factors:

$$\frac{71 \text{ grams of } Cl_2}{6 \times 10^{23} \text{ molecules}} \quad \text{or} \quad \frac{6 \times 10^{23} \text{ molecules}}{71 \text{ grams of } Cl_2}$$

Step 3: Our problem asks the following question:

$$7.1 \text{ grams of } Cl_2 = (?) \text{ molecules}$$

In order that our answer appear in the correct units, we use the second of the two unit factors shown in Step 2.

$$7.1 \text{ grams of } Cl_2 \times \frac{6 \times 10^{23} \text{ molecules}}{71 \text{ grams of } Cl_2}$$

$$= \frac{7.1 \times 6 \times 10^{23} \text{ molecules}}{71}$$

$$= 0.1 \times 6 \times 10^{23} \text{ molecules}$$

$$= 6 \times 10^{22} \text{ molecules}$$

6.15 Chemical Equations

Chemical equations are a shorthand way of representing what occurs in a chemical reaction. A chemical equation contains the formulas of the starting materials, or **reactants,** separated by an arrow from the resulting materials, or **products.**

$$A + B \longrightarrow C + D$$
$$\text{Reactants} \qquad \text{Products}$$

Atoms are neither created nor destroyed in a chemical reaction, so the chemical equation representing the reaction must be **balanced.** That is, for each element involved in the reaction, the equation must show the same number of atoms on the reactant side as are found on the product side (Figure 6.15).

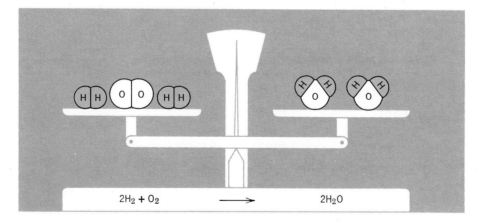

$$2H_2 + O_2 \qquad \longrightarrow \qquad 2H_2O$$

Figure 6.15 When an equation is balanced, the number of atoms of each element on the reactant side will equal the number of atoms of each element on the product side. Therefore, the total mass of the reactants will equal the total mass of the products.

Consider the reaction for the production of sodium chloride.

$$Na + Cl_2 \longrightarrow NaCl$$

This is not a balanced equation; there are two chlorine atoms on the reactant side of the equation and only one on the product side. Placing a coefficient of 2 in front of the formula for sodium chloride will balance the chlorine atoms in this equation.

$$Na + Cl_2 \longrightarrow 2NaCl$$

But now there are two sodium atoms on the product side, and only one on the reactant side. Placing a coefficient of 2 in front of the sodium on the reactant side will balance the equation.

$$2Na + Cl_2 \longrightarrow 2NaCl$$

Balancing equations is not really a complicated procedure. The following steps should help you quickly master this skill.

1. First, write the correct chemical formula for each reactant and each product. Once this is done you should never change the formula of a substance; only the coefficients can be changed. Any change of subscript would change the nature of the compound. For example, CO and $2CO$ represent one and two molecules of the compound carbon monoxide, respectively. But CO and CO_2 are entirely different substances, and you wouldn't want to confuse the two. Carbon monoxide is deadly, but carbon dioxide is produced within and exhaled from our bodies.

2. Since the balancing of equations involves juggling coefficients, it is often easiest to begin by assigning the compound having the most complicated formula the coefficient 1.

3. You must then start balancing each of the elements. You will often find it easiest if you leave oxygen to the last.

4. Treat complex ions as one unit (that is, just as if they were a single element) if they remain unchanged in the reaction.

5. When you think the equation is balanced, it is a good idea to check each element again to verify that it really is balanced.

The following examples illustrate these five steps for balancing equations. After you have read through the examples, try them yourself to see if you can get the same answer.

Example 1: _____
A small amount of the pollutant sulfur trioxide is released into the air when the sulfur in coal and petroleum is burned.

Step 1:

$$S + O_2 \longrightarrow SO_3$$

Steps 2, 3, 4: When the equation is written as above, the sulfurs balance, but not the oxygen. In order for the oxygen to balance, we must have the same number of atoms on both sides, and the lowest possible number is 6.

$$S + 3O_2 \longrightarrow 2SO_3$$

Now to balance the sulfur,

$$2S + 3O_2 \longrightarrow 2SO_3$$

Step 5: We now have two sulfur atoms on each side and six oxygen atoms on each side (Figure 6.16).

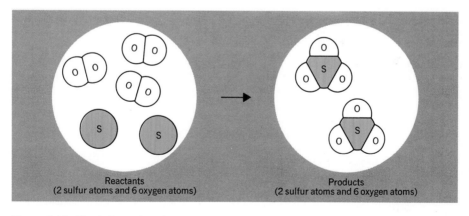

Reactants
(2 sulfur atoms and 6 oxygen atoms)

Products
(2 sulfur atoms and 6 oxygen atoms)

Figure 6.16 The balanced equation for the formation of sulfur trioxide is $2S + 3O_2 \longrightarrow 2SO_3$.

Example 2: _____
Reacting aluminum with sulfuric acid produces hydrogen gas and aluminum sulfate.

Step 1:

$$Al + H_2SO_4 \longrightarrow H_2 + Al_2(SO_4)_3$$

Steps 2, 3, 4: Start with one $Al_2(SO_4)_3$. First balance the aluminum.

$$2Al + H_2SO_4 \longrightarrow H_2 + Al_2(SO_4)_3$$

Next balance the sulfate, SO_4.

$$2Al + 3H_2SO_4 \longrightarrow H_2 + Al_2(SO_4)_3$$

Finally, balance the hydrogens.

$$2Al + 3H_2SO_4 \longrightarrow 3H_2 + Al_2(SO_4)_3$$

Step 5: There are now two atoms of aluminum on each side, six atoms of hydrogen on each side, and three sulfate ions on each side (Table 6.7).

Table 6.7   A Tally of Atoms for the Following Balanced Equation
$$2Al + 3H_2SO_4 \longrightarrow 3H_2 + Al_2(SO_4)_3$$

Symbol of Atom	Number of Atoms on the Reactant Side	Number of Atoms on the Product Side
Al	2	2
H	6	6
S	3	3
O	12	12

6.16 Calculations Using Balanced Equations

A balanced chemical equation contains a great deal of information for the chemist. Not only does it indicate the identity of the reactants and products, but it tells the relative number of these substances involved in the reaction. Consider the following reaction:

$$2H_2 + O_2 \longrightarrow 2H_2O$$

This equation can be read: "Two hydrogen molecules will react with one oxygen molecule to produce two water molecules," or on a scale with which we can work: "Two moles of hydrogen will react with one mole of oxygen to produce two moles of water."

Example 1: _____
Imagine that you are working in a laboratory and want to obtain 2.4 moles of magnesium chloride, a compound that is produced (along with water) when magnesium oxide is heated with hydrochloric acid, HCl. How many grams of each reactant would you need?

Step 1: First you need to write a balanced chemical equation for this reaction.

Hydrochloric acid + Magnesium oxide $\longrightarrow$ Magnesium chloride + Water

Unbalanced: $HCl + MgO \longrightarrow MgCl_2 + H_2O$

Balanced: $2HCl + MgO \longrightarrow MgCl_2 + H_2O$

We have now written a chemical equation representing the production of one mole of magnesium chloride.

Step 2: No matter how much magnesium chloride we might want to produce, the important fact to remember is that the proportion of reactants and products involved in the reaction will always remain the same. For example, we will always need two moles of hydrochloric acid for every mole of magnesium chloride produced, and will always obtain one mole of water for every mole of magnesium oxide that reacts. Using the coefficients of the balanced equation, we can write a series of reaction ratios expressing the proportions that will hold between any two of the reactants or products:

(a) $\dfrac{2 \text{ moles of HCl}}{1 \text{ mole of MgO}}$ or $\dfrac{1 \text{ mole of MgO}}{2 \text{ moles of HCl}}$

(b) $\dfrac{2 \text{ moles of HCl}}{1 \text{ mole of MgCl}_2}$ or $\dfrac{1 \text{ mole of MgCl}_2}{2 \text{ moles of HCl}}$

(c) $\dfrac{2 \text{ moles of HCl}}{1 \text{ mole of H}_2\text{O}}$ or $\dfrac{1 \text{ mole of H}_2\text{O}}{2 \text{ moles of HCl}}$

(d) $\dfrac{1 \text{ mole of MgO}}{1 \text{ mole of MgCl}_2}$ or $\dfrac{1 \text{ mole of MgCl}_2}{1 \text{ mole of MgO}}$

(e) $\dfrac{1 \text{ mole of MgO}}{1 \text{ mole of H}_2\text{O}}$ or $\dfrac{1 \text{ mole of H}_2\text{O}}{1 \text{ mole of MgO}}$

(f) $\dfrac{1 \text{ mole of MgCl}_2}{1 \text{ mole of H}_2\text{O}}$ or $\dfrac{1 \text{ mole of H}_2\text{O}}{1 \text{ mole of MgCl}_2}$

Step 3: Now we need to refer back to the problem to see which of the above reaction ratios will help us determine the correct answer. The problem asks how many grams of each reactant is needed to produce 2.4 moles of $MgCl_2$. Therefore we need to select reaction ratios that will show the relationships between the reactants HCl and MgO, and the product $MgCl_2$. We can then use these ratios to calculate the number of moles of each reactant that would be needed.

$$2.4 \text{ moles of MgCl}_2 \times \frac{2 \text{ moles of HCl}}{1 \text{ mole of MgCl}_2} = 4.8 \text{ moles of HCl}$$

$$2.4 \text{ moles of MgCl}_2 \times \frac{1 \text{ mole of MgO}}{1 \text{ mole of MgCl}_2} = 2.4 \text{ moles of MgO}$$

Step 4: We now know how many moles of each reactant are required, but we have been asked to express the answer in grams. Therefore, we must use the appropriate unit factors to convert moles of each reactant into grams.

(a) 4.8 moles of HCl = (?) grams.

$$1 \text{ mole of HCl} = 1 + 35.5 = 36.5 \text{ grams}$$

Therefore,

$$4.8 \text{ moles of HCl} \times \frac{36.5 \text{ grams}}{1 \text{ mole of HCl}} = 175.2 \text{ grams}$$

(b) 2.4 moles of MgO = (?) grams.

$$1 \text{ mole of MgO} = 24 + 16 = 40 \text{ grams}$$

Therefore,

$$2.4 \text{ moles of MgO} \times \frac{40 \text{ grams}}{1 \text{ mole of MgO}} = 96 \text{ grams}$$

So, to produce 2.4 moles of $MgCl_2$ we would need 175.2 grams of HCl and 96 grams of MgO.

Example 2: ─────────────────────────────────

Antacid tablets are taken by millions of Americans to reduce the discomfort of an upset stomach. The active ingredient in some commercial antacid tablets is magnesium hydroxide, $Mg(OH)_2$, which will react with stomach acid (HCl) to produce magnesium chloride ($MgCl_2$) and water. One popular tablet contains 0.1 grams of $Mg(OH)_2$. How many grams of stomach acid will this tablet neutralize?

Step 1: Write the balanced equation.

Unbalanced: $Mg(OH)_2 + HCl \longrightarrow MgCl_2 + H_2O$
Balanced: $Mg(OH)_2 + 2HCl \longrightarrow MgCl_2 + 2H_2O$

Step 2: Write the reaction ratios that will be needed to solve the problem. In this case we are interested in the relationship between $Mg(OH)_2$ and HCl.

$$\frac{1 \text{ mole of } Mg(OH)_2}{2 \text{ moles of HCl}} \quad \text{or} \quad \frac{2 \text{ moles of HCl}}{1 \text{ mole of } Mg(OH)_2}$$

Step 3: In order to use the above reaction ratios, we need to determine how many moles of $Mg(OH)_2$ are found in one tablet.

$$1 \text{ mole of } Mg(OH)_2 = 24 + [2 \times (16 + 1)] = 58 \text{ grams}$$

Therefore,

$$0.1 \text{ gram} \times \frac{1 \text{ mole of } Mg(OH)_2}{58 \text{ grams}} = \frac{0.1}{58} \text{ moles of } Mg(OH)_2$$

$$= 0.0017 \text{ moles of } Mg(OH)_2$$

Step 4: Using a reaction ratio from Step 2, we can now compute the number of moles of HCl that will be neutralized.

$$0.0017 \text{ moles of } Mg(OH)_2 \times \frac{2 \text{ moles of HCl}}{1 \text{ mole of } Mg(OH)_2} = 0.0034 \text{ moles of HCl}$$

Step 5: To solve the problem, then, all we need do is convert moles of HCl into grams of HCl.

$$1 \text{ mole of HCl} = 1 + 35.5 = 36.5 \text{ grams}$$

$$0.0034 \text{ moles of HCl} \times \frac{36.5 \text{ grams}}{1 \text{ mole of HCl}} = 0.124 \text{ grams}$$

Therefore, 0.1 grams of $Mg(OH)_2$ will neutralize 0.124 grams of HCl.

Additional Reading

Book

1. J. J. Lagowski, *The Chemical Bond*, Houghton Mifflin, Boston, 1966.

Articles

1. R. T. Sanderson, "What Is Bond Polarity and What Difference Does It Make?" *Chemistry*, September 1973, page 12.

2. "Shaping the Chemical Key to Unlock Hallucination," *New Scientist*, April 12, 1973, page 71.

Questions and Problems

1. Define the following terms:

    (a) Octet rule
    (b) Ionic bond
    (c) Ionic compound
    (d) Electrovalence
    (e) Covalent bond
    (f) Diatomic molecule
    (g) Double bond
    (h) Triple bond

    (i) Covalence number
    (j) Complex ion
    (k) Polar covalent bond
    (l) Electronegativity
    (m) Hydrogen bond
    (n) Mole
    (o) Avogadro's number
    (p) Formula weight

2. Write the formula for the following ionic compounds.

    (a) Lithium fluoride
    (b) Potassium sulfide
    (c) Magnesium bromide

    (d) Aluminum oxide
    (e) Copper(I) sulfide
    (f) Iron(II) chloride

3. State the name of the following ionic compounds.

   (a) AgCl
   (b) BaI_2
   (c) Fe_2S_3

4. State the difference between a single, double, and triple covalent bond. Give examples of each.

5. Draw the electron dot diagrams for the following molecules.

   (a) Hydrogen iodide, HI
   (b) Fluorine, F_2
   (c) Chloromethane, CH_3Cl (a carbon bonded to three hydrogens and one chlorine)
   (d) Hydrogen cyanide, HCN
   (e) Oxygen difluoride, OF_2 (an oxygen bonded to two fluorines)

6. Draw the bond diagram for each of the following compounds.

   (a) Ammonia, NH_3
   (b) Carbon tetrachloride, CCl_4
   (c) Hydrogen chloride, HCl

7. Write the formulas for the following compounds.

   (a) Sodium bicarbonate
   (b) Ammonium phosphate
   (c) Potassium sulfate
   (d) Aluminum carbonate
   (e) Potassium permanganate
   (f) Magnesium nitrite

8. Predict whether the bonds formed between the following atoms will be polar or nonpolar. If the bond is polar, indicate which is the more electronegative atom.

   (a) Cl and Cl
   (b) H and Br
   (c) N and O
   (d) P and O

9. Predict whether the following molecules will be polar or nonpolar.

   (a) Oxygen difluoride, OF_2
   (b) Fluorine, F_2
   (c) Hydrogen iodide, HI
   (d) Methane, CH_4
   (e) Chloromethane, CH_3Cl

10. Would you expect hydrogen bonds to form between molecules of hydrogen fluoride, HF? Why or why not?

11. Calculate the formula weight of the following substances.

    (a) Nitrogen, N_2
    (b) Methane, CH_4
    (c) Sodium phosphate, Na_3PO_4
    (d) Aluminum hydroxide, $Al(OH)_3$

12. What is the weight of 0.5 moles of each of the substances in question 11?

13. How many moles of nitrogen would you have if you had the following masses of nitrogen?

(a) 2.8 g   (c) 0.14 g
(b) 56 g   (d) 140 g

14. Calculate the weight of one mole of the following substances.

(a) Aluminum, Al
(b) Oxygen, O_2
(c) Sulfuric acid, H_2SO_4

15. How much would each of the following weigh?

(a) 0.5 moles of aluminum
(b) 2 moles of oxygen
(c) 3×10^{23} molecules of sulfuric acid

16. Write a balanced equation for the following reactions.

(a) Hydrogen reacts with bromine to form hydrogen bromide.
(b) Silver nitrate + copper reacts to form copper(II) nitrate + silver.
(c) Hydrogen + nitrogen reacts to form ammonia.
(d) Methane + chlorine reacts to form carbon tetrachloride (CCl_4) + hydrogen chloride

17. Calculate the number of moles in each of the following.

(a) 36 grams of water
(b) 36×10^{23} molecules of oxygen
(c) 0.98 g of sulfuric acid

18. Balance the following equations.

(a) $Na + H_2O \longrightarrow NaOH + H_2$
(b) $KClO_3 \longrightarrow KCl + O_2$
(c) $MnO_2 + HCl \longrightarrow Cl_2 + MnCl_2 + H_2O$
(d) $C_3H_8 + O_2 \longrightarrow CO_2 + H_2O$
(e) $NH_3 + O_2 \longrightarrow NO + H_2O$

19. The pollutant sulfur trioxide is removed from the atmosphere by rain, which reacts with the SO_3 to form sulfuric acid. Along major freeways, rain water has been found to be quite acidic and, as a result, is damaging to plant life. For every kilogram of SO_3 that reacts with rain, how many grams of sulfuric acid will be formed?

20. Sodium lauryl sulfate, a detergent, can be prepared by the following reactions:

1. $C_{12}H_{25}OH$ + H_2SO_4 $\longrightarrow$ $C_{12}H_{25}OSO_3H + H_2O$
   lauryl alcohol
2. $C_{12}H_{25}OSO_3H$ + NaOH $\longrightarrow$ $C_{12}H_{25}OSO_3Na$ + H_2O
   Detergent

If a day's production of detergent is 11 tons, how many tons of lauryl alcohol would be required?

chapter 7

Reaction Rates and Chemical Equilibrium

Learning Objectives

By the time you have finished this chapter, you should be able to:

1. Define "activation energy."

2. Draw a potential energy diagram for an endothermic and exothermic reaction.

3. State four factors that will affect the rate of a chemical reaction.

4. Define "catalyst."

5. Define "chemical equilibrium," and give two examples.

6. Given the equilibrium constant for a reaction, predict whether reactants or products will predominate at equilibrium.

7. State two ways in which a chemical equilibrium can be disrupted.

8. State Le Chatelier's principle, and predict the changes that will occur in an equilibrium when a stress is applied.

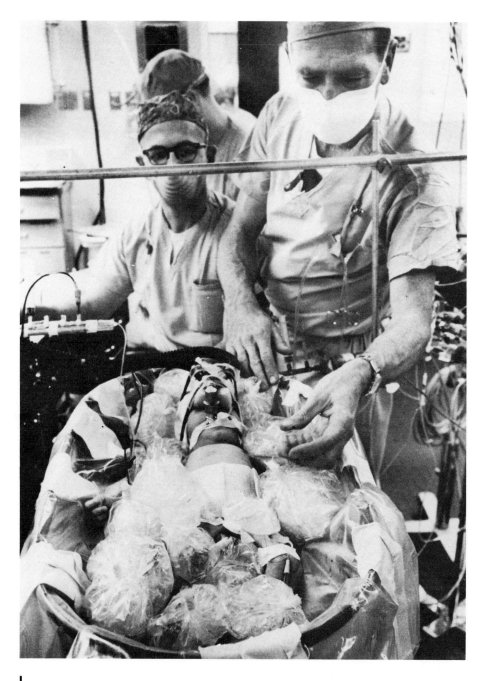

In the delivery room of a large Boston hospital, a baby is born to Jan and Tom Harper. The baby, a boy named Tim, is found to be suffering from a fatal heart defect. In order to survive, Tim must have surgery soon.

His chances for survival under conventional open-heart surgery techniques are small. The heart-lung machine, which would ordinarily be used to maintain blood circulation during such an operation, is risky to use on a child so young; the machine damages red blood cells, and the tubes and other equipment that would have to fit into such a small chest cavity would make it difficult for the surgeon to see what he was doing. But Tim's chances for life are still good, thanks to a new surgical technique called deep hypothermia.

In Chapter 2 we saw that lowering the body temperature could result in death. But doctors now use hypothermia under controlled conditions to save human lives. Just 36 hours old, Tim is taken to surgery and is anesthetized. His body is covered with a plastic blanket, which is then covered with cracked ice. When his body temperature has dropped to 77°F (25°C), the surgeon opens Tim's chest and hooks him up to a heart-lung machine, which pumps cooled blood through Tim's body until his temperature is lowered to 68°F (20°C). At this point the pump is turned off. Tim's heart is motionless, and his blood circulation has stopped. The surgeons then begin the delicate process of rearranging the blood vessels between the heart and the lungs. Less than an hour later, the task is complete. The pump is turned on, now pumping warmed blood through Tim's body. As his temperature increases, his heart again begins to pump blood through the repaired vessels. Tim recovered quickly from the surgery, and has since grown to be a very healthy and active toddler.

But how can the body, especially the brain, survive without any blood circulating for an hour? Under normal conditions the brain undergoes irreparable damage if it is cut off from its oxygen supply for as little as three minutes. However, by lowering the body temperature, doctors are able to slow down the chemical reactions occurring within the brain cells. Under controlled conditions, these chemical processes can be slowed to a point where blood circulation can be stopped for up to an hour.

Controlling the rate of chemical reactions can, therefore, mean the difference between life and death to a young child. It can also mean the difference between a nuclear holocaust and the peaceful use of nuclear power to generate electricity. In order to understand how scientists control chemical reactions for our benefit, it is important to study the various factors that influence the rate at which chemical reactions occur.

Rates of Chemical Reactions

7.1 Activation Energy

For a reaction to occur between two particles, they must be brought close enough together for their outer shell electrons to interact. In fact, they must collide. Not only must they collide, however, but they must do so with sufficient energy to overcome the repelling forces set up between the electrons surrounding the two nuclei. The amount of energy necessary for a successful collision is called the **activation energy, E_a.** In order for a reaction to occur, molecules must collide with energy at least equal to the activation energy. It may help to picture the activation energy as a hill the particles must get over in order to complete the reaction. Each reaction is associated with a hill of a different height, or different activation energy. Imagine a person trying to roll a bowling ball up such a hill: in most cases the bowling ball will roll only part way up the hill and then roll down again. This is true of a reaction—most collisions between molecules occur with insufficient energy to overcome the activation barrier, and the molecules bounce off each other unreacted. Just as it will be only occasionally that the bowler can give the ball enough energy to get up over the hill, so it is only on occasion that the particles in a reaction collide with sufficient energy to overcome the activation energy barrier and form products (Figure 7.1).

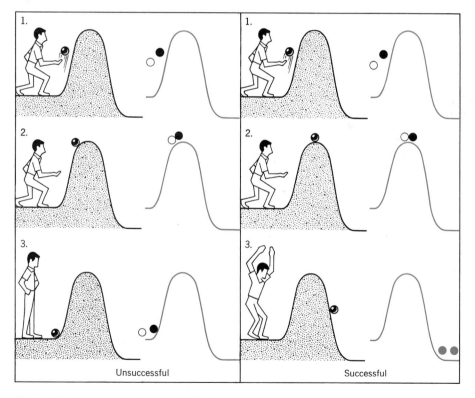

Figure 7.1 Just as a bowling ball will roll over the hill only when it has been given enough energy, so a reaction will occur between two molecules only when they collide with sufficient energy to get over the activation energy barrier.

You know that a candle will not spontaneously burst into flame at room temperature, but you may wonder what allows it to keep burning once it is lit. The reaction between candle wax and oxygen has a high activation energy, and very few molecules have enough energy at room temperature to react. The heat of a match is required if a significant number of molecules are to overcome the activation energy barrier and cause the wax to burn. Now, the reaction between candle wax and oxygen is **exothermic,** meaning that the products of the reaction have less potential energy (energy stored in the chemical bonds) than the reactants. Such reactions result in the release of energy as the reaction occurs. Therefore, once these molecules begin to react, this reaction releases enough energy to boost additional molecules over the activation energy barrier. Hence the reaction becomes self-sustaining, and the candle continues to burn (Figure 7.2).

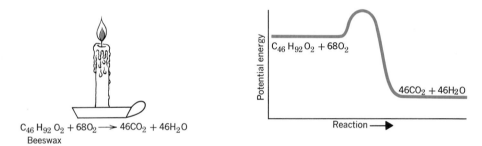

$$C_{46}H_{92}O_2 + 68O_2 \longrightarrow 46CO_2 + 46H_2O$$
Beeswax

Figure 7.2 Burning candle wax is an exothermic reaction in which the products have less potential energy than the reactants.

Some highly exothermic reactions can be explosive. Dynamite, for example, must be set off with a percussion cap. This very small explosion releases enough energy for a small number of dynamite molecules to react, and the energy released by these molecules is enough to cause the remainder of the molecules to simultaneously react, releasing a tremendous amount of energy.

Other reactions such as the decomposition of water into hydrogen and oxygen, or the formation of sugar molecules in the process of photosynthesis require the application of energy for the reaction to occur. Reactions requiring a constant input of energy are called **endothermic.** In such reactions the potential energy of the products is greater than that of the reactants; energy has been absorbed. In endothermic reactions, reactants require an initial input of energy to get over the activation energy barrier, and then a continuing supply of energy to keep the reaction going. Stop the supply of electricity, and the decomposition of water will come to a halt; keep a green plant in the dark, and photosynthesis will cease, causing the plant to die (Figures 7.3 and 7.4).

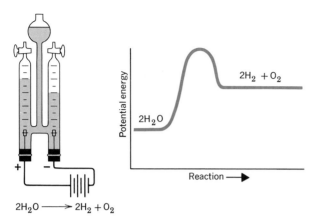

$$2H_2O \longrightarrow 2H_2 + O_2$$

Figure 7.3 Electrolysis (the breaking apart of water molecules with electrical energy) is an endothermic reaction in which the products, hydrogen and oxygen, have more potential energy than the reactant, water.

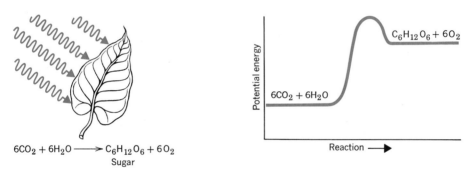

$$6CO_2 + 6H_2O \longrightarrow C_6H_{12}O_6 + 6O_2$$
$$\text{Sugar}$$

Figure 7.4 Photosynthesis is an endothermic process requiring a constant supply of energy from the sun.

Figure 7.5 contains potential energy diagrams for an exothermic and an endothermic reaction. The energy required or released is called the **heat of reaction, ΔH,** and appears on these diagrams as the difference in height between the potential energy of the products and the potential energy of the reactants. Each reaction has its own characteristic activation energy and heat of reaction.

Factors Affecting Reaction Rates

7.2 The Nature of the Reactants
The nature of the reactants can influence the rate of the chemical reaction.

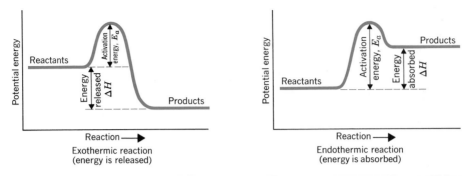

Figure 7.5 Potential energy diagrams for an exothermic and endothermic reaction. E_a = activation energy. ΔH = heat of reaction, or energy released or absorbed.

For example, when colorless nitric oxide escapes from a test tube into the air, reddish-brown nitrogen dioxide forms very quickly.

$$2NO + O_2 \longrightarrow 2NO_2 \qquad \text{Fast reaction at 25°C}$$

But when carbon monoxide from automobile exhaust fumes is released into the air, the reaction with oxygen to form carbon dioxide is quite slow, much to the detriment of inhabitants of large urban areas.

$$2CO + O_2 \longrightarrow 2CO_2 \qquad \text{Very slow reaction at 25°C}$$

These two balanced equations look quite a bit alike, but the large difference in their rates of reaction stems from the nature of the carbon monoxide and nitric oxide molecules themselves.

7.3 The Concentration of Reactants

The concentration of reactants can greatly influence the rate of a chemical reaction. We have seen that particles must collide in order for them to react, and it makes sense that the more reactant particles there are in a given space, the more collisions will occur. (We see the same phenomenon on our highways; the more crowded the highways, the higher the probability of a fatal collision. This is especially clear when you consider the high death tolls on holiday weekends.) Increasing the concentration of the reactants will increase the number of collisions and, therefore, will increase the reaction rate (Figure 7.6).

Physicians make use of this principle when prescribing medicines to treat specific diseases. The appropriate treatment for tonsillitus or pneumonia may involve a dose of 250 mg of penicillin, but meningitis would call for two to five times that dosage. This higher concentration of penicillin increases the rate of absorption into the blood stream and, therefore, increases the effective concentration in the blood.

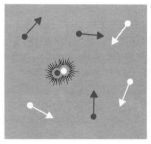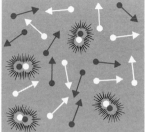

Figure 7.6 Increasing the concentration of the reactants will increase the frequency of collisions and thereby increase the rate of the reaction.

7.4 The Surface Area of a Solid Reactant

Increasing the surface area of a solid reactant will increase the rate of reaction. An increase in the surface area increases the number of solid particles exposed, thereby increasing the number of collisions possible between the reactants. Increasing the surface area can sometimes increase the rate of reaction to explosive levels. For example, lumber mills seldom worry about their log piles spontaneously catching fire, but sawdust piles can burst into flame and must, therefore, be kept wet. You probably don't think of flour as being potentially dangerous, but finely divided flour dust can explode. Because of its increased surface area, a crushed aspirin will get to your headache faster than a whole tablet.

7.5 The Temperature of the Reaction

Increasing the temperature of the reactants will increase the kinetic energy of the particles. This will not only increase the frequency of collisions, but will also increase the likelihood that the colliding particles will have enough energy to overcome the activation energy barrier. (This is similar to the intuitive notion that increasing speeds on the highway will increase the likelihood that any collision will be a fatal one.) (See Figure 7.7.)

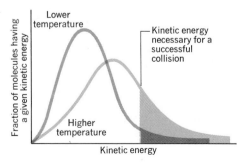

Figure 7.7 Increasing the temperature of a reaction increases the average kinetic energy of the molecules, increasing the number of molecules with enough energy to collide successfully and produce products.

Changes in temperature have profound effects on living organisms. A fever increases the rate of chemical reactions in the body, as is evidenced by the increased pulse rate, increased breathing rate, and abnormalities in digestive and nervous systems. When a person has a fever, his basal metabolism rate goes up by 5% for each degree rise in body temperature. The effects on living organisms from an increase in external temperatures is illustrated by the release of warm water from power plants into streams and lakes, causing an increase in the metabolic rate in the aquatic life. In order to support their new higher rate of metabolism, these organisms require more oxygen. But the warming of the water simultaneously decreases the concentration of oxygen in the water, contributing to the death of fish in these streams and lakes. Another detrimental effect of these higher water temperatures is the resulting increase in the sensitivity of fish to pollutants in the water (Figure 7.8).

Figure 7.8 This fish kill resulted when warm waste water from a nuclear power plant was discharged into a nearby stream. (Wide World)

Decreasing the body's temperature slows down bodily reaction rates. During open heart surgery, the patient's body temperature is usually lowered four to five degrees Fahrenheit to decrease his metabolism rate and his oxygen requirements. At the beginning of this chapter we saw an extreme example of the decrease in bodily reactions from a lowering of body temperatures. As another example, a trainer might spray a surface anesthetic called ethyl chloride on the injured knee of a basketball player to relieve the pain. The rapid evaporation of the ethyl chloride will lower the temperature of the tissue enough to block the reactions responsible for the transmission of nerve impulses.

Many animals hibernate during the cold winter months. Their body temperatures fall to a few degrees above freezing, and all the chemical reactions in their bodies slow down. Their breathing and heart rate become very slow, and much less energy is needed to sustain life. Therefore, the animal can live on stored body fat alone. To illustrate the extent of this slowdown, consider that a woodchuck's heart beats about 80 times a minute while he is active, but when hibernating his heart will beat only four times a minute (Figure 7.9). The intriguing notion of keeping human beings

Figure 7.9 The body temperature of this hibernating woodchuck is only a few degrees above freezing; thus all chemical reactions in its body have slowed down. This means much less energy is used to sustain life, allowing the animal to survive the winter on stored body fat alone. (Leonard Lee Rue/ Photo Researchers)

in suspended animation by lowering their body temperatures has been the subject of many science fiction stories. In 1974 two scientists from the Darwin Research Institute discovered that bacteria which had been frozen in the cold rock of Antarctica 10,000 to 1 million years ago could be revived in the warmth of the laboratory and, incredibly, could even reproduce.

7.6 Catalysts

Many reactions that proceed slowly can be made to take place at a more rapid rate by the introduction of substances called catalysts. A **catalyst** is a substance that increases the rate of a chemical reaction without being consumed in the reaction. The effect of the catalyst is to lower the activation energy required for the reaction. In our illustration of the man and the bowling ball, the action of a catalyst would resemble that of a bulldozer cutting a much lower path over the hill, allowing many more balls rolled by the bowler to reach the top and roll down the other side (Figure 7.10).

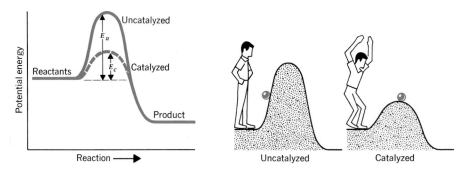

Figure 7.10 A catalyst lowers the activation energy, increasing the chances for a successful collision.

Industry makes wide use of catalysts, especially those that allow companies to produce significant amounts of a product at a lower temperature, thereby saving on energy costs. An example of this occurs in the production of sulfuric acid, H_2SO_4, the so-called "king of chemicals." Over 20 million tons of sulfuric acid are used each year in the United States, mostly by the steel, fertilizer, and petroleum industries. One step in the production of sulfuric acid involves the production of sulfur trioxide from sulfur dioxide and oxygen. This reaction has a very high activation energy, and is quite slow even at high temperatures. But the reaction has been made economically feasible by the introduction of a catalyst, such as finely divided platinum, which greatly increases the reaction rate (Figure 7.11). Similarly, the cost of producing gasoline from much larger molecules of crude petroleum is kept low through the use of catalysts in the cracking process—the process by which large molecules are broken into small fractions.

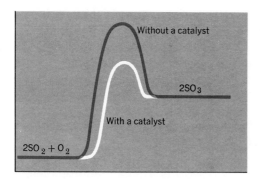

Figure 7.11 The production of sulfuric acid involves a reaction between sulfur dioxide and oxygen which has a very high activation energy. The addition of platinum as a catalyst makes the process economically feasible.

Most of the chemical reactions that occur in our bodies would not ordinarily occur at a significant rate at body temperature. But the body produces special compounds called **enzymes** that act as catalysts, permitting these reactions to occur readily. The removal or damaging of body enzymes can have disastrous effects. For example, the poison cyanide prevents enzymes from catalyzing reactions essential to the life of the cell, thereby causing the death of the organism. "Everybody needs milk" does not apply to nearly 90% of non-Caucasian adults, whose digestive tracts lack the enzyme that breaks down lactose, the sugar found in milk. For these individuals, drinking milk results in nausea, gas, and diarrhea.

Chemical Equilibrium

7.7 What Is Chemical Equilibrium?

At one time or another you may have added sugar to a glass of iced tea, stirred, and watched the sugar dissolve. You may have noticed that additional sugar added to this solution continues to dissolve until you reach a certain point, after which any more sugar added just settles to the bottom of the glass. Although you may continue stirring, or might let the glass sit for a period of time, neither the amount of sugar on the bottom of the glass nor the amount of sugar dissolved in the iced tea will change.

Under these circumstances the iced tea now contains all the sugar molecules it can possibly hold, and we will see that the molecules of sugar dissolved in the iced tea and the molecules of sugar in the crystals sitting on the bottom of the glass are in equilibrium. A molecular equilibrium such as this is not a static condition such as the equilibrium reached when two children are equally balanced on a seesaw. Rather, **a chemical equilibrium is a dynamic state in which events are constantly occurring, but at exactly equal rates.** In our glass of iced tea, for example, new sugar molecules are constantly dissolving and other sugar molecules are constantly crystallizing out of solution, but the rate of each of these processes is the same. Therefore, there is no overall change in the number of sugar molecules dissolved in the iced tea or lying on the bottom of the glass. In other words, for every sugar molecule that dissolves, one sugar molecule crystallizes out of solution (Figure 7.12).

Let's examine this iced tea example in chemical terms. We said that when we first add sugar to the iced tea, it dissolves. We can represent this process by the following equation.

$$\text{Sugar}_{(s)} \longrightarrow \text{Sugar}_{(aq)}$$

$(s) =$ solid
$(aq) =$ aqueous, dissolved in water

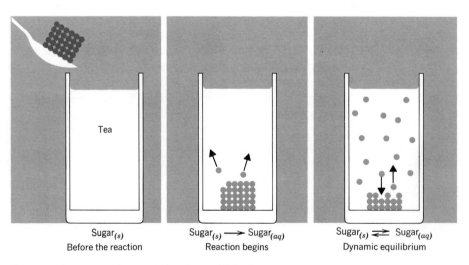

| Sugar$_{(s)}$ | Sugar$_{(s)} \longrightarrow$ Sugar$_{(aq)}$ | Sugar$_{(s)} \rightleftharpoons$ Sugar$_{(aq)}$ |
| Before the reaction | Reaction begins | Dynamic equilibrium |

Figure 7.12 A dynamic equilibrium is established when the rate of the forward reaction equals the rate of the reverse reaction.

As more and more sugar molecules are added to the iced tea, the number dissolved in the tea will increase to the point that the reverse reaction (that is, the crystallization of sugar from the tea) begins to occur.

$$\text{Sugar}_{(s)} \rightleftharpoons \text{Sugar}_{(aq)}$$

Notice that in this equation we use two arrows pointing in different directions to indicate that this reaction is reversible. As the number of sugar molecules in the tea continues to increase, the rate of the reverse reaction will increase until the rate of the reverse reaction equals the rate of the forward reaction. If we denote the rate of the forward reaction by $rate_1$ and the rate of the reverse reaction by $rate_2$, we then have $rate_1 = rate_2$.

$$Sugar_{(s)} \underset{Rate_2}{\overset{Rate_1}{\rightleftarrows}} Sugar_{(aq)}$$

At this point, a state of equilibrium has been reached, and no net change will occur in our glass of sugared iced tea. But it is again important to emphasize that the forward reaction and reverse reaction still continue to occur, only at exactly equal rates. We can define **chemical equilibrium,** then, as **a dynamic state in which the rate of the forward reaction equals the rate of the reverse reaction.**

7.8 The Equilibrium Constant

We often need to discuss chemical equilibrium in precise terms and, therefore, need an indication of how far a reaction proceeds before equilibrium is established. Consider the following representation of a general reaction:

$$A + B \rightleftharpoons C + D$$

A principle of reaction kinetics called the Law of Mass Action states that the rate of the forward reaction can be determined by the following equation:

$$Rate_1 \quad \propto \quad [A] \times [B]$$

Therefore,

$$Rate_1 = k_1 \times [A] \times [B]$$

The notation $[A]$ indicates the concentration of reactant A in moles per liter, $[B]$ indicates the concentration of reactant B in moles per liter, and k_1 is a rate constant. The rate of the reverse reaction can be similarly expressed:

$$Rate_2 = k_2 \times [C] \times [D]$$

Since at equilibrium we will have $rate_1 = rate_2$, at equilibrium we can write the equation

$$k_1 \times [A] \times [B] = k_2 \times [C] \times [D]$$

If we solve this equation for the equilibrium constants, we get the following:

$$\frac{k_1}{k_2} = \frac{[C] \times [D]}{[A] \times [B]} = \frac{[C][D]}{[A][B]}$$

We can then define a new constant, $K_{eq} = k_1/k_2$; therefore, at equilibrium we now have the relationship

$$K_{eq} = \frac{[C][D]}{[A][B]}$$

We call K_{eq} the **equilibrium constant** for that reaction. The value of K_{eq} for any reaction will be a constant at a given temperature; varying the temperature of the system will change the value of K_{eq} (Table 7.1).

Table 7.1 Some Equilibrium Constants

Reaction	K_{eq}	Value of K_{eq} at Specific Temperatures
$HSO_4^- \rightleftharpoons H^+ + SO_4^{2-}$	$K = \dfrac{[H^+][SO_4^{2-}]}{[HSO_4^-]}$	1.3×10^{-2} at 25°C
$CH_3COOH \rightleftharpoons H^+ + CH_3COO^-$	$K = \dfrac{[H^+][CH_3COO^-]}{[CH_3COOH]}$	1.8×10^{-5} at 25°C
$N_2O_{4(g)} \rightleftharpoons 2NO_{2(g)}$	$K = \dfrac{[NO_2]^2}{[N_2O_4]}$	8.3×10^{-1} at 55°C
$Ag^+ + 2NH_3 \rightleftharpoons Ag(NH_3)_2^+$	$K = \dfrac{[Ag(NH_3)_2^+]}{[Ag^+][NH_3]^2}$	1.7×10^7 at 25°C

Note. In the equilibrium constant expression, the coefficient becomes the power to which the concentration is raised. In general, $aA \rightleftharpoons bB$

$$K_{eq} = \frac{[B]^b}{[A]^a}$$

The numerical value of K_{eq} yields valuable information about the reaction. If the value of $K_{eq} > 1$, then $[C][D]$ will be greater than $[A][B]$ at equilibrium. If K_{eq} is a very large number, this indicates that the forward reaction goes almost to completion before the equilibrium is established. If $K_{eq} < 1$, then $[A][B]$ is greater than $[C][D]$ at equilibrium, so the forward reaction does not proceed very far before equilibrium is established.

7.9 Altering the Equilibrium

Altering a chemical equilibrium is an important technique in industrial or biological chemistry, especially when it is desirable for a reaction having a $K_{eq} < 1$ to go to completion. Various factors affecting the rate of a chemical reaction can alter a chemical equilibrium. Let's review each of the factors we discussed previously as causing a change in the rate of chemical reactions.

7.10 Changes in Concentration

Changing the concentration of a reactant or product in a reaction at equilibrium can affect the rate of the forward or reverse reaction. Increasing the concentration of a reactant increases the number of collisions between reactant molecules, thereby increasing the rate of the forward reaction. When this occurs, the system is no longer in equilibrium:

$$Rate_1 > Rate_2$$

The forward reaction will proceed at a higher rate, producing more product molecules until the rate of the reverse reaction again increases enough to equal the rate of the forward reaction.

One colorful example of the effect on a chemical equilibrium of changing reactant concentrations can be seen in the equilibrium between the yellow chromate ion, CrO_4^{2-}, and the orange dichromate ion, $Cr_2O_7^{2-}$.

$$2CrO_4^{2-}{}_{(aq)} + 2H^+{}_{(aq)} \underset{Rate_2}{\overset{Rate_1}{\rightleftharpoons}} Cr_2O_7^{2-}{}_{(aq)} + H_2O$$

Yellow   Orange

Increasing the concentration of hydrogen ions on the reactant side of the equation will increase the number of collisions between the chromate ions and the hydrogen ions, causing an increase in the rate of the forward reaction. As more dichromate forms, the rate of the reverse reaction also increases until it again equals the rate of the forward reaction. At that point a new equilibrium will be established. In this new equilibrium there will be a higher concentration of dichromate ions and a lower concentration of chromate ions than at the previous equilibrium. Therefore the solution will appear more orange (Figure 7.13). It is a common procedure for chemists to add more reactant, or to remove a product in a reaction, in order to alter a chemical equilibrium and to drive a reaction to completion.

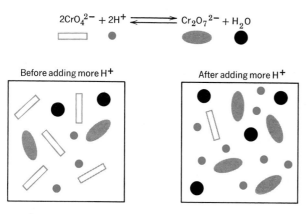

$$2CrO_4^{2-} + 2H^+ \rightleftharpoons Cr_2O_7^{2-} + H_2O$$

Before adding more H^+ After adding more H^+

Figure 7.13 A molecular view of the chromate-dichromate equilibrium.

In the plasma of the blood, an equilibrium exists between carbon dioxide and carbonic acid.

$$CO_2 \quad + \quad H_2O \underset{Rate_2}{\overset{Rate_1}{\rightleftharpoons}} H_2CO_3$$

Carbon dioxide   Carbonic acid

In the tissues, carbon dioxide is a waste product that enters the blood. As the concentration of carbon dioxide in the blood increases, the above reaction is driven to the right (rate$_1$ > rate$_2$), and carbonic acid, which can be transported by the blood stream, is formed. When the blood reaches the lungs, we exhale carbon dioxide. This lowers the concentration of carbon dioxide in the blood, driving the reaction to the left and allowing us to rid the blood stream of the waste product carbonic acid (Figure 7.14).

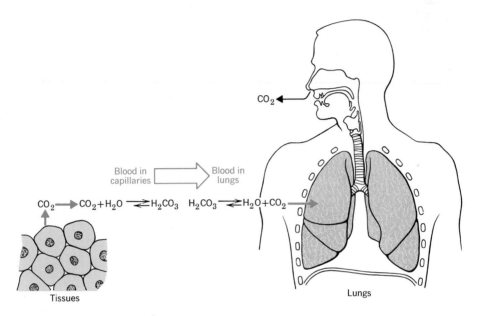

Figure 7.14   Changes in the levels of carbon dioxide in the body affect the carbon dioxide/carbonic acid equilibrium

$$CO_2 + H_2O \rightleftarrows H_2CO_3$$

Carbon dioxide is produced in the tissues, driving the reaction toward the formation of carbonic acid. This substance is carried in the blood to the lungs, where the constant removal of carbon dioxide in exhaled breath drives the reaction toward the formation of more carbon dioxide.

7.11 Changes in Temperature

Changing the temperature of a system in equilibrium will alter the equilibrium. Increasing the temperature will favor the endothermic reaction (the reaction requiring energy), and lowering the temperature will favor the exothermic reaction (the reaction in which energy is released). Let's return to our example of iced tea. The process of dissolving is an endothermic reaction. Therefore, increasing the temperature will increase the number of sugar molecules that will dissolve.

$$Sugar_{(s)} + Energy \underset{Rate_2}{\overset{Rate_1}{\rightleftarrows}} Sugar_{(aq)}$$

Warming our tea will increase rate$_1$, since that is the endothermic reaction. Therefore we will have rate$_1$ > rate$_2$, and more sugar will be dissolving than is crystallizing. If we raise the temperature and hold it constant, rate$_2$ will increase until it equals rate$_1$, establishing a new equilibrium with a greater number of sugar molecules dissolved in the tea (Figure 7.15). Obviously, if you like your tea sweet you should drink it warm!

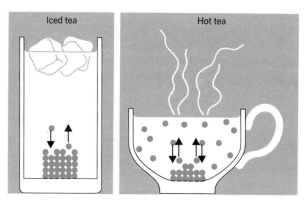

$$Sugar_{(s)} + Heat \rightleftarrows Sugar_{(aq)}$$

Figure 7.15 Increasing the temperature of a system at equilibrium will increase the rate of the endothermic reaction. As the temperature of the iced tea is increased, more sugar will dissolve.

7.12 Catalysts

We have discussed the fact that a catalyst will increase the rate of a chemical reaction, but you may be surprised to learn that a catalyst will have no effect on the equilibrium concentrations of the reactants and products in an equilibrium system. Why is this? Remember that the way in which a catalyst increases the rate of a chemical reaction is by lowering the activation energy of the reaction (Figure 7.10). At the same time, however, the activation energy for the reverse reaction will be lowered by an equal amount, so the rate of the reverse reaction will also be increased. A catalyst, then, increases the rate of both the forward and the reverse reaction by the same amount and, therefore, will have no effect on the equilibrium system.

7.13 Le Chatelier's Principle

We have seen that a chemical system is in equilibrium when the rates of the forward and reverse reactions are equal, and that changes made in the system can disrupt the equilibrium and thereby alter the equilibrium concentrations of reactants and products. After studying a great number of equilibrium systems, the French chemist Le Chatelier was able to formulate a principle that helps to predict whether the reactants or products will be favored in a given change. **Le Chatelier's principle states that when a stress is placed on a system in equilibrium, the system will change in a direction that will tend to remove the stress.**

When we talk of applying a stress to an equilibrium, we are referring to such actions as changing the concentrations of the reactants or products, changing the temperature of the system, or changing the pressure in a gaseous system. Le Chatelier's principle tells us that if we increase the concentration of a reactant, the system will move in a direction to remove that increase; that is, $rate_1$ will be greater than $rate_2$. If we increase the temperature, the reaction which uses up that increase in energy will be favored; that is, the rate of the endothermic reaction will be greater than the rate of the exothermic reaction.

Example:_____

We might consider the effect of various stresses on the following reaction when it is at equilibrium.

$$2CO_{(g)} + O_{2(g)} \rightleftarrows 2CO_{2(g)} + \text{Heat}$$

$(g) = \text{gas}$

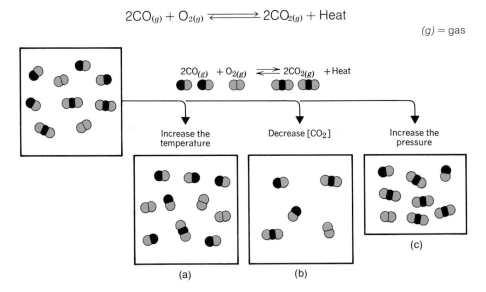

Increase the temperature

Decrease $[CO_2]$

Increase the pressure

(a) (b) (c)

a. *Increasing the temperature.* The endothermic reaction, which is the reverse reaction, would be favored. Therefore, $rate_2$ will be greater than $rate_1$.

b. *Decreasing the concentration of CO_2.* Since the stress is a decrease in the amount of product CO_2, the forward reaction will be favored so as to counteract that change. Another way of reaching this conclusion is to realize that there will be fewer CO_2 molecules to react, so $rate_2$ will decrease.

c. *Increasing the pressure.* The reaction that will be favored is the one which produces a system that exerts less pressure; that is, a system with fewer molecules. The forward reaction produces a system with two molecules, while the reverse reaction produces a system with three molecules. Therefore, to remove the stress, $rate_1$ will be greater than $rate_2$.

We can very conveniently use Le Chatelier's principle to predict the changes that will occur in an equilibrium system when a stress is applied. All that we need to remember is that whenever a change occurs in an equilibrium system, the system will move so as to counteract that change.

Additional Reading

Articles

1. V. Haensel and R. L. Burwell, Jr., "Catalysis," *Scientific American,* December 1971, page 46.

2. D. L. Morris, "Stress, Collisions, and Constants," *Chemistry,* April 1971, page 10.

3. _____, "The Carbonates: An Entertainment in Equilibrium," *Chemistry,* July–August 1974, page 6.

4. G. Porter, "Equilibrium," *Chemistry,* September 1968, page 16.

Questions and Problems

1. Define the following terms:

   (a) Activation energy
   (b) Exothermic reaction
   (c) Endothermic reaction
   (d) Catalyst

   (e) Enzyme
   (f) Chemical equilibrium
   (g) Equilibrium constant
   (h) Le Chatelier's principle

2. Using the concept of activation energy, explain why a dropped glass might remain intact on one occasion and shatter on another.

3. The heat of reaction for the conversion of graphite to diamond is quite small, yet this reaction will take place only under extremely high temperatures and pressure. Why is this? Draw a potential energy diagram for the reaction

$$0.45 \text{ kcal} + C_{(graphite)} \longrightarrow C_{(diamond)}$$

4. Draw potential energy diagrams for a general endothermic reaction and a general exothermic reaction. Label the reactants, the products, the activation energy, and the heat of reaction.

5. Which of the following reactions are endothermic and which are exothermic?

    (a) $2H_{2(g)} + O_{2(g)} \longrightarrow 2H_2O_{(g)} + 115.6$ kcal
    (b) $N_{2(g)} + 2O_{2(g)} + 16.2$ kcal $\longrightarrow 2NO_{2(g)}$
    (c) $2NH_{3(g)} + 22$ kcal $\longrightarrow N_{2(g)} + 3H_{2(g)}$
    (d) $3C_{(s)} + 2Fe_2O_{3(s)} + 110.8$ kcal $\longrightarrow 4Fe_{(s)} + 3CO_{2(g)}$

6. The following is the potential energy curve for the reaction

    $$CO + NO_2 \longrightarrow CO_2 + NO$$

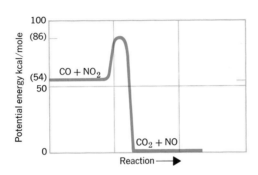

    (a) Is this reaction endothermic or exothermic?
    (b) What is the heat of reaction?
    (c) What is the activation energy for the forward reaction?
    (d) What is the activation energy for the reverse reaction

    $$CO_2 + NO \longrightarrow CO + NO_2?$$

7. Explain why special industrial procedures must be designed for the handling of large amounts of finely divided, dry, combustible materials.

8. (a) Compare the difference in reaction rates for the following reaction if the carbon is supplied (1) in chunks and (2) as a finely divided powder.

    $$C_{(s)} + O_{2(g)} \longrightarrow CO_{2(g)} + Energy$$

    (b) Draw the potential energy diagrams for the reactions in (1) and (2).

9. You have cut up peaches for dessert. What could you do to retard the browning of the peach slices?

10. Why do labels of certain antibiotic drugs state that the drugs must be kept under refrigeration?

11. What is a catalyst?

12. An important industrial process is the production of ammonia through the following reaction.

$$N_{2(g)} + 3H_{2(g)} \rightleftharpoons 2NH_{3(g)} + 24 \text{ kcal}$$

This process uses a catalyst.

(a) Draw the potential energy diagram for the uncatalyzed and the catalyzed reaction.
(b) Describe three steps that manufacturers could take to maximize the yield of ammonia in this reaction.

13. Is an equilibrium established in a can of ether

(a) with the lid off?
(b) with the lid on?

Give a reason for your answers.

14. What is "equal" in a reaction at equilibrium?

15. Why is a chemical equilibrium described as dynamic rather than static?

16. Equilibrium constants are given for the following three reactions. In each case state whether the reactants or products predominate at equilibrium.

(a) $CH_3COOH_{(aq)} \rightleftharpoons H^+_{(aq)} + CH_3COO^-_{(aq)}$ $K_{eq} = 1.8 \times 10^{-5}$
(b) $CdS_{(s)} \rightleftharpoons Cd^{2+}_{(aq)} + S^{2-}_{(aq)}$ $K_{eq} = 7.1 \times 10^{-28}$
(c) $H^+_{(aq)} + HS^-_{(aq)} \rightleftharpoons H_2S_{(aq)}$ $K_{eq} = 1 \times 10^7$

17. For reaction (a) in question 16, state two ways for increasing the concentration of CH_3COO^-.

18. What is the effect of a catalyst on a reaction at equilibrium?

19. State Le Chatelier's principle in your own words.

20. What effect would the following changes have on the equilibrium concentration of $O_{2(g)}$ in the following reaction? Give a reason for each of your answers.

$$4HCl_{(g)} + O_{2(g)} \rightleftharpoons 2H_2O_{(g)} + 2Cl_{2(g)} + 27 \text{ kcal}$$

(a) Increasing the temperature of the reaction.
(b) Increasing the pressure.
(c) Decreasing the concentration of Cl_2.
(d) Increasing the concentration of HCl.
(e) Adding a catalyst.

chapter 8

Water and Solution Chemistry

Learning Objectives

By the time you have finished this chapter, you should be able to:

1. List six unique properties of water.

2. Define the following terms:

   (a) Solution   (d) Electrolyte
   (b) Solute   (e) Nonelectrolyte
   (c) Solvent   (f) Saturated solution

3. State the difference between a strong and weak electrolyte.

4. Explain how to prepare the following solutions:

   (a) 100 ml of a 0.5% NaCl solution.
   (b) 250 ml of a 3M HCl solution.

5. Define osmosis, hypertonic, isotonic, and hypotonic.

6. Give the Brønsted-Lowry definition of an acid and base, and identify the conjugate acid-base pairs in a chemical equation.

7. Given the equilibrium constant for an acid, state whether the acid is strong or weak.

8. Given the pH of a solution, calculate the hydrogen ion concentration and the hydroxide ion concentration.

9. Describe conditions in the body that would result in acidosis.

10. Give an example of the way in which the blood plasma protects against sharp changes in pH.

11. Define a colloidal dispersion and state three properties that are unique to colloids.

12. Describe a laboratory procedure for separating cellular proteins from the crystalloid contents of the cell.

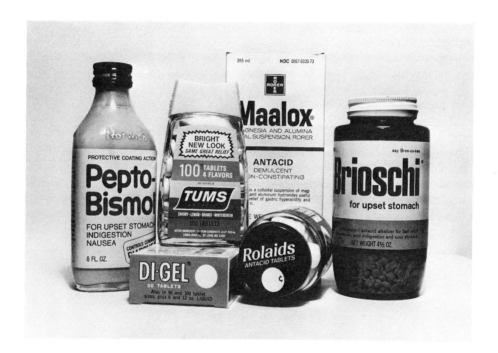

Fred Masterson was a middle-aged insurance agent with a very sensitive stomach. He kept a roll of antacids handy in his pocket to ward off the acid indigestion from which he constantly suffered; he also drank large amounts of milk in hopes of preventing an ulcer from forming. One afternoon Fred began to suffer some very strange symptoms. He was extremely dizzy, had high and erratic blood pressure, felt nauseous, and had an acid stomach that was worse than usual. His doctor put Fred into the hospital for a series of tests, but was unable to determine what was wrong. The symptoms soon disappeared, and Fred was released. But not long after he returned to work, the symptoms reappeared. Again he was hospitalized, and again the doctors could find nothing wrong. On Fred's third visit to the hospital, his doctor even went so far as to call in psychiatrists for consultation on Fred's case. Finally, after long discussions with Fred about his daily habits, a very determined intern found a reference to the "milk-alkali" syndrome in one of his medical textbooks. This syndrome, which included the symptoms suffered by Fred, is caused by a high intake of calcium-containing antacids and large quantities of milk.

To understand what was happening to Fred we must take a closer look at the work done by the stomach. Cells in our stomach lining produce hydrochloric acid, which is very corrosive. In fact, at the concentrations secreted by the stomach (0.1 moles/liter), hydrochloric acid is able to dissolve a piece of zinc, and is deadly to cells. In the stomach, this acid functions to kill bacteria in foods, to soften foods, and to convert the inactive enzyme pepsinogen into its active form, pepsin, to begin the digestion of protein.

Hydrochloric acid reacts with water in the stomach fluid to form hydronium ions, H_3O^+, and chloride ions.

$$HCl + H_2O \longrightarrow H_3O^+ + Cl^-$$

It is the presence of the hydronium ion that makes the stomach contents acidic. Sometimes the stomach produces excess acid, resulting in indigestion and heartburn. The various products sold as antacids all contain compounds that can react with hydronium ions and thereby neutralize the excess acid.

The "active ingredient" in the antacid tablets that Fred was taking is calcium carbonate. Dissolved in the water mixture in the stomach, calcium carbonate breaks apart to form the calcium ion and the carbonate ion.

$$CaCO_{3(s)} + H_2O \longrightarrow Ca^{2+}_{(aq)} + CO_3^{2-}{}_{(aq)}$$

It is the carbonate ions that react with and neutralize hydronium ions in the stomach, relieving the indigestion and heartburn.

$$CO_3^{2-} + 2H_3O^+ \longrightarrow 3H_2O + CO_2$$

Calcium carbonate is an excellent antacid for many reasons: it acts rapidly, can neutralize a large amount of acid, and has a prolonged effect at low cost. However, as Fred discovered, it is not without serious side effects when used over long periods of time. Recent research has revealed that calcium carbonate can stimulate increased secretion of stomach acid and, in effect, become self-defeating. Calcium carbonate also tends to cause constipation. A more important side effect, however, results from the high levels of calcium ions in the blood that can occur when a person takes calcium-based antacids repeatedly. This possibility of high blood calcium is greatly increased in a person who also consumes large amounts of milk, as was the case with Fred, or with a person who has kidney problems. While calcium ions are essential for normal body functioning, excessively high blood calcium levels will disrupt reactions in the body, and will cause the strange symptoms suffered by Fred. Such elevated calcium levels can also cause impaired kidney function, and lead to the formation of kidney stones. Fred's symptoms cleared up about a week after doctors limited his intake of milk and had him switch to an aluminum-magnesium antacid to relieve his indigestion. Most medical authorities now recommend antacids that contain aluminum or magnesium compounds, since these have less severe side effects.

The interaction between stomach acid and commercial antacids is only one example of the many reactions taking place in water solutions in the body. Water is the solvent for all molecules of life, so before we begin our study of the molecules of life, it is important that we first study in detail the nature of water and the properties of aqueous (water) solutions.

Water

8.1 Water and the Living Organism

Three-quarters of the surface of the earth is covered with water, either as a liquid in lakes, streams, and oceans, or as ice in glaciers and the polar ice caps. Water is critical to the survival of living organisms. It is second only to oxygen in importance for human survival; we can survive for several weeks without food, for a few days without water, and only minutes without oxygen. The water content of living organisms varies from less than 50% in some bacterial cells, to 96 and 97% in some marine invertebrates. Water is the most abundant chemical compound in the human body, making up 60 to 70% of the adult body weight. This water is distributed within the cells, in the extracellular fluid that bathes the cells, and in the blood plasma (Figure 8.1).

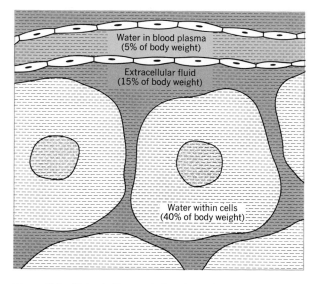

Water in blood plasma
(5% of body weight)

Extracellular fluid
(15% of body weight)

Water within cells
(40% of body weight)

Figure 8.1 Water is the most abundant compound in the human body. It is distributed within the cells, in the extracellular fluid that bathes the cells, and in the blood plasma.

Water performs many biological functions: it is the fluid found throughout living organisms, transporting food and oxygen to the cells, and carrying away wastes. It is the medium in which digestion takes place, and is a lubricant for the cells and tissues. Water plays a major role in the regulation of body temperature, and in the acid-base balance of body fluids and cells. Water is a reactant and product in many chemical reactions that take place in living organisms.

Unique Properties of Water

8.2 Solvent Properties

Water is the most abundant liquid in the world, yet knowledge of its structure is far from complete. It is the unique properties of water that make it so vital to life. Water, H_2O, is a molecule containing an oxygen atom covalently bonded to two hydrogen atoms. The shape of the molecule is bent.

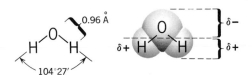

The oxygen atom is much more electronegative than the hydrogen atom, and each covalent bond is polar. In addition, the bent shape of the molecule makes the entire molecule polar, with the oxygen at the negative end and the hydrogens at the positive end. The polarity of water gives this molecule its solvent properties. Water is called the "universal solvent" since it is a better solvent than most other liquids. It will dissolve ionic compounds and molecular compounds that are polar or that contain polar groups.

8.3 High Freezing and Boiling Points

Other properties of water can be explained by the hydrogen bonding that exists between water molecules (Chapter 6, Figure 6.13). Water has a high freezing point and a high boiling point when compared to other molecules of comparable formula weight. Most compounds of comparable formula weight are gases at the temperature that water is a liquid. For example, water has a formula weight of 18 and a boiling point of 100°C, whereas ammonia (with a formula weight of 17) and methane (with a formula weight of 16) have boiling points of −33 and −164°C, respectively. The high freezing point and boiling point result from the attraction of hydrogen bonding between the molecules of water, requiring more energy to pull these molecules apart.

8.4 Density of Ice

The density of a substance is a measure of the substance's mass per unit volume [density = mass (g)/volume (cc)]. When heated, nearly all substances will increase in volume. Since the mass of the substance remains constant, its density will therefore decrease as the temperature increases (Figure 8.2a). As water cools, it contracts until it reaches its maximum density at 4°C. As it continues to cool, however, and then freeze at 0°C it *expands* by 9% (Figure 8.2b). It is the open lattice structure of ice, in which each molecule of water is hydrogen bonded to four other water molecules, that makes ice less dense than the more compact liquid water.

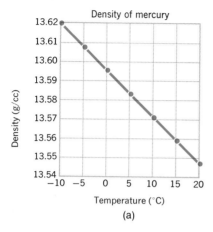

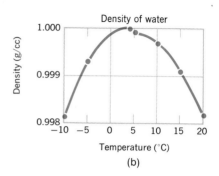

Figure 8.2 The density of most substances, such as mercury, gradually increases as the temperature decreases. The density of water, however, increases to a maximum at 4°C and then gradually decreases. As a result, ice is less dense than water and will float in water.

This is the reason why pipes break when the water in them freezes. The fact that ice is less dense than water is critical to aquatic life. When the surface of a lake freezes, the ice floats and the most dense water (at 4°C) sinks to the bottom, giving plants and animals a place to survive. If ice were more dense than water, lakes and oceans would freeze from the bottom up, and would probably never completely thaw out in summer.

8.5 High Heat of Vaporization

Water has a high heat of vaporization. The strong attractive forces between water molecules cause this liquid to boil away very slowly. It takes nearly seven times as much energy to boil off a pint of water at 100°C as it does to heat it from 21°C (room temperature) to 100°C. Evaporation is essentially the same process as boiling; it involves changing liquid water to gas, only at a slower rate. As with boiling, evaporation uses up a great deal of energy or heat. Animals make use of this principle to rid their bodies of excess heat through the evaporation of sweat.

8.6 High Heat of Fusion

Water also has a high heat of fusion. That is, a great deal of heat must be released before ice can be formed. To convert one kilogram of water at 0°C to one kilogram of ice requires four times as much refrigeration as is needed to cool this much water from 21 to 0°C. This explains why ice forms slowly in the winter and melts slowly in the spring. Water stored as snow in the mountains will, therefore, run off relatively slowly, preventing disastrous floods in the spring.

8.7 High Specific Heat

The specific heat of a liquid is the amount of heat required to raise the temperature of one gram of that liquid from 15 to 16°C. Water has a high specific heat compared to other liquids (Table 8.1). The higher the specific

Table 8.1   Specific Heat in Calories per Gram

Compound	Specific Heat	Compound	Specific Heat
Water	1.0	Chloroform	0.23
Ethanol	0.58	Ethyl acetate	0.46
Methanol	0.60	Liquid ammonia	1.12
Acetone	0.53		

heat of a substance, the less its temperature will change when it absorbs a given amount of heat. The high specific heat of water enables this fluid to keep the temperature of an organism relatively constant in the face of fluctuating internal or external heat levels. On a larger scale, the water in lakes and oceans will absorb and store large quantities of solar energy, explaining why such large bodies of water have moderating effects on local climates.

Solutions

We mentioned previously that water is often called the universal solvent because its highly polar nature allows a large number of substances to dissolve in it. You will remember that we defined a **solution** to be a homogeneous or uniform mixture of two or more substances whose particles are of atomic or molecular size. The substance being dissolved is called the **solute,** and the **solvent** is the substance in which the solute is being dissolved. If water is the solvent, the resulting solution is called an **aqueous solution.**

8.8 Electrolytes and Nonelectrolytes

Solute particles can take two forms: they may be charged ionic particles, or uncharged molecular particles. Substances that form uncharged molecular particles when dissolved in a solvent are called **nonelectrolytes.** Substances that dissolve to form charged ionic particles in solution are called **electrolytes.** The term "electrolytes" refers to the fact that solutions of electrolytes will conduct electricity.

Most ionic compounds are soluble in water because the attraction of the polar water molecules for the ions making up the compound is greater than the attraction between the ions themselves. An ionic crystal will break apart, or dissociate, in water to form an electrolytic solution. For example, table salt (NaCl) will dissolve in water because the attraction of the water molecules for the positive sodium and the negative chloride ions is greater than the attraction between these ions in the crystal lattice (Figure 8.3).

$$NaCl_{(s)} + H_2O \longrightarrow Na^+{}_{(aq)} + Cl^-{}_{(aq)}$$

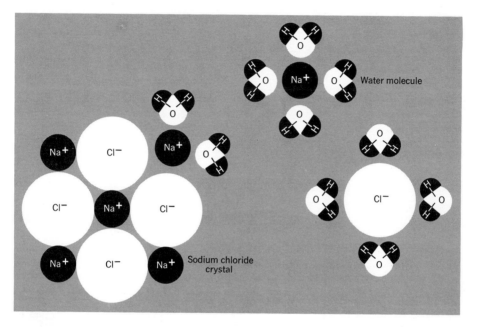

Figure 8.3 The ionic compound sodium chloride will dissolve in water because the attraction of the water molecules for the sodium and chloride ions is greater than the attraction between these atoms in the crystal.

Some covalently bonded molecules that are highly polar will also be pulled apart, or ionized, by water molecules and will form electrolytes in solution (Figure 8.4).

$$HCl_{(g)} + H_2O \longrightarrow H_3O^+{}_{(aq)} + Cl^-{}_{(aq)}$$

Electrolytes may be classified as either strong or weak. A **strong electrolyte** is a substance that completely ionizes, or dissociates into ions, in solution. A **weak electrolyte** is a substance that only partially ionizes in solution (Figure 8.5).

Electrolytes perform many important regulatory roles in our bodies, and are responsible for maintenance of the acid-base and water balance. The major positive ions, or **cations,** found in living tissue are Na^+, K^+, Ca^{2+}, and Mg^{2+}. The major negative ions, or **anions,** are HCO_3^-, Cl^-, HPO_4^{2-}, SO_4^{2-}, organic acids, and proteins.

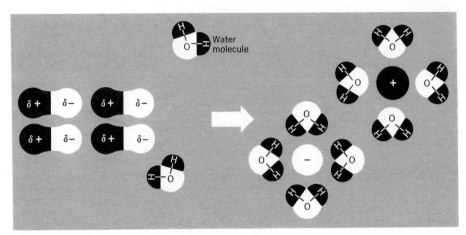

Figure 8.4 When placed in water, highly polar covalent compounds will be ionized or pulled apart and will form electrolytes in solution.

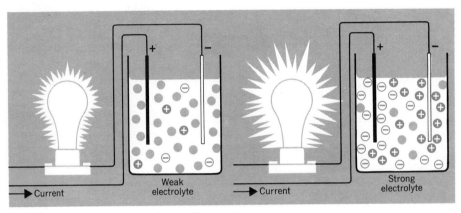

Figure 8.5 Solutions of electrolytes will conduct electricity. The stronger the electrolyte, the better the solution conducts electricity.

8.9 Solubility of a Solute

Many factors affect the solubility of a solute. A primary factor is the nature of the solvent and the solute; polar solvents will dissolve polar solutes, and nonpolar solvents will dissolve nonpolar solutes. The surface area of a solid solute will affect the rate of dissolving; the larger the surface area, the more rapidly the solute will dissolve. Stirring will also increase the rate of dissolving. For most solid substances, increasing the temperature will increase the solubility of the solute. However, there are a few compounds such as $CaCr_2O_7$, $CaSO_4$, and $Ca(OH)_2$ whose solubility decreases with an increase in temperature.

The solubility of a gas in a liquid decreases with an increase in temperature. For example, the warming of river water caused by heat discharge from a nuclear power plant decreases the supply of oxygen available for the river's fish. And, as you recall (Chap. 2), the solubility of a gas in a liquid increases with an increase in pressure (Henry's law).

8.10 Concentration

We may say that a solution is **dilute** if there are only a few solute particles dissolved in it, or **concentrated** if there are many solute particles dissolved. But the terms "dilute" and "concentrated" are not particularly precise, so various methods have been developed to more accurately describe the concentration of a solute in solution. A **saturated** solution is one which contains all the solute particles the solvent can possibly hold at that temperature; a state of equilibrium will exist between the dissolved solute and any undissolved solute in the container (Figure 8.6).

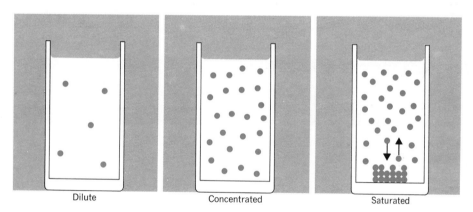

Dilute   Concentrated   Saturated

Figure 8.6 A dilute solution has very few solute particles. A concentrated solution has many solute particles. A saturated solution has all the solute particles the solvent can hold at that temperature.

8.11 Weight/Volume Percent

One way to precisely define the concentration of solute in a solution is by weight/volume percent. **The concentration of a solution in weight/volume percent is defined to be the number of grams of solute in 100 milliliters (ml) of solution.** For example, a 10% NaCl solution will contain 10 grams of NaCl in enough water to make 100 ml of solution. Intravenous solutions often contain a 0.9% NaCl solution (called normal, or physiological, saline), or a 5% glucose solution (Figures 8.7 and 8.8). Note that weight-volume percent solutions do not take into account the differences in formula weights of solutes.

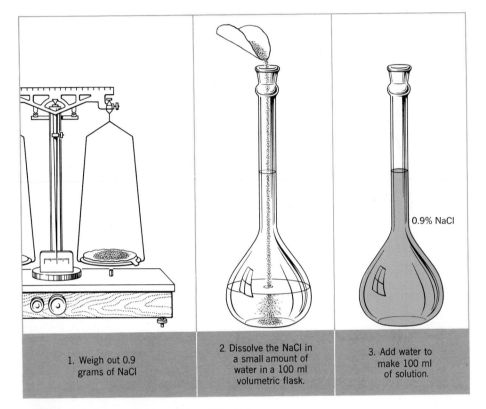

1. Weigh out 0.9 grams of NaCl

2. Dissolve the NaCl in a small amount of water in a 100 ml volumetric flask.

3. Add water to make 100 ml of solution.

0.9% NaCl

Figure 8.7 The procedure for making 100 ml of physiological saline (0.9% NaCl).

8.12 Molarity

A second way to describe the concentration of a solute — one that does take into account formula weights — is by means of molarity. **The molarity of a solution is defined as the number of moles of solute in a liter of solution.** A one molar, or 1M solution, will contain one mole of solute in enough water to make a liter of solution.

Example: —————————————————————————————

How would you make a 0.1M NaCl solution?

1. First of all you would weigh out 0.1 moles of NaCl.
   Molecular weight of NaCl $= 23 + 35.5 = 58.5$
   1 mole of NaCl $= 58.5$ grams

   $$0.1 \text{ moles NaCl} \times \frac{58.5 \text{ grams}}{1 \text{ mole NaCl}} = 5.85 \text{ grams}$$

2. Second, you would dissolve 5.85 grams of NaCl in a small amount of water, and then add enough water to make one liter of solution (Figure 8.9).

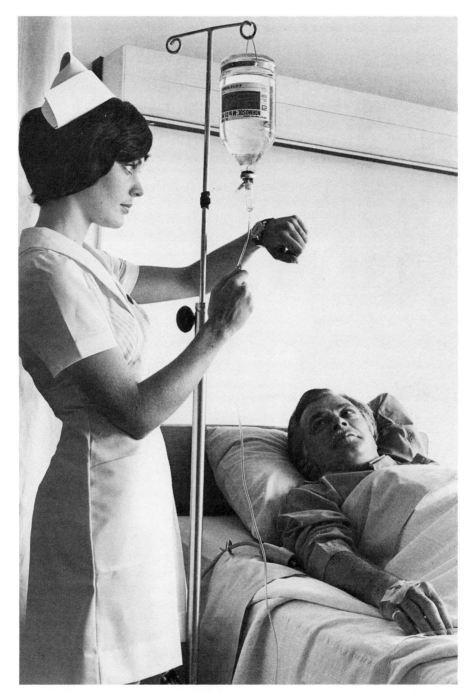

Figure 8.8 Drugs dissolved in 0.9% NaCl (physiological saline) can be administered to a patient through this intravenous apparatus. (Courtesy Abbott Laboratories)

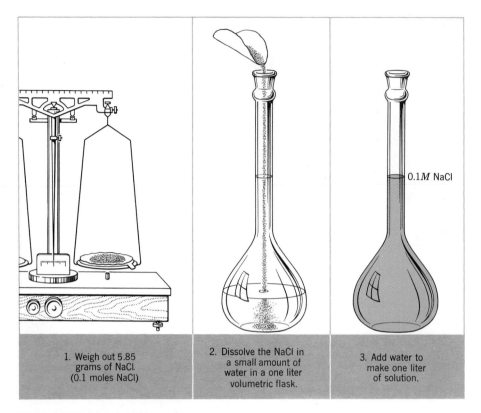

Figure 8.9 The procedure for making one liter of 0.1M NaCl.

1. Weigh out 5.85 grams of NaCl. (0.1 moles NaCl)

2. Dissolve the NaCl in a small amount of water in a one liter volumetric flask.

3. Add water to make one liter of solution.

0.1M NaCl

8.13 Colligative Properties of Water

The colligative properties of water are properties that depend only on the number of solute particles dissolved in the water, regardless of the chemical nature of the solute. Two important colligative properties of water are the lowering of the freezing point of water and the raising of the boiling point by solute particles. The more solute particles, the lower the freezing point and the higher the boiling point. For example, automotive antifreeze lowers the freezing point of the water in a car's radiator so that it doesn't freeze in the winter, and raises the boiling point so that it doesn't boil over in the summer.

8.14 Osmosis

Another colligative property of water is osmotic pressure. **Osmosis** is the flow of water molecules through a differentially permeable membrane from a region of lower solute concentration to a region of higher solute concentration. A differentially permeable membrane is a barrier that will allow the passage of water, but not solute particles.

For example, if pure water and a sugar solution are separated by a differentially permeable membrane, the water molecules will flow through the membrane into the sugar solution at a higher rate than they return. Therefore, the volume of the sugar solution will increase (Figure 8.10).

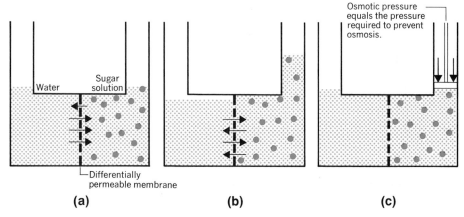

Figure 8.10 Osmosis. (a) A differentially permeable membrane will let water molecules pass through, but not sugar molecules. The water will move from the region of lower concentration to the region of higher concentration. (b) When equilibrium is established, the side with the sugar solution ends up with more material. (c) The osmotic pressure equals the pressure that must be applied to the sugar solution to prevent the movement of water, or osmosis, into the solution.

Osmotic pressure is defined as the amount of pressure that must be applied to prevent the flow of water, or osmosis, through the membrane. Osmotic pressure is dependent only upon the number of solute particles in the solution—the greater the concentration of solute, the greater the pressure that must be applied to prevent osmosis.

The passage of water in and out of living cells is an important biological process. If the concentration of solute is equal on both sides of the cell membrane, the osmotic pressures will be equal and the solutions are said to be **isotonic.** The usual concentration of the salts in blood plasma is isotonic to a 0.9% NaCl solution. This concentration of NaCl is called normal saline or physiological saline.

A normal saline solution is isotonic to red blood cells; when red blood cells are placed in the saline, no change will be observed. If red blood cells are placed in a solution of lower solute concentration, called a **hypotonic solution** (for example, distilled water), water will move into the cells. The cells will swell and rupture in a process called hemolysis. If red blood cells are placed in a solution of higher solute concentration, called a **hypertonic solution** (for example, blood plus 3% NaCl solution), the

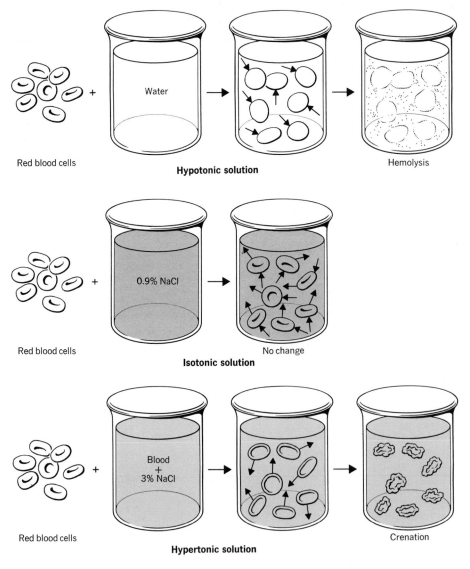

Figure 8.11   The changes that occur in red blood cells when they are placed in solutions of different concentrations.

water will move from the cell to the solution. The cell will then shrink in a process called crenation (Figure 8.11). As you can see, solutions cannot be safely introduced into the blood stream unless they are isotonic. All intravenous preparations are made with isotonic solutions.

Some of the solutes in the blood stream are not capable of passing through the walls of the capillaries, giving the blood a higher osmotic pressure than the surrounding extracellular fluid. As a result, water moves from the extracellular fluid to the capillaries, keeping the veins full and preventing the collapse of the blood vessels. Sometimes the introduction of a hypertonic or hypotonic solution into the body can have therapeutic effects. Saline laxatives such as epsom salts act by forming hypertonic solutions in the intestines. The resulting osmotic pressure draws water through the intestinal wall, thus moistening the contents of the intestines and making evacuation easier.

Acids and Bases

8.15 What Are Acids and Bases?

Acids and bases play significant roles in living organisms. Strong acids and bases can be quite harmful; they will destroy tissue by dissolving protein material and extracting water. For example, concentrated sulfuric acid is a strong dehydrating (water-removing) agent, and will rapidly injure tissues on contact. Concentrated bases will react with the fats that make up the protective membranes of cells, destroying such membranes and causing even more extensive destruction to tissues than acids. Strong laundry soaps and detergents contain bases. Clothes containing wool and silk (which are animal proteins) cannot be washed in such soaps because the base in the soap will cause the fibers to shrink and partially dissolve.

In general, **acids** are compounds which, when dissolved in water, produce solutions that conduct electricity, react with metals such as zinc or magnesium to produce hydrogen gas, taste sour, and turn paper containing litmus dye from blue to red. **Bases** also form solutions that conduct electricity; however, they taste bitter, feel slippery to the touch, and turn litmus paper from red to blue. Bases will react with acids to neutralize their properties. Medicines for the relief of pain, indigestion, and constipation often contain the base bicarbonate, which will neutralize the hydrochloric acid in the stomach. A critical balance between acids and bases must be maintained in the living organism to permit normal physiological reactions to occur (Figure 8.12).

8.16 Acids and Bases: The Brønsted-Lowry Definition

What makes a compound an acid or a base? Although there are several definitions that can be used, we will find it convenient for our study of

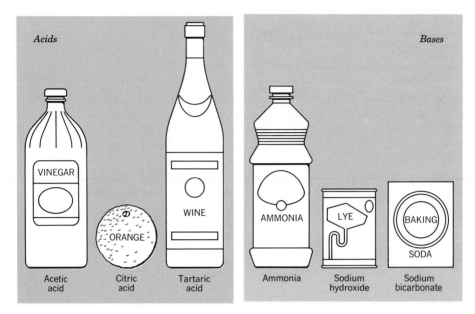

Figure 8.12 Some household acids and bases.

biochemistry to use the definitions proposed in 1923 by the Danish chemist J. N. Brønsted and the English chemist T. M. Lowry:

An acid is a substance that can donate a proton, H^+.
A base is a substance that can accept a proton, H^+.

Let's look at some examples. Consider the following reaction:

1. HCl + H_2O $\longrightarrow$ H_3O^+ + Cl^-
     Hydrochloric   Hydronium
     acid   ion

In this reaction, hydrochloric acid is donating a proton to water. Therefore, hydrochloric acid is the acid (the proton donor), and water acts as the base (the proton acceptor).

2. NH_3 + H_2O $\longrightarrow$ NH_4^+ + OH^-
     Ammonia   Ammonium   Hydroxide
                                       ion   ion

In this reaction ammonia, NH_3, accepts a proton from a water molecule. Therefore, ammonia is the base and water, as the proton donor, is the acid. You can see that water may function either as an acid or as a base depending upon the nature of the substance with which it is reacting.

Consider for a moment the reverse reaction in equation 1. Here the hydronium ion, H_3O^+, is the acid, and the chloride ion is the base. Likewise, in the reverse of equation 2, the ammonium ion, NH_4^+, is the acid and the hydroxide ion, OH^-, is the base.

$$\text{HCl} \ + \ \text{H}_2\text{O} \ \rightleftharpoons \ \text{H}_3\text{O}^+ \ + \ \text{Cl}^-$$

Acid₁   Base₂   Acid₂   Base₁

$$\text{NH}_3 \ + \ \text{H}_2\text{O} \ \rightleftharpoons \ \text{NH}_4^+ \ + \ \text{OH}^-$$

Base₁   Acid₂   Acid₁   Base₂

You might notice that HCl and Cl^-, and H_2O and H_3O^+, differ only by a proton; they are known as **conjugate acid-base pairs.** Chloride ion is the conjugate base of hydrochloric acid, and hydronium ion is the conjugate acid of water.

8.17 The Strength of Acids and Bases

We will define strong and weak acids in much the same way that we defined strong and weak electrolytes. A **strong acid** is one that completely, or almost completely, ionizes to donate all of its protons. A **weak acid** only partially ionizes to donate protons. Similarly, a **strong base** will have a very large attraction for protons; a **weak base** will have a weak attraction, and only a small percentage of its molecules will accept protons. If an acid is strong and has a large tendency to donate protons, then its conjugate base will be weak and will have a low attraction for protons. The reverse will be true for a weak acid; a weak acid will have a conjugate base with a strong attraction for protons.

8.18 Equilibrium Constant, K_A

The following is a general equation for the ionization of an acid, HA.

$$\text{HA} \rightleftharpoons \text{H}^+ + \text{A}^-$$

We can write the equilibrium constant for this reaction as follows:

$$K_A = \frac{[\text{H}^+] \times [\text{A}^-]}{[\text{HA}]} = \frac{[\text{H}^+][\text{A}^-]}{[\text{HA}]}$$

where, as you remember, $[\text{H}^+]$ denotes the concentration of H^+ in moles per liter. The value of the equilibrium constant K_A gives us an indication of the strength of the acid. If K_A is a large number (that is, $[\text{H}^+][\text{A}^-] > [\text{HA}]$), the acid is mostly ionized and is a strong acid. If K_A is very small (that is, $[\text{H}^+][\text{A}^-] < [\text{HA}]$), the acid is only partially ionized and is a weak acid (Table 8.2).

8.19 Ion Product of Water, K_W

Water has a slight tendency to ionize.

$$\text{H}_2\text{O} \ + \ \text{H}_2\text{O} \ \rightleftharpoons \ \text{H}_3\text{O}^+ \ + \ \text{OH}^-$$

Hydronium   Hydroxide
ion   ion

Table 8.2   The Relative Strength of Some Acids and Bases

Conjugate Acid-Base Pair Acid (HA)	Base (A⁻)	Relative Strength of the Acid	Relative Strength of the Base	$K_A = \dfrac{[H^+][A^-]}{[HA]}$
HCl	Cl⁻	Very strong	↑	Very large
HNO_3	NO_3^-			Very large
H_2SO_4	HSO_4^-		Very weak	Large
H_3PO_4	$H_2PO_4^-$			7.1×10^{-3}
CH_3COOH	CH_3COO^-			1.8×10^{-5}
H_2CO_3	HCO_3^-			4.4×10^{-7}
$H_2PO_4^-$	HPO_4^{2-}			6.3×10^{-8}
NH_4^+	NH_3			5.7×10^{-10}
HCO_3^-	CO_3^{2-}	Very weak		4.7×10^{-11}
HPO_4^{2-}	PO_4^{3-}			4.4×10^{-13}
H_2O	OH^-	↓	Strong	1.0×10^{-14}

One molecule of water out of every 550 million will be found to be ionized. That means that there is one mole of H_3O^+ in every 10 million liters of water. Therefore, one liter of water will contain $\frac{1}{10,000,000}$, or 10^{-7} moles of H_3O^+. For convenience, we often write the equation for the ionization of water as follows:

$$H_2O \rightleftharpoons H^+ + OH^-$$

Although we use the terms proton or hydrogen ion when talking about aqueous solutions, it is important to remember that solitary protons never exist in water. They are always associated with a molecule of water in the form of the hydronium ion, H_3O^+.

We can write the equilibrium constant for the ionization of water as follows:

$$K_{eq} = \frac{[H^+][OH^-]}{[H_2O]}$$

The concentration of water is 55.5 moles per liter, and will remain essentially constant. Therefore, we can incorporate this constant along with the equilibrium constant, and call the resulting value the **ion product of water, K_W.**

$$K_W = K_{eq} \times 55.5 = [H^+][OH^-]$$

For every mole of H^+ produced by the ionization of water, one mole of OH^- will be produced. We have already stated that the concentration of H^+ will be 10^{-7} moles per liter.

$$[H^+] = [OH^-] = 1 \times 10^{-7}M$$

Therefore we see that

$$K_W = [H^+][OH^-] = 10^{-7} \times 10^{-7} = 10^{-14} \quad (\text{or} \quad 1 \times 10^{-14})$$

From this equation we can calculate either the concentration of H^+ or OH^- if we know the concentration of the other ion.

$$[H^+] = \frac{1 \times 10^{-14}}{[OH^-]} \qquad\qquad [OH^-] = \frac{1 \times 10^{-14}}{[H^+]}$$

8.20 pH Scale

In 1909, a chemist named Sorenson developed the pH scale as a convenient way of indicating the hydrogen ion concentration in aqueous solutions within a range of $1M$ H^+ to $10^{-14}M$ H^+. The **pH** of a solution is defined as follows:

$$pH = -\log_{10} [H^+]$$

Since $[H^+] = [OH^-]$ in water, water is neutral; therefore, the pH of a neutral solution will be seven. An acidic solution is one in which $[H^+] > [OH^-]$, and the pH is less than seven. A basic solution is one in which $[OH^-] > [H^+]$, and the pH is greater than seven (Figure 8.13 and Table 8.3).

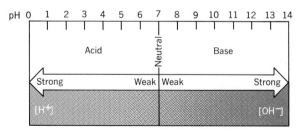

Figure 8.13   The pH scale. Acidic solutions have a range of pH from 0 to 7, and basic solutions a pH range from 7 to 14. A solution with a pH of 7 is neutral, and will have equal concentrations of hydrogen ions and hydroxide ions.

Table 8.3 The pH Scale and the Corresponding Concentrations of
Hydrogen and Hydroxide Ions

$[H^+]$	pH	$[OH^-] = \dfrac{1 \times 10^{-14}}{[H^+]}$
$10^0 = 1$	0	10^{-14}
$10^{-1} = 0.1$	1	10^{-13}
$10^{-2} = 0.01$	2	10^{-12}
$10^{-3} = 0.001$	3	10^{-11}
$10^{-4} = 0.0001$	4	10^{-10}
$10^{-5} = 0.00001$	5	10^{-9}
$10^{-6} = 0.000001$	6	10^{-8}
$10^{-7} = 0.0000001$	7	10^{-7}
$10^{-8} = 0.00000001$	8	10^{-6}
$10^{-9} = 0.000000001$	9	10^{-5}
$10^{-10} = 0.0000000001$	10	10^{-4}
$10^{-11} = 0.00000000001$	11	10^{-3}
$10^{-12} = 0.000000000001$	12	10^{-2}
$10^{-13} = 0.0000000000001$	13	10^{-1}
$10^{-14} = 0.00000000000001$	14	10^0

8.21 Measuring pH

The measurement of pH is an important laboratory procedure since the pH
of a solution affects the activity of biological molecules and, therefore, will
influence the behavior of cells and organisms. For example, bacteria will
grow best in a very narrow range of pH; therefore, the pH of culture media
must be carefully controlled. Enzymes, the biological catalysts, work best in
a very narrow range of pH that can vary from an optimum pH range of 1 to
4 for pepsin (an enzyme in the stomach), to an optimum pH range of 8 to 9
for trypsin (an enzyme in the small intestine). Most body fluids maintain a
very narrow range of pH that, if changed, can be toxic to the organism
(Tables 8.4 and 8.5).

Table 8.4   The Normal pH Range of Some Body Fluids

Fluids	pH
Blood	7.35–7.45
Urine	5.5 –7.0
Saliva	6.5 –7.5
Gastric juice	1.0 –3.0
Bile	7.8 –8.6
Pancreatic juice	8

Table 8.5   The pH of Some Common Foods

Foods	pH
Lemons	2.8–3.4
Oranges	3.0–4.0
Peaches	3.4–3.6
Bananas	4.5–4.7
Crackers	6.5–8.5
Eggs	7.6–8.0
Beans	5.0–6.0
Peas	5.8–6.4
Drinking water	6.5–8.0
Wine	2.8–3.8
Soft drinks	2.0–4.0
Milk	6.3–6.6

The pH of a solution can be measured using a pH meter (Figure 8.14). These instruments make use of the fact that the voltage of an electric current passing through a solution will depend upon the pH of the solution. A second, but less accurate, method of measuring pH is by a colorimetric determination. Such methods use chemical dyes called acid-base indicators, which will change color at certain hydrogen ion concentrations (Table 8.6). For example, paper on which the dye nitrazine has been placed

Figure 8.14 A pH meter. (Courtesy Beckman Instruments)

Table 8.6 Colors of Some Acid-Base Indicators at Various pH Levels

Indicator	pH														
	0	1	2	3	4	5	6	7	8	9	10	11	12	13	14
Thymol blue*	Red	Transition	Yellow												
Methyl orange			Red	Transition	Yellow										
Methyl red				Red	Transition	Yellow									
Litmus					Red	Transition		Blue							
Bromothymol blue						Yellow	Transition		Blue						
Metacresol purple							Yellow	Transition	Purple						
Thymol blue*								Yellow	Transition	Blue					
Phenolphthalein								Colorless	Transition	Red					

*Thymol blue indicator undergoes two color changes—one in the acid range and one in the base range.

is yellow at a pH of 4.5, and blue at a pH of 7.5. Such paper is used in hospitals to test the pH of urine. Acidic urine—urine with a pH of 4.5 or lower—will turn the paper yellow, and is often an indication of a serious disorder.

8.22 Titrations

In the Brønsted-Lowry definition of acids and bases, the transfer of protons from acids to bases is called a **neutralization reaction.** We can use a neutralization reaction to measure the amount of acid or base in a solution by means of a procedure called **titration** (Figure 8.15). In a titration, a solution having a known molar concentration of acid or base is added from a buret to a base or acid solution of unknown molar concentration. The neutralization reaction can be monitored by a pH meter or by an acid-base indicator. Titrations can be used to determine the alkali (or basic) constituents of the blood, the acidity of the stomach, or the acidity of the urine.

Example:
Titration of the contents of the stomach can often be of diagnostic value. What is the concentration of acid in 100 ml of stomach fluid if complete neutralization occurs when 27 ml of 0.1M base has been added? 0.1M base will contain 0.1 moles of base in 1000 ml. Thus, 27 ml will contain

$$27 \text{ mL} \times \frac{0.1 \text{ moles}}{1000 \text{ mL}} = 0.0027 \text{ moles of base}$$

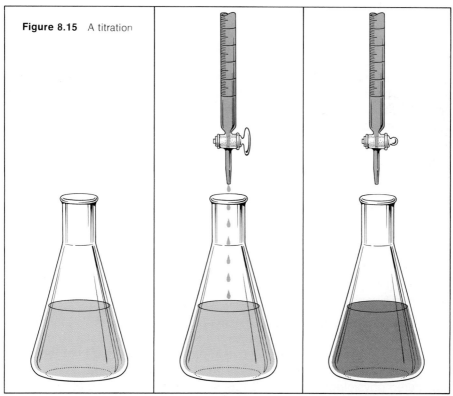

Figure 8.15 A titration

1. An acidic solution of unknown concentration is placed in a flask with an indicator.

2. A basic solution of known concentration is added slowly from the buret.

3. The flow of base from the buret is stopped when the indicator changes color, showing that just enough base has been added to react with all the acid in the flask.

If this many moles of base completely neutralized the acid in the stomach fluid, then there were 0.0027 moles of hydrogen ions in the 100 ml of stomach fluid. Therefore, the concentration of acid in moles per liter (1000 ml) would be $0.0027 \times 10 = 0.027M$. What is the pH of the stomach contents?

$$pH = -\log_{10} [H^+] = -\log_{10} [0.027]$$
$$= -\log_{10} (2.7 \times 10^{-2})$$
$$= 1.57$$

In general, a person is considered to be suffering from hyperacidity when 44 ml or more of $0.1M$ base are required to neutralize 100 ml of stomach fluid (pH < 1.4), and from hypoacidity when 10 ml or less is required (pH > 2).

8.23 Buffers

Buffers are substances that, when present in solution, resist sudden changes in pH. In particular, they protect against drastic changes in pH when acids or bases are added to the system. Living cells are extremely sensitive to even very slight changes in pH. As we stated earlier, the reason for this sensitivity is that the enzymes that catalyze metabolic reactions operate in only a small range of pH; altering the pH will slow down or stop the action of the enzyme. Fortunately, the contents of cells, the extracellular fluid, and the blood have developed buffer systems that protect against pH changes.

The best buffer systems consist of a weak acid and its conjugate base, or a weak base and its conjugate acid. Such systems have their highest buffering capacity at a pH where the (concentration of acid) = (concentration of conjugate base), or (concentration of base) = (concentration of conjugate acid). The following are some acid-base pairs that can be used in buffer systems.

H_2CO_3	$\rightleftharpoons$	$HCO_3^- + H^+$
Carbonic acid		Bicarbonate ion
CH_3COOH	$\rightleftharpoons$	$CH_3COO^- + H^+$
Acetic acid		Acetate ion
$H_2PO_4^-$	$\rightleftharpoons$	$HPO_4^{2-} + H^+$
Dihydrogen phosphate ion		Hydrogen phosphate ion
$NH_4OH + H^+$	$\rightleftharpoons$	$NH_4^+ + H_2O$
Ammonium hydroxide		Ammonium ion

In the human body the blood plasma has a normal pH of 7.4; if the pH should fall below 7.0 or rise above 7.8 the results would be fatal. The buffer systems in the blood are very effective in protecting this fluid from large changes in pH. For example, if 1 ml of $10.0M$ HCl were added to unbuffered physiological saline ($0.15M$ NaCl) at a pH of 7, the pH would fall to 2. But if 1 ml of $10.0M$ HCl is added to blood plasma at pH 7.4, the pH will drop only to 7.2 (Figure 8.16)!

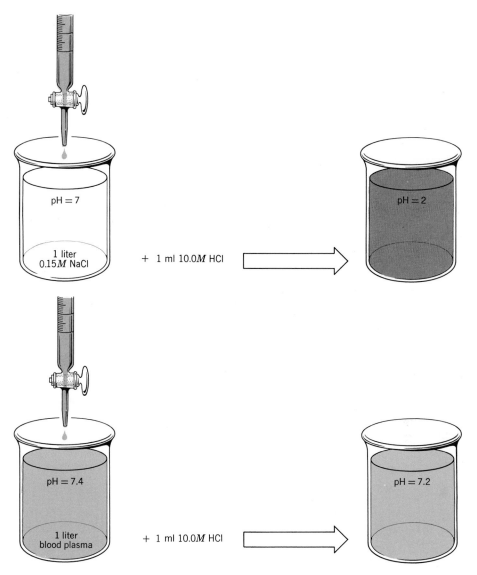

Figure 8.16 The buffer systems in the blood are
very effective in preventing large changes in pH,
which would be fatal.

How do such buffer systems protect the blood? The major buffer
system in the blood is the carbonic acid–bicarbonate system. Consider
the following equilibrium equation:

$$H_2CO_3 \rightleftharpoons HCO_3^- + H^+$$

Adding a strong acid to the system will increase the concentration of H^+, driving the reaction to the left and forming more carbonic acid.

$$H_2CO_3 \rightleftharpoons HCO_3^- + H^+$$

Carbonic acid is unstable, and will decompose to form carbon dioxide and water.

$$H_2CO_3 \rightleftharpoons CO_2 + H_2O$$

The carbon dioxide so formed can be removed from the blood and exhaled by the lungs. The buffer system will continue to protect against the pH change until all the carbonate has reacted.

Various factors can cause abnormal increases in acid levels in the blood: hypoventilation caused by emphysema, congestive heart failure, or bronchopneumonia; an increase in the production of metabolic acids, such as occurs in diabetes or some low-carbohydrate/high-fat diets; ingestion of excess acids; excess loss of bicarbonate in severe diarrhea; or decreased excretion of hydrogen ions in kidney failure. Each of these conditions will cause an increase in the hydrogen ion level in the blood, and a lowering of the concentration of basic components (such as bicarbonate), known as the alkaline reserves. The pH of the blood can drop to 7.1 or 7.2, resulting in a condition known as **acidosis.** However, the body has compensatory mechanisms to restore the blood pH to normal: first by expelling the excess carbon dioxide (formed from the carbonic acid) through an increase in the rate of respiration and, second, by the increased excretion of H^+ and the retention of HCO_3^- by the kidneys, resulting in acidic urine.

The bicarbonate buffer system also protects against an addition of strong base to the system. A base will react with the hydrogen ions to produce water, decreasing the concentration of hydrogen ions in the system. This will drive the reaction to the right.

$$H_2CO_3 \longrightarrow \rightleftharpoons HCO_3^- + H^+$$

An increase in base in the blood will occur in cases of hyperventilation during extreme fevers or hysteria, excessive ingestion of basic substances such as antacids, and in severe vomiting. The pH of the blood can increase to a pH of 7.5, resulting in a condition known as **alkalosis.** Alkalosis is not as common as acidosis. The body's compensatory mechanisms for returning the pH to normal are a decrease in expulsion of carbon dioxide by the lungs and an increase in excretion of HCO_3^- by the kidneys, resulting in an alkaline urine (Figure 8.17).

Another buffer system, active mainly within the cells, is the phosphate buffer system, which has a maximum buffering action at a pH of 7.2.

$$H_2PO_4^- \rightleftharpoons HPO_4^{2-} + H^+$$

Adding strong acid to this system will drive the reaction to the left, increasing the concentration of $H_2PO_4^-$, which is only weakly acidic. Large amounts of $H_2PO_4^-$ will result in acidosis, but the body will eliminate the excess in the urine as NaH_2PO_4. Adding strong base to the system will

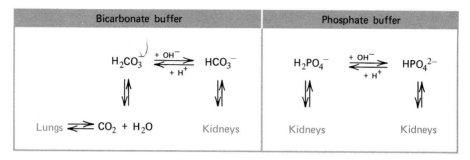

Figure 8.17 Two buffer systems which work to prevent changes of pH in the human body.

drive the reaction to the right, as the hydrogen ions react with the base to form water. Large amounts of HPO_4^{2-} would be found in alkalosis, but under normal kidney function the HPO_4^{2-} is excreted in the urine as Na_2HPO_4.

Normal metabolic reactions result in the continuous production of acids. Cells produce an average of about 10 to 20 moles of carbonic acid each day, which is equivalent to 1 or 2 liters of concentrated HCl. This acid must be removed from the cells and carried to the organs of excretion without disrupting the pH of the blood. It is through the action of the buffer systems in the cells and extracellular fluid that our bodies are protected from changes in pH that would otherwise be caused by these acids.

Colloids

8.24 The Nature of Colloids

The aqueous systems of cells are not composed entirely of true solutions, but also contain colloidal dispersions, or colloids. **Colloids** are mixtures whose chief characteristic is the size of the particles of the dissolved substance. Colloidal chemistry is important in the study of biological systems; tissues and cells are colloidal in nature, and reactions occurring within them involve colloidal chemistry. For example, food digestion involves the formation of colloids before the food can be digested; contraction of muscles can be explained by colloidal chemistry; and the body's proteins are of colloidal size. Particles larger than colloidal size are visible to the eye, and when placed in water will form suspensions that will settle out in time. Particles smaller than colloids are atomic or molecular in size (called **crystalloids**), and when placed in water form true solutions. Although there are not sharp boundaries, Table 8.7 will help define the nature of colloids.

Table 8.7   A Comparison of Some Properties of True Solutions, Colloids, and Suspensions

Property	Solution	Colloids	Suspension
Particle size	Less than 1 mμ*	1 to 1000 mμ	More than 1000 mμ
Filtration	Will pass through filters and membranes	Will pass through filters but not membranes	Stopped by filters and membranes
Visibility	Invisible	Visible in an electron micro- scope	Visible to the eye or in a light microscope
Motion	Molecular motion	Brownian movement	Movement only by gravity
Passage of light	Transparent	Transparent, Tyndall effect	Translucent

* 1 mμ = 10^{-9} meters = 1 nanometer.

Colloids can be classified by the solvent (or dispersing medium), and by the colloidal matter (or the dispersed medium). Eight classes result, and are shown in Table 8.8.

Table 8.8   Classes of Colloids

Class	Example
1. Solid in solid	Colored glass; certain alloys
2. Solid in liquid	Gelatin in water; protein in water
3. Solid in gas	Aerosols; dust in air; smoke
4. Liquid in solid	Water in gems, such as opals and pearls; jellies
5. Liquid in liquid	Emulsions; egg yolk; mayonnaise; protoplasm
6. Liquid in gas	Aerosols; fog, mist
7. Gas in solid	Activated charcoal; styrofoam
8. Gas in liquid	Foam; whipped cream
(Gas in gas)	(True solution)

Properties of Colloids

8.25 Brownian Movement

Have you ever watched particles of dust dancing in a sunbeam, or moving randomly about in the light from a movie projector? The random movement of the particles in a colloid is caused by the bombardment of these particles by the solvent molecules. The resulting movement is called **Brownian movement** after Robert Brown, who first observed this erratic movement of particles.

8.26 Tyndall Effect

When you watch dust dancing in a sunbeam, you are not seeing the actual dust particles — they are too small to be seen with the naked eye. What you are seeing is the reflection of light from these particles. A similar effect can be observed when a strong light is passed through a colloid. The path of the beam will be clearly visible because of the reflection of light from the dispersed colloidal particles (Figure 8.18). This property of colloids is known as the **Tyndall effect,** and can be used to distinguish true solutions from colloids.

Figure 8.18 The Tyndall effect. The container nearest the light source holds a true solution, and the second container holds a colloid. The light beam is visible in the second container, but it passes through the first container unscattered. (From C. W. Keenan, W. E. Bull and J. H. Wood, *Fundamentals of College Chemistry,* 3rd ed., p. 218, Harper & Row, New York, 1972. Used by permission.)

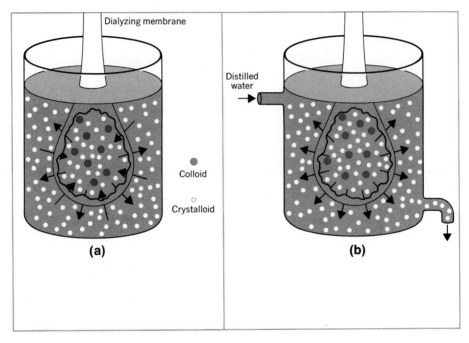

Figure 8.19 A dialysis system. (a) Crystalloid particles will readily pass through the membrane, while the colloidal particles cannot. An equilibrium will eventually be established between the crystalloid particles inside and outside the bag. (b) Complete separation of the crystalloid particles from the colloidal particles can be accomplished if distilled water is kept flowing slowly through the system. An equilibrium between the crystalloid particles inside and outside of the bag can never be reached.

8.27 Adsorption

Colloidal particles range in size from 1 to 1000 millimicrons (nanometers) in diameter. Particles of this size have very large surface areas compared to their weight. This large surface area gives colloidal particles the ability to take up, or **adsorb,** substances on their surface. For example, powdered charcoal, whose particles are of colloidal size, has many practical uses. It is put in gas masks to adsorb poisonous gases in the air, is used to remove gases and odors from city water supplies, is used to remove colored impurities from solutions in the laboratory and industry, and is used as an antidote for ingested poisons. A newly developed device is saving lives by using a charcoal filter to remove toxic substances from the blood of persons who have swallowed poisons or overdoses of drugs.

8.28 Dialysis

Some membranes will allow crystalloids to pass through them, but not colloids. Such membranes, called **dialyzing membranes,** can separate particles by size. Most animal membranes are dialyzing membranes. For example, only particles of crystalloid size can pass through the intestinal wall. The process of dialysis is a useful laboratory procedure for separating cell contents that are colloidal in nature from small molecules found in the cell (Figure 8.19).

Additional Reading

Books

1. F. Franks, ed., *Water, A Comprehensive Treatise,* Volume 1, Chapter 1 "Water, the Unique Chemical," Plenum Press, New York, 1972.

2. W. R. Frisell, *Acid-Base Chemistry in Medicine,* Macmillan, New York, 1968.

Articles

1. T. A. Boyd, "The Wonder of Water," *Chemistry,* June 1974, page 7.

2. H. W. Davenport, "Why the Stomach Does Not Digest Itself," *Scientific American,* January 1972, page 86.

3. C. Hall, "Water," *Chemistry,* September 1971, page 6.

4. W. D. Hobey, "Stomach Upset Caused by Aspirin," *Journal of Chemical Education,* March 1973, page 212.

5. "Indigestion Aids: Which Should You Use?," *Consumers Reports,* September 1973, page 584.

6. D. L. Morris, "Stress, Collisions, and Constants. Part II: Buffers," *Chemistry,* May 1971, page 15.

7. A. K. Solomon, "The State of Water in Red Cells," *Scientific American,* February 1971, page 88.

Questions and Problems

---

1. Define the following terms:

   (a) Solution
   (b) Solute

   (c) Solvent
   (d) Aqueous solution

(e) Nonelectrolyte
(f) Electrolyte
(g) Cation
(h) Anion
(i) Dilute
(j) Concentrated
(k) Saturated
(l) Weight/volume percent
(m) Molarity
(n) Osmosis
(o) Isotonic
(p) Hypotonic
(q) Hypertonic

(r) Acid
(s) Base
(t) Conjugate acid and base
(u) K_W
(v) pH
(w) Titration
(x) Buffer
(y) Acidosis
(z) Alkalosis
(aa) Colloid
(bb) Brownian movement
(cc) Tyndall effect
(dd) Dialysis

2. Explain the reason for each of the following properties of water in terms of the hydrogen bonding that exists in water.

   (a) High boiling point   (c) High heat of vaporization
   (b) Density of ice

3. Which of the following are cations and which are anions?

   (a) PO_4^{3-} (c) Br^-
   (b) H^+ (d) Ba^{2+}

4. Explain why a soft drink will go "flat" faster at room temperature than in the refrigerator.

5. How would you prepare 500 ml of 5% glucose solution?

6. How would you make up 200 ml of a 0.5M NaCl solution?

7. If you dissolve 4.9 g of H_2SO_4 in 250 ml of solution, what is the molarity of the solution?

8. Quite often celery stored for long periods in a refrigerator goes limp. It can be made crisp again by putting it in a container of water. Explain the principle behind the celery's return to its original crispness.

9. A pharmaceutical company recalled a batch of intravenous glucose solution after a patient died from introduction of this solution into her blood stream. It was discovered that the solution was 15% glucose. Explain how the introduction of this solution might lead to death.

10. What would happen to red blood cells if they would be placed in the following solutions?

   (a) 1.5% NaCl solution
   (b) 0.154M NaCl solution
   (c) 0.15% NaCl solution

11. Identify each of the solutions in question 10 as isotonic, hypertonic, or hypotonic to red blood cells.

12. Label the conjugate acid-base pairs in the following reactions.

(a) $HF + H_2O \rightleftharpoons H_3O^+ + F^-$
(b) $CO_3^{2-} + H_2O \rightleftharpoons OH^- + HCO_3^-$
(c) $CH_3COOH + H_2O \rightleftharpoons H_3O^+ + CH_3COO^-$

13. The K_A for a particular acid is 1.5×10^{-7}. Is this a strong or weak acid? Give a reason for your answer.

14. Which of the following acids is the strongest, and which is the weakest?

Hydrofluoric acid, HF $\qquad K_A = 6.7 \times 10^{-4}$
Sulfurous acid, $H_2SO_3 \qquad K_A = 1.7 \times 10^{-2}$
Hydrogen sulfide, $H_2S \qquad K_A = 9 \times 10^{-8}$

15. What is the pH of a solution having an $[H^+]$ of

(a) 0.001?   (b) 1×10^{-8}?   (c) $[OH^-] = 1 \times 10^{-4}$?

16. What is the $[H^+]$ and the $[OH^-]$ of a solution having a pH of

(a) 1?   (b) 6?   (c) 12?

17. Indicate whether each of the solutions in question 15 and question 16 above is acidic or basic.

18. Solution A has a pH of 3 and Solution B a pH of 5. Which solution is more acidic, A or B? What is the hydrogen ion concentration in each solution, and by what factor do the two hydrogen ion concentrations differ?

19. What is the concentration of acid, in moles per liter, in a sample of waste water from a chemical plant if 52 ml of 0.1M base completely neutralizes the acid in 100 ml of the sample?

20. A large amount of lactic acid is produced in the muscles during strenuous exercise. This lactic acid must be transported by the blood to the liver, where it is broken down. Why doesn't the pH of the blood change drastically after strenuous exercise?

21. What compensatory mechanisms does the body use to correct the conditions of (a) acidosis and (b) alkalosis?

22. People with impaired kidney function cannot rid their body of low formula weight waste products produced by the cells. They require cleaning of their blood by dialysis. Design a machine for cleansing the blood using dialysis, and explain how your machine works.

section III
the elements
necessary for life

chapter 9

Carbon and Hydrogen

Learning Objectives

By the time you have finished this chapter, you should be able to:

1. Define "organic chemistry."

2. Describe the difference between molecular and structural formulas of a compound.

3. Describe the difference between the alkanes, alkenes, and alkynes, and give examples of molecules found in each class.

4. Given the structural formula of a hydrocarbon, identify whether it is saturated or unsaturated.

5. Define "structural isomer" and give two examples.

6. Define "geometric isomer" and give an example of the *cis* and *trans* forms of an unsaturated hydrocarbon and a cyclic hydrocarbon.

7. Describe the polymerization reaction of an alkene.

8. Compare the chemical reactivity of the alkanes, alkenes, and alkynes.

9. Give several examples of cyclic hydrocarbons.

10. Give several examples of aromatic hydrocarbons.

11. Describe the structure of the benzene molecule.

12. Define "carcinogen."

13. Give several examples of compounds belonging to the halogenated hydrocarbons, and explain their uses.

14. Using specific examples, discuss the benefits and risks to society of misusing synthetic organic chemicals.

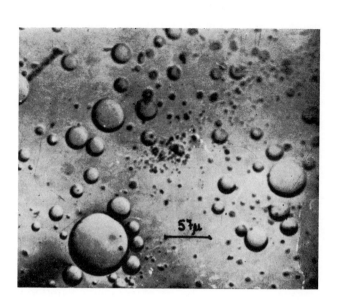

Four and one-half billion years ago the planet earth was in its infancy, having formed as a mass of steaming vapors from huge accumulations of gases. Over endless centuries the surface of the earth slowly cooled and solidified to form the rocks known as the earth's mantle. As this cooling continued, steam condensed to form water that filled the low places in the mantle, forming the oceans of the earth. This primordial earth was a lifeless expanse of rock and water, constantly racked by violent storms and torrential rains. The first atmosphere that may have formed, enriched by gases spewing from the interior of the earth in volcanic eruptions, consisted mostly of methane and nitrogen, with hydrogen, ammonia, carbon monoxide, and water vapor in smaller quantities. There was no free oxygen.

In this primordial atmosphere, chemical reactions were initiated by ultraviolet radiations, ionizing radiations, and lightning. Many of the products of these reactions were simple carbon compounds. These molecules were brought to earth with the rains, and settled in the pools and oceans, forming a warm, dilute, aqueous solution of dissolved minerals and organic compounds—a primordial soup rich in the building blocks for basic life processes. Many theories have been proposed describing how life originated and developed, and the following narrative has been proposed by the Russian biochemist A. I. Oparin.

As the centuries passed, the carbon compounds floating in the primordial soup began to join together to form larger and larger molecules, forming a film on the water. Once these molecules reached a certain critical size, they began to congregate and collect into droplets distinctly separate from the surrounding medium. These droplets, or coacervates, had the extraordinary ability to absorb materials selectively from the surrounding environment and to form a surface layer, or boundary, having properties different from the droplet itself. Various molecules could be held in close proximity to one another within the coacervate, allowing their concentrations to be increased to levels much higher than the surrounding environment. This represented the first major step toward the emergence of life.

The coacervates that survived over time were those in which chemical reactions guaranteeing their preservation, such as repairing or replacing the surface molecules, took place at a faster rate than in the outside primordial soup. As the ages passed, coacervate systems developed that were capable of growth as well as self-preservation. This capability required the development of complicated chemical pathways that tapped the chemical energy stored in the bonds of molecules found in the primordial soup. While all coacervate systems multiplied by fragmentation due to wave action or to internal stress, the droplets that survived such fragmentation were in large part the ones with improved chemical systems having specific chemical pathways for the types of reactions most fundamental to the reactions of life. The coacervate droplets possessing this improved internal organization are called protobionts, and represent systems intermediate between the original droplets and the most primitive of living organisms. The evolution of the protobiont into a primitive living organism occurred about 3.5 billion years ago, when the protobionts were able to develop mechanisms for controlling the chemical reactions taking place within themselves. This led to a more efficient use of nutrients and a much more rapid growth, and allowed the development of reproduction processes that insured the survival of the organism. It was with the development of mechanisms to control the organism's own replication that we can truly say life began.

Elements Necessary for Life

9.1 Elements That Are Abundant in Living Organisms

Living organisms, like all other matter on earth, are composed of atoms of the 90 naturally occurring elements. However, not all 90 of these elements are found in such organisms. The periodic table in Figure 9.1 shows the 24 elements that have thus far been established as essential to life. Hydrogen, carbon, nitrogen, and oxygen are the most abundant elements in the living organism. They make up 99.3% of all the atoms in your body, while the remaining 20 elements account for only 0.7%.

Why are these 24 elements always found in living organisms, while others are not? Perhaps we can identify some special properties possessed by these elements that make them important in living systems.

Abundance in Sea Water

According to the Oparin coacervate theory, life originated in the oceans. Therefore, we might expect that the elements most prevalent in sea water should also be abundant in living organisms. Table 9.1 shows that many of the elements that are abundant in sea water are indeed also abundant in the human body.

Table 9.1 The Chemical Elements of Life

Composition of Universe		Composition of Earth's Crust		Composition of Sea Water		Composition of the Human Body	
H	91	O	47	H	66	H	63
He	9.1	Si	28	O	33	O	25.5
O	0.057	Al	7.9	Cl	0.33	C	9.5
N	0.042	Fe	4.5	Na	0.28	N	1.4
C	0.021	Ca	3.5	Mg	0.033	Ca	0.31
Si	0.003	Na	2.5	S	0.017	P	0.22
Ne	0.003	K	2.5	Ca	0.006	Cl	0.03
Mg	0.002	Mg	2.2	K	0.006	K	0.06
Fe	0.002	Ti	0.46	C	0.0014	S	0.05
S	0.001	H	0.22	Br	0.0005	Na	0.03
		C	0.19			Mg	0.01
All others 0.01		All others 0.1		All others 0.1		All others 0.01	

Percent of Total Number of Atoms

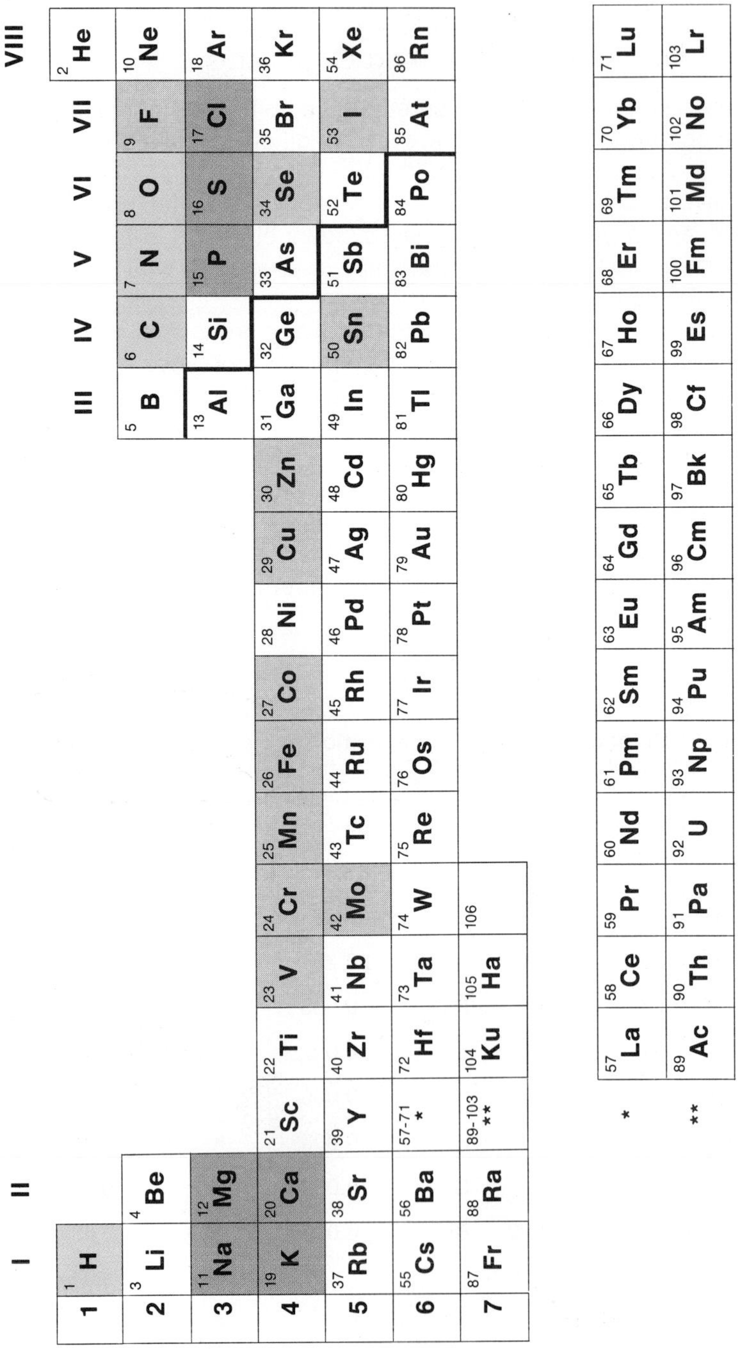

Figure 9.1 The elements that are essential to life. The elements most abundant in the living organism, making up 99.3% of all the atoms in the human body, are shown in the brightest shading. The seven elements shown in medium shading make up only about 0.7% of the atoms in the body. The remaining elements that are essential to life are called the trace elements and make up less than 0.01% of the atoms in the body.

Interaction with Water

Water is by far the most abundant compound found in every living organism. Not only is water the major constituent of all living systems, it is also the substance in which all the compounds of life are either dissolved or suspended. Therefore, the way in which an element or compound responds to water will determine its usefulness in the living system. Many of the elements found in living organisms are dissolved in water in the form of hydrated ions, which are ions bound to water molecules. In fact, most elements found in living organisms are present either as hydrated ions or in chemical combination with other elements. Neutral átoms of an element (that is, the element in unreacted form) can often be quite harmful to an organism. Such atoms may be highly unstable and may react chemically with the living organism, damaging or killing the surrounding tissue. As we saw in Chapter 6, the element chlorine is potentially very harmful to living organisms, but when reacted with sodium to form sodium chloride, the stable chloride ion becomes an essential nutrient.

The Size of the Atom

The size of the atom of an element seems to influence the role that the element will play in the living system, for living organisms are composed largely of atoms that have small atomic radii and low atomic weights. As evidence of this, 21 of the lightest 34 elements have been established as essential to life, while there are only three heavier elements (molybdenum, tin, and iodine) that have been verified as being necessary. We have seen that the four most abundant elements in the human body are hydrogen, carbon, nitrogen, and oxygen. Notice that atoms of each of these elements are small; they are all found on the first and second periods of the periodic table.

9.2 The Role of Carbon

In the previous section we described a primordial soup rich in molecules containing carbon atoms. What is so special about carbon-containing molecules that they should comprise the building blocks of life?

Look at the periodic table in Figure 9.1. Carbon is the first member of group IV, a family also containing silicon, germanium, tin, and lead. Each member of this family has four valence electrons. To attain a stable octet of electrons these elements can lose four electrons, gain four electrons, or share four electrons. Under ordinary conditions carbon has a strong tendency to share four electrons and, thus, to form four covalent bonds. The most stable arrangement for these four electron pairs occurs when they are geometrically as far from the other pairs as possible. (Remember that each covalent bond is made up of a pair of electrons, and that these negatively charged electron pairs repel each other.) Therefore, these bonds will be found directed toward the corners of a tetrahedron with the carbon atom in the center, as if the carbon atom were at the joint of a stable tripod with the

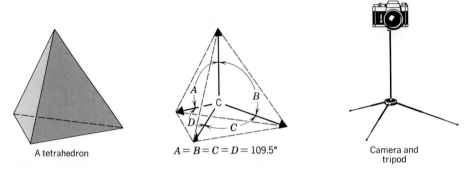

A tetrahedron $A = B = C = D = 109.5°$ Camera and tripod

Figure 9.2 The tetrahedral carbon atom. The most stable arrangement for the four covalent bonds formed by carbon is achieved when these bonds are directed toward the four corners of a tetrahedron.

four bonds represented by the three legs and the camera stand, each of equal length (Figure 9.2).

A carbon atom can form four single bonds with four different atoms or with other carbon atoms. In this second case the other carbon atoms can, in turn, be bonded to up to three more carbon atoms, and so on. The very stable compound that results when every carbon atom is bonded to four other carbon atoms is diamond (Figure 9.3). A major feature, then, making carbon unique among the elements is its ability to form strong stable bonds with up to four other atoms identical to itself.

Figure 9.3 A model of the bonding in diamond. Each carbon atom sits at the center of a tetrahedron and is bonded to four other carbon atoms.

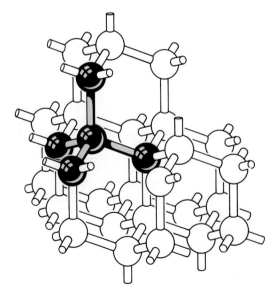

We have mentioned that diamond is the compound formed when every carbon is bonded to four other carbon atoms. But all four bonds need not be to other carbon atoms, and the molecules found in living organisms are rarely of this type. Rather, in such molecules we find carbon atoms bound to other carbon atoms in the form of long chains, branched chains, or rings (Figure 9.4). As the number of carbon atoms found in a molecule increases,

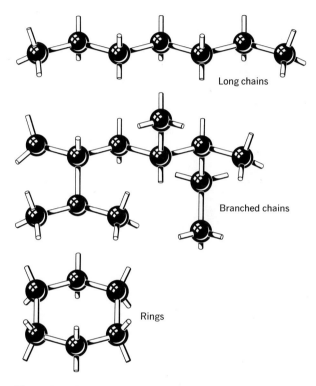

Long chains

Branched chains

Rings

Figure 9.4 Some possible arrangements of carbon atoms found in organic compounds. (Only the carbon atoms are shown.)

the number of ways in which they can be arranged increases, yielding compounds with the same chemical composition (the same number of atoms of each element), but with different structures and with correspondingly different chemical and physical properties. Compounds having the same molecular formulas but with different geometric structures are called **isomers** (Figure 9.5).

As we just mentioned, not only can carbon form stable bonds with other carbon atoms, but it can also form equally strong bonds with a number of other elements: hydrogen, oxygen, nitrogen, sulfur, phosphorus, and the halogens. The resulting diversity of compounds has given rise to a field of chemistry devoted entirely to the study of just such carbon compounds; this is the field of **organic chemistry.** The number of known

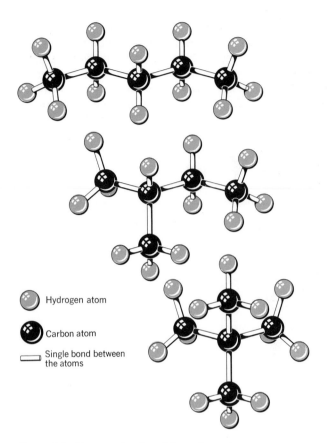

Figure 9.5 These molecules all have the molecular formula C_5H_{12}. They have different structural formulas and, therefore, are isomers.

Hydrogen atom

Carbon atom

Single bond between the atoms

organic compounds, both natural and synthetic, is greater than two million, while the number of known inorganic compounds (those formed from combinations of all the other known elements) is only about 500,000. The reason for this comparatively small number of inorganic compounds is that molecules of covalent inorganic compounds are composed of only a few atoms. Two or three atoms will form a very stable inorganic molecule, but as more atoms are added the molecule becomes unstable and becomes more likely to fall apart. Inorganic molecules containing more than 12 atoms are quite rare. However, it is not uncommon for a large organic molecule such as a protein to contain more than a million atoms!

You might not relish the thought of studying a field of chemistry covering more than two million compounds, but fortunately the chemical properties of many of these compounds are similar, allowing them to be conveniently grouped into several classes or series of compounds. In that way, by looking at one or two examples of the chemistry of organic compounds belonging to a given class, we can obtain an understanding of the chemistry of all compounds belonging to that class.

Formulas and Nomenclature

Before beginning our study of organic chemistry, we will pause to take a brief look at two concepts that often cause a bit of confusion. These are the areas of formula writing and nomenclature, and mastering a basic understanding of these areas should help you overcome any initial confusion.

9.3 Writing Structural Formulas

First of all, let us take a look at formula writing. A few pages ago we mentioned that in organic chemistry it is common to find several different compounds having the same chemical formula. Therefore, we will often find it important to include the structural formula of the molecule. Structural formulas are no more than chemical diagrams. Many things require diagrams to be completely understood; you might be hard pressed to put together an unassembled bicycle or to sew a dress without some helpful diagrams accompanying the instructions. Structural formulas make organic chemistry easier to follow just as wiring diagrams, say, make electronic repairs easier to perform. Although all chemical compounds have three-dimensional structures, these are often very difficult to draw; so we will use two-dimensional representations for these three-dimensional structures.

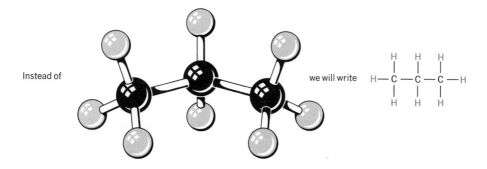

In the structural formula of a compound, the two electrons in a covalent bond are represented by a line drawn between the two atoms connected by that bond. Each atom is represented by the symbol of its element. For example, there are three compounds (that is, three isomers) having the molecular formula C_5H_{12}.

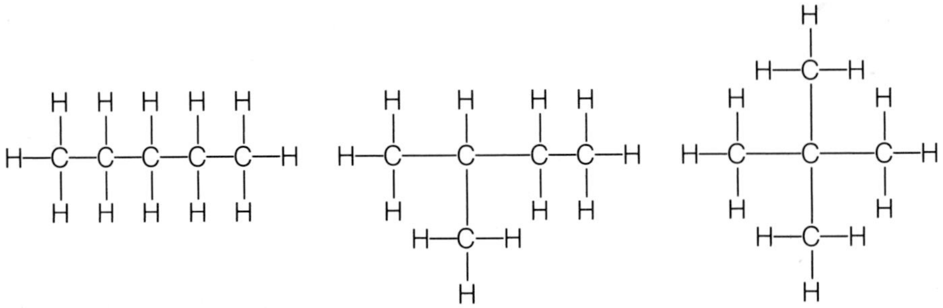

This subject would be pretty straightforward if we could stop our discussion right here, but drawing the structural formulas of large organic molecules can be quite time-consuming. Therefore, the organic chemist has devised several ways of shortening the procedure. One method commonly used is to indicate the atoms that are bonded to the carbon by writing their symbols after the carbon symbol, but without bothering to use a dash for the bonds.

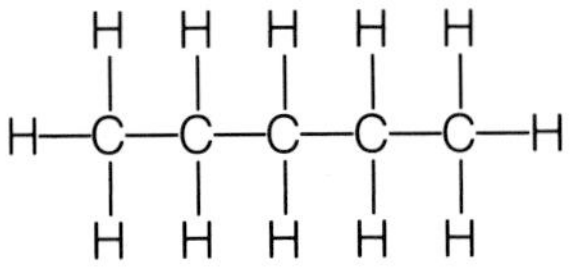

becomes $CH_3CH_2CH_2CH_2CH_3$

or

$CH_3(CH_2)_3CH_3$

Condensed structural formulas

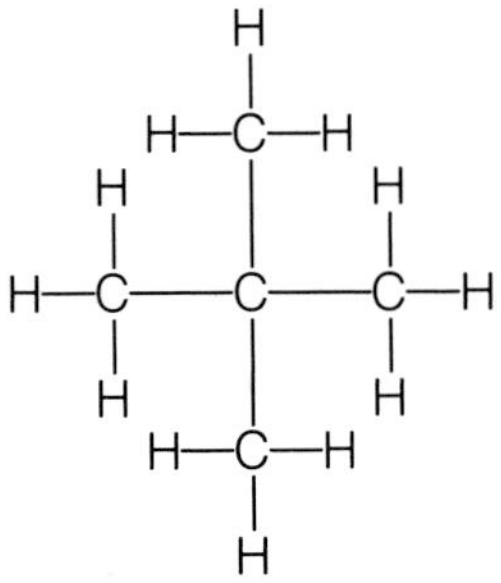

becomes $CH_3-\overset{\underset{\displaystyle CH_3}{|}}{\underset{|}{\overset{CH_3}{C}}}-CH_3$

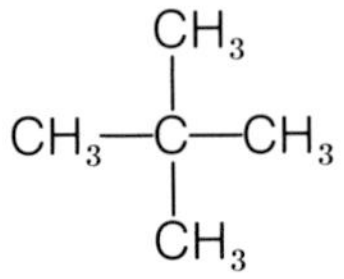

or

$CH_3C(CH_3)_2CH_3$

Or, quite often the part of the molecule in which we are interested is drawn out in detail, while the rest of the molecule is represented in shorthand.

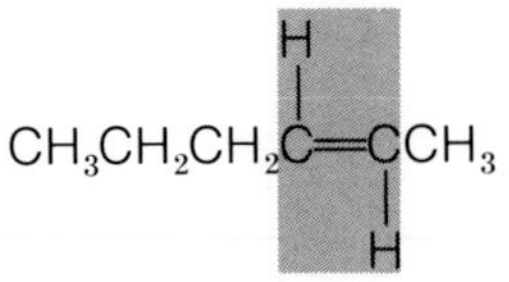

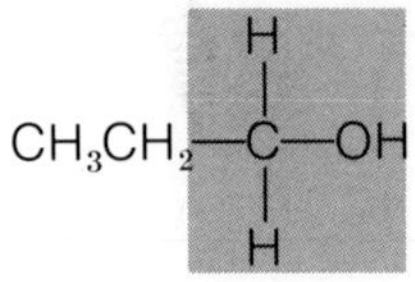

$CH_3CH_2CH_2C=CCH_3$

CH_3CH_2-C-OH

As you read this chapter and encounter more structural formulas, you will find that they soon become quite easy to read, and help make chemical discussions much easier to follow.

9.4 IUPAC Nomenclature

Nomenclature is an equally important concept for the beginner to understand, for it is virtually impossible to talk about any of the molecules of life without having a basic knowledge of the way in which these compounds are named. Before a standard procedure was finally established for naming compounds, most organic compounds were known by common names, usually derived from the source of that compound rather than from its chemical structure. For example, lactic acid is the common name of a compound found in sour milk, getting its name from the Latin word for milk, *lactis*. However, the name "lactic acid" does not give us any clue as to the structure of this compound, seemingly requiring us to memorize the structure of the lactic acid molecule. Given that there are more than two million organic compounds known, the study of organic chemistry would clearly be impossible without some system of nomenclature that could be standardized and that would indicate the structure of the molecule being named. The International Union of Pure and Applied Chemistry (IUPAC) began meeting in Geneva in 1892 to establish the rules for such a naming system.

A complete discussion of the IUPAC conventions for naming compounds must be left for other textbooks, but we will briefly describe some of the basic rules.

Rule 1 Prefixes are used to indicate the number of carbon atoms in the main carbon chain of the compound.

Table 9.2 Prefixes Used to Indicate the Number of Carbon Atoms in an Organic Compound

Number of Carbon Atoms	Prefix	Number of Carbon Atoms	Prefix
1	meth-	6	hex-
2	eth-	7	hept-
3	prop-	8	oct-
4	but-	9	non-
5	pent-	10	dec-

Rule 2 Organic compounds can be categorized into groups or classes. Each group is given a characteristic suffix that identifies the group.

Table 9.3   Some IUPAC Suffixes That Identify Classes of Organic
Compounds

Suffix	Class of Compounds	Example
-ane	Alkanes	Prop**ane**
-ene	Alkenes	Prop**ene**
-yne	Alkynes	Prop**yne**
-ol	Alcohols	Propan**ol**
-al	Aldehydes	Propan**al**
-one	Ketones	Propan**one**
-oic acid	Organic acids	Propan**oic acid**

Rule 3 (a)   Atoms or groups of atoms (other than hydrogen)
attached to the carbon chain are listed in front
of the name of the compound. For example,
bromomethane is a compound containing an
atom of bromine in a methane molecule, and
chloroethane is a compound containing an atom
of chlorine in an ethane molecule.

(b)   Groups of atoms that are attached to the main
carbon chain and that contain only carbon and
hydrogen atoms are called **alkyl groups.** They
are named using the prefixes shown in Table 9.2:

$-CH_3$ is the methyl group
$-CH_2CH_3$ is the ethyl group
$-CH_2CH_2CH_3$ is the propyl group, etc.

(c)   If the organic compound contains more than one
of the same type of these "other" atoms, the
number is indicated by a prefix: di- indicates
two, tri- indicates three, and tetra- indicates four.
For example, dibromoethane is an ethane
molecule containing two atoms of bromine, and
tetrachloromethane is a methane molecule
containing four atoms of chlorine.

(d)   To indicate the carbon atoms to which these
"other" atoms are attached, numbers
corresponding to the carbon atoms precede the
name of the compound. The carbon atoms are
numbered from the end of the chain nearest the
attached group. For example, 1,2-dichloroethane

is an ethane molecule containing two atoms of chlorine—one attached to carbon number 1 and one attached to carbon number 2.

To see how these rules work in actual practice, let's return to our example of lactic acid. The IUPAC name for lactic acid is 2-hydroxypropanoic acid. The structure of lactic acid is as follows:

$$H-\underset{\underset{H}{|}}{\overset{\overset{H}{|}}{C_3}}-\underset{\underset{OH}{|}}{\overset{\overset{H}{|}}{C_2}}-\overset{\overset{O}{\|}}{C_1}-OH$$

In the IUPAC name, the prefix "prop-" tells us this compound has three carbon atoms in its main carbon chain. The suffix "-oic acid" tells us that this compound belongs to the class of organic acids, which are molecules containing the following group of atoms:

$$-\overset{\overset{O}{\|}}{C}-OH$$

And finally, "2-hydroxy" tells us that this compound contains a hydroxyl group of atoms, the combination —OH, and that this group is attached to carbon number 2. To complete the structure of the molecule, any other possible carbon bonds not already specified are assumed to lead to hydrogen atoms. Thus, this IUPAC name allows us to figure out the entire structure of the lactic acid molecule. Let's face it, though, the name lactic acid is certainly easier to say and to remember than 2-hydroxypropanoic acid. This is the reason that common names persist, and we will usually use the common name throughout this book except in instances when it is important to establish the structure of a compound.

Hydrocarbons

9.5 Petroleum

The **hydrocarbons** are a large class of organic compounds, each of which contains atoms of only carbon and hydrogen. These compounds are the basic building blocks from which all other organic compounds can be derived. A major source of hydrocarbons in commercial quantities is petroleum, which is a mixture of hundreds of different hydrocarbons. Petroleum was formed over the course of millions of years as vast regions of plant life died, partially decayed, and were covered over as the earth slowly changed. After crude petroleum is pumped out of the ground, it is transported to refineries where the hydrocarbons are separated by a process called **fractional distillation.** This process makes use of the fact that compounds composed of long hydrocarbon chains require more heat to vaporize than do compounds composed of shorter chains. In other words, a hydrocarbon with a long carbon chain has a higher boiling point than one with a shorter carbon chain. In fractional distillation, the petroleum is heated slightly; this causes the molecules with short carbon chains to vaporize, and they are drawn off. As the heat is increased, molecules with longer and longer carbon chains vaporize, are drawn off, and are collected. The actual contents of each fraction that is drawn off may vary with the source of the petroleum. Each fraction may then be further processed to produce the particular hydrocarbons that are desired (Figure 9.6).

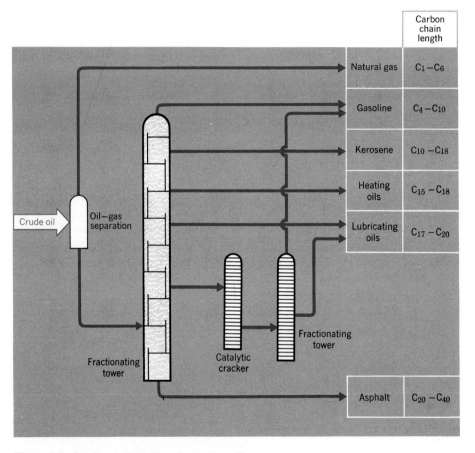

Figure 9.6 Fractional distillation of petroleum. The molecules found in petroleum (crude oil) are separated on the basis of their boiling points. The petroleum is heated and the molecules with short carbon chains are drawn off. As the heat is increased, molecules with longer and longer carbon chains are drawn off. The molecules that remain are called heavy bottoms; they have very long carbon chains and are used to make asphalt and coke.

Saturated Hydrocarbons

9.6 Alkanes

The simplest of all hydrocarbons is the gas methane, CH_4. In this molecule the carbon is bonded to four hydrogen atoms that are located at the four corners of a tetrahedron. This makes the methane molecule symmetrical and nonpolar. It is not soluble in water, and has a very low boiling point, $-161.5°C$ (Figure 9.7).

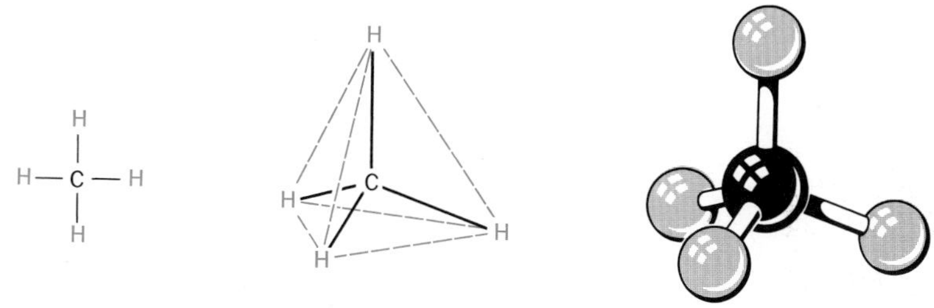

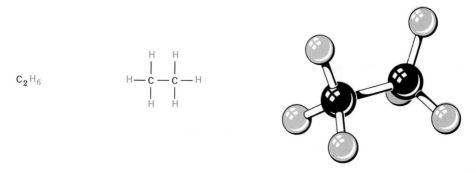

Figure 9.7 Methane

Methane is a major component of the natural gas used to heat homes and cook food. It is occasionally found concentrated in pockets in fields of coal, and is one of the causes of explosions in coal mines. Methane can be formed by the action of bacteria on decaying matter, and is found among the gases that bubble out of marshes and swamps and, for that reason, it was once known as marsh gas. In the Middle Ages, the fires that resulted when these marsh gases were ignited were thought to be the spirits of the dead.

The second member of this group is ethane, C_2H_6, which is formed when a hydrogen on methane is replaced by a methyl group ($-CH_3$) (Figure 9.8). Methane and ethane are members of the hydrocarbon class of

Figure 9.8 Ethane

compounds called the **alkanes,** whose formulas all fit the general pattern of C_nH_{2n+2} (Table 9.4).

Table 9.4   Some Straight Chain Alkanes

Number of Carbons	Molecular Formula	IUPAC Prefix	Name	Structural Formula	Boiling Point in °C
1	CH_4	meth-	methane	CH_4	−162
2	C_2H_6	eth-	ethane	CH_3CH_3	−89
3	C_3H_8	prop-	propane	$CH_3CH_2CH_3$	−42
4	C_4H_{10}	but-	butane	$CH_3CH_2CH_2CH_3$	0
5	C_5H_{12}	pent-	pentane	$CH_3CH_2CH_2CH_2CH_3$	36
6	C_6H_{14}	hex-	hexane	$CH_3CH_2CH_2CH_2CH_2CH_3$	69
7	C_7H_{16}	hept-	heptane	$CH_3CH_2CH_2CH_2CH_2CH_2CH_3$	98
8	C_8H_{18}	oct-	octane	$CH_3CH_2CH_2CH_2CH_2CH_2CH_2CH_3$	126
9	C_9H_{20}	non-	nonane	$CH_3CH_2CH_2CH_2CH_2CH_2CH_2CH_2CH_3$	151
10	$C_{10}H_{22}$	dec-	decane	$CH_3CH_2CH_2CH_2CH_2CH_2CH_2CH_2CH_2CH_3$	174

The identifying characteristic of the alkanes is that in each molecule the carbons are bonded singly to four other atoms; that is, each carbon forms four single bonds. Molecules having this property are said to be **saturated.** Single bonds between carbon atoms are strong and stable, making the alkanes the least reactive class of hydrocarbons.

Propane, C_3H_8, and butane, C_4H_{10}, are the next two members of the alkane series. They have progressively higher boiling points and can be liquefied in tanks under pressure. This allows them to be stored and transported easily, and they find wide use as fuels for lighters, torches, and furnaces in rural homes.

9.7 Structural Isomers

There are two different compounds that have the molecular formula C_4H_{10}. These two compounds have their carbon and hydrogen atoms arranged differently in the three-dimensional structure of the molecule, and are known as **structural isomers.** The two isomers having the molecular formula C_4H_{10} are butane (common name, n-butane) whose carbons all lie in a straight chain, and 2-methylpropane (common name, isobutane) whose carbons are arranged in a branched chain (Figure 9.9 and Table 9.5).

While there are only two structural isomers of butane, the number of possible isomers increases as the number of carbon atoms in the molecule increases. Octane, which is the eight-carbon alkane, has 18 different structural isomers. Each octane isomer behaves a little differently. One of

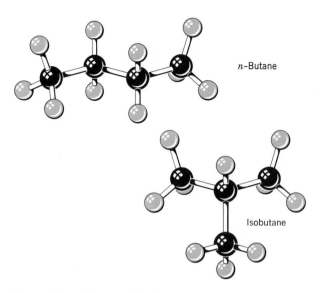

n-Butane

Isobutane

Figure 9.9 *n*-Butane and isobutane are structural isomers having the molecular formula C_4H_{10}.

Table 9.5   Some Structural Isomers

Molecular Formula	Common Name (IUPAC Name)	Structural Formula	Boiling Point °C
C_4H_{10}	*n*-Butane (Butane)	$CH_3CH_2CH_2CH_3$	0
	Isobutane (2-Methylpropane)	CH_3CHCH_3 with CH_3	−12
C_6H_{14}	*n*-Hexane (Hexane)	$CH_3CH_2CH_2CH_2CH_2CH_3$	69
	(2-Methylpentane)	$CH_3CHCH_2CH_2CH_3$ with CH_3	60
	(3-Methylpentane)	$CH_3CH_2CHCH_2CH_3$ with CH_3	63
	(2,3-Dimethylbutane)	$CH_3CH-CHCH_3$ with CH_3 CH_3	58
	Neohexane (2,2-Dimethylbutane)	$CH_3CCH_2CH_3$ with CH_3 above and CH_3 below	50

them, isooctane, burns very well in car engines, and is used as a standard in determining the octane rating of a gasoline.

$$CH_3\text{—}\overset{\overset{\displaystyle CH_3}{|}}{\underset{\underset{\displaystyle CH_3}{|}}{C}}\text{—}CH_2\text{—}\overset{\overset{\displaystyle CH_3}{|}}{\underset{\underset{\displaystyle H}{|}}{C}}\text{—}CH_3$$

Isooctane (2,2,4-Trimethylpentane)

Someone once calculated that a 40-carbon compound would have more than 60 trillion possible isomers!

Reactions of Alkanes

9.8 Oxidation

You are probably most familiar with alkanes as the fuels that heat your home, run your car, and power your camping stove and lantern. Although we mentioned earlier that the saturated hydrocarbons are fairly unreactive, one of the reactions that they readily undergo is the reaction with oxygen known as burning. In this reaction, oxygen atoms from the air combine with the carbon and hydrogen atoms in the hydrocarbon until the carbon and hydrogen atoms are bonded to as many oxygens as possible. The reaction is exothermic, and the end products of the reaction are carbon dioxide and water. This reaction with oxygen, by the way, is occurring all the time even at room temperature; but it usually occurs at an imperceptible rate. However, if the compound is heated in the presence of air, this process will speed up until a point is reached, called the ignition point, where the reaction becomes rapid enough that the heat generated can be seen and felt. It is when we actually see and feel the reaction occurring that we say the substance is burning, and refer to the reaction as "combustion." The general name for any reaction with oxygen, no matter what the rate, is **oxidation.** While the process of oxidation occurring in a forest fire is easily seen and impressive in scale, even the results of very slow oxidation can be observed by noticing the yellowing of pages in old books (Figure 9.10). The equation for the particular oxidation reaction that occurs when methane is burned is

$$CH_4 + 2O_2 \longrightarrow CO_2 + 2H_2O + Energy$$

Oxidation is occurring continuously in our bodies and, in fact, is the source of all the energy for our cells. If this oxidation occurred in the same way as methane burns, our cells would be destroyed. However, oxidation occurs in the body at a very controlled rate, allowing cells to trap and use the energy released. Our bodies actually do not use alkanes as their main fuel supply, but rather use derivatives of alkanes—carbohydrates and fats. The oxygen needed for oxidation is supplied to the cells by the blood, which picks up oxygen in the lungs and distributes it throughout the body, returning the waste product carbon dioxide to the lungs to be exhaled. The

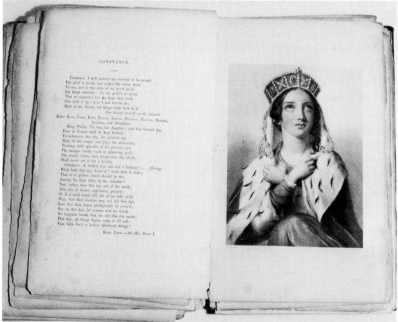

Figure 9.10 Both of these photographs illustrate oxidation reactions, but the reactions are occurring at vastly different rates. (Top, Courtesy U.S. Forest Service; bottom, Ron Nelson)

water produced in the oxidation reaction either remains in the cells and tissues, or is excreted through sweat and urine. Just as a fire can be smothered by a blanket cutting off the oxygen supply, so our lives can be smothered by anything that cuts off our supply of oxygen for as short a period as five minutes.

If there is an insufficient amount of oxygen present when alkanes are burned, incomplete burning (or incomplete combustion) occurs, producing end products of carbon monoxide and water. For example, if methane is burned in an oxygen-poor environment, the following reaction occurs.

$$2CH_4 + 3O_2 \longrightarrow 2CO + 4H_2O$$

The carbon monoxide produced by incomplete combustion in factories and cars represents the largest source of carbon monoxide pollution in our society. It was computed that in 1965, 91 million tons of carbon monoxide entered the atmosphere from automobile exhaust alone. Recent studies have shown that the carbon monoxide from automobile exhaust can accelerate the formation of photochemical smog.

While carbon dioxide is a normal part of the environment of living cells, carbon monoxide is toxic. When inhaled, carbon monoxide greatly reduces the oxygen-carrying ability of our red blood cells. At high levels, this impairment can be so great as to produce coma and death; but even at low levels, disruption of the central nervous system can be measured. Cigarette smokers inhale carbon monoxide into their lungs along with the cigarette smoke, producing elevated levels of carbon monoxide in their bloodstreams. These levels are often sufficient to produce measurable effects on the central nervous system. Urban drivers who smoke can easily

Table 9.6   The Effects of Carboxyhemoglobin* Blood Levels on the Human Body

Blood Levels of Carboxyhemoglobin	Effects
2 to 5%	Impairment of the central nervous system.
5%	Impairment of perception and psychomotor performance.
10%	Oxygen transport significantly impaired.
15%	Headaches, dizziness, and lassitude.
15 to 40%	Ringing ears, nausea, vomiting, heart palpitations, difficulty breathing, muscular weakness, apathy.
40% and above	Collapse, coma, and death.

*Carboxyhemoglobin is a hemoglobin–carbon monoxide complex that is formed when carbon monoxide enters the blood.

elevate their carbon monoxide level above that necessary to produce headaches, dizziness, and weariness—the first symptoms of carbon monoxide poisoning (Tables 9.6 and 9.7).

Table 9.7   Median Carboxyhemoglobin Levels in Smokers and Nonsmokers*

Location	Cigarette smokers	Nonsmokers
Anchorage	4.7%	1.5%
Chicago	5.8%	1.7%
Denver	5.5%	2.0%
Houston	3.2%	1.2%
Los Angeles	6.2%	1.8%
Vermont and New Hampshire	4.8%	1.2%
Washington, D.C.	4.9%	1.2%

*The maximum level recommended by the Clean Air Act of 1971 is 1.5%.
Reprinted from the Journal of the American Medical Association, August 26, 1974, Volume 229. Copyright © 1974. American Medical Association.

Our bodies have several different mechanisms for responding to elevated carbon monoxide levels, all of which result in increased strain on the heart and increased risk of heart disease. Short-term high levels cause the heart to pump faster, and long-term low levels cause the body to increase the number of red blood cells that carry oxygen, thereby thickening the blood and increasing the work load on the heart. Studies by John Goldsmith of the California State Department of Public Health have shown a direct relationship between levels of carbon monoxide and death from heart attack. Carbon monoxide levels in the air in excess of 10 parts per million (ppm) may be associated with increased death rates in patients with heart disease. (Note that the air in Los Angeles and New York normally averages more than 10 ppm carbon monoxide on a daily basis!) The difference in death rate monitored in these studies occurred only during periods of increased carbon monoxide pollution, showing no correlation with other air pollutants.

9.9 Substitution Reactions

Under the proper chemical conditions, alkanes can also react with nitric acid and with the elements of group VII (the halogens) in reactions called **substitution reactions.** These reactions are so named because another atom or group of atoms is substituted for one or more of the hydrogens on the alkane. For example, when methane reacts with nitric acid, nitromethane is formed.

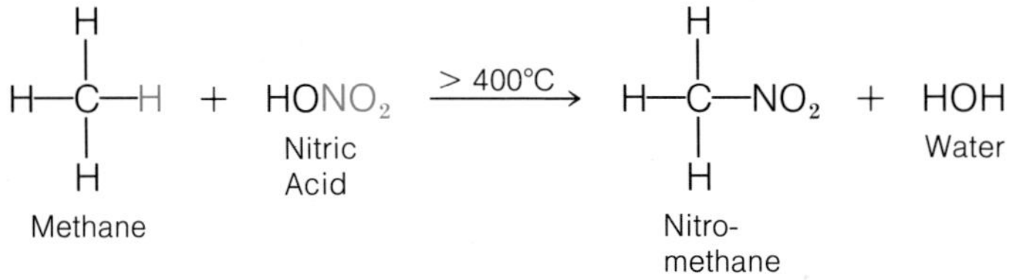

$$
\begin{array}{c}
\overset{\displaystyle H}{\underset{\displaystyle H}{H-C-H}} + HONO_2 \xrightarrow{\;>400°C\;} \overset{\displaystyle H}{\underset{\displaystyle H}{H-C-NO_2}} + HOH
\end{array}
$$

Methane   Nitric Acid   Nitro-methane   Water

In the above chemical equation you will notice that > 400°C has been written above the arrow. Whenever special chemical conditions are required for a reaction to occur, they are quite often indicated in this way. In this reaction, then, the temperature must be greater than 400°C for the reaction to occur. Nitromethane, the product of this reaction, is used as a solvent, as an important chemical in the production of other organic compounds, and as a high energy fuel for racing cars.

When methane and chlorine are exposed to sunlight, a reaction occurs that produces a mixture of products.

$$
CH_4 + Cl_2 \xrightarrow{\;Sunlight\;}
\begin{cases}
HCl, \\
CH_3Cl, \\
CH_2Cl_2, \\
CHCl_3, \\
CCl_4
\end{cases}
$$

Of the products from this reaction, methyl chloride (CH_3Cl) is used as a solvent and refrigerant; trichloromethane, or chloroform ($CHCl_3$), is an anesthetic; and tetrachloromethane, or carbon tetrachloride (CCl_4), was used as a dry cleaning solvent until it was shown to cause liver damage.

Unsaturated Hydrocarbons

9.10 Alkenes

A study of the alkanes just begins to hint at the diverse chemistry of carbon. Not only can carbon form strong single bonds with other carbon atoms, it can also share four or six electrons with another carbon atom, thereby forming a double or triple bond (Figure 9.11). We have seen that carbon is commonly found with four bonds in a tetrahedral arrangement, since this structure keeps the negative pairs of electrons as far apart as

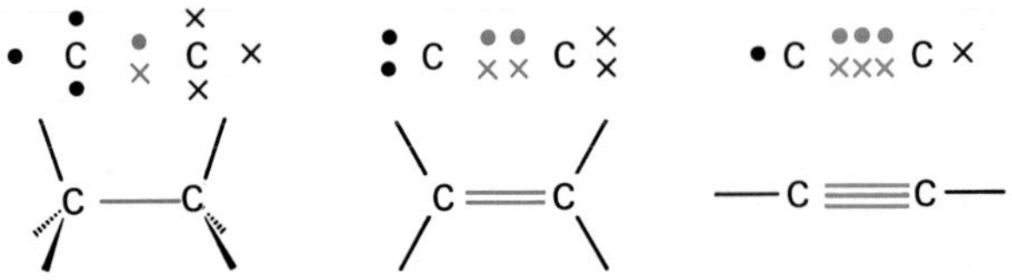

Figure 9.11 A carbon atom can form a single, double, or triple bond with another carbon atom.

possible. Understandably, it takes more energy to form a double bond than a single bond because there are now two pairs of negative electrons located between the carbon atoms. Even more energy is required to hold the three electron pairs in a triple bond. This circumstance makes double and triple bonds progressively more reactive than the single bond, and multiple bonds form an unstable, or reactive, spot in the organic molecule. Compounds with double or triple bonds are said to be **unsaturated.** If a compound has several double or triple bonds, then it is **polyunsaturated** (Figure 9.12). You probably have heard that word before on television commercials, and we will deal with the question of polyunsaturated fats and oils in the chapter on lipids.

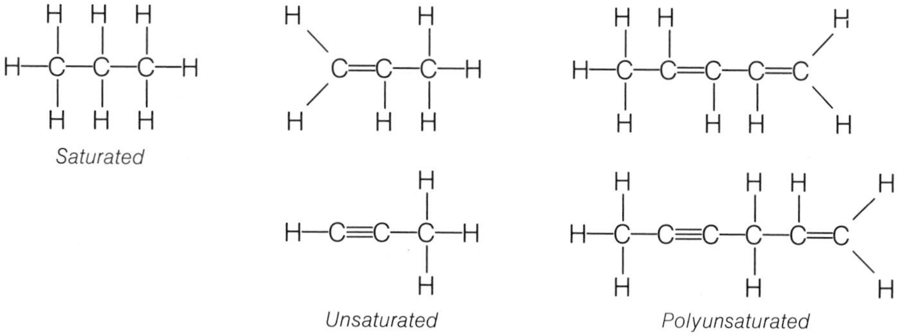

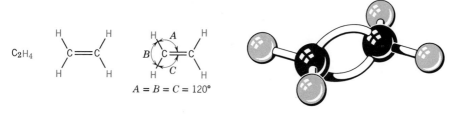

Saturated

Unsaturated

Polyunsaturated

Figure 9.12 The degree of saturation of an organic molecule is determined by the number of carbon-to-carbon double or triple bonds in the molecule.

The class of compounds containing double bonds between carbon atoms is known as the **alkenes.** The simplest alkene is ethene, or ethylene, C_2H_4 (Figure 9.13). Ethylene is a flammable, anesthetic gas that is nontoxic to tissues even in high concentrations. Its anesthetic effects are rapid; two to four minutes after administration, a patient is ready for surgery. A commercially useful property of ethylene gas is its ability to shorten the ripening time of citrus fruit.

C_2H_4

$A = B = C = 120°$

Figure 9.13 Ethene (ethylene)

Notice that the structure of ethylene is no longer tetrahedral. The most stable arrangement for this molecule results when the bonds are as far apart as possible, and this occurs when all the atoms in the ethylene molecule are located in the same plane. In general, alkene molecules can have two or more double bonds, which can be located anywhere in the molecule. Large alkene molecules can have dozens of double bonds, leading to millions of possible isomers (Table 9.8).

Table 9.8 Some Alkenes

Number of Carbons	Number of Double Bonds	Molecular Formula	Common Name (IUPAC Name)	Structural Formula	Boiling Point °C
2	1	C_2H_4	Ethylene (Ethene)	$CH_2{=}CH_2$	−104
3	1	C_3H_6	Propylene (Propene)	$CH_3CH{=}CH_2$	−47
4	1	C_4H_8	α-Butylene (1-Butene)	$CH_2{=}CHCH_2CH_3$	−6
			β-Butylene (2-Butene)	$CH_3CH{=}CHCH_3$	cis 4 trans 1
			Isobutylene (2-Methyl-1-propene)	$CH_2{=}\overset{\underset{\mid}{CH_3}}{C}{-}CH_3$	−7
4	2	C_4H_6	Bivinyl (1, 3-Butadiene)	$CH_2{=}CHCH{=}CH_2$	−4
			Methylallene (1, 2-Butadiene)	$CH_2{=}C{=}CHCH_3$	11
6	3	C_6H_8	Divinylethylene (1, 3, 5-Hexatriene)	$CH_2{=}CHCH{=}CHCH{=}CH_2$	78

9.11 Geometric Isomers

The existence of a carbon-to-carbon double bond gives rise to another type of isomerism, **geometric isomerism.** A single bond between carbon atoms,

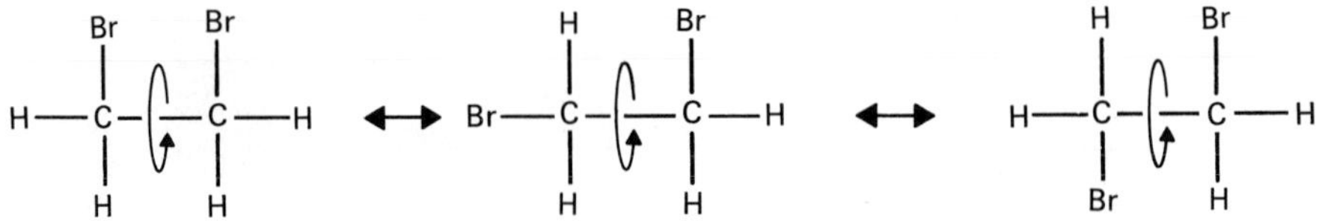

Figure 9.14 *Cis-trans* isomerism does not exist in 1,2-dibromoethane because the carbon-to-carbon single bond is free to rotate; as a result, the atoms bonded to the carbons can be in any position.

as is found in the alkanes, does not restrict the rotation of atoms around that bond; the carbon atoms can twist freely around their single bonds just as two spheres can twist freely when connected by a string (Figure 9.14). However, a double or triple bond between two carbon atoms is structurally rigid, preventing free rotation of the carbon atoms. This results in two possible arrangements of atoms on either side of the unsaturated bond (Figure 9.15). In terms of our analogy, replácing a single bond between two carbon atoms by a double or triple bond is equivalent to replacing the

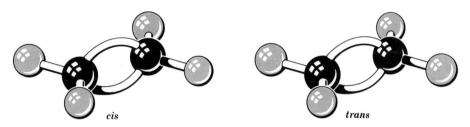

cis   *trans*

Figure 9.15 *Cis-trans* isomerism does exist in 1,2-dibromoethene. The double bond is structurally rigid, preventing free rotation of the carbon atoms.

string with a rigid pole between the two spheres. When specified atoms or groups of atoms attached to the doubly bonded carbons appear on the same side of the bond, the molecule is called a **cis isomer.**

$$CH_3 \diagdown \qquad \diagup CH_3$$
$$C{=}C$$
$$\diagup \qquad \diagdown$$
$$H \qquad H$$
cis-Butene

$$Br \diagdown \qquad \diagup Br$$
$$C{=}C$$
$$\diagup \qquad \diagdown$$
$$H \qquad H$$
cis-Dibromoethene

A **trans isomer** is formed when specified atoms or groups of atoms appear on opposite sides of the double bond.

$$CH_3 \diagdown \qquad \diagup H$$
$$C{=}C$$
$$\diagup \qquad \diagdown$$
$$H \qquad CH_3$$
trans-Butene

$$Br \diagdown \qquad \diagup H$$
$$C{=}C$$
$$\diagup \qquad \diagdown$$
$$H \qquad Br$$
trans-Dibromoethene

You may wonder why we are bothering to point out these different types of isomers. A major reason is that isomers are critically important to the workings of living cells. Although you may have trouble telling the difference between the *cis* and *trans* isomer of the following complicated compound, your body can—especially the cells in the retina of your eye.

Vitamin A.
(all *trans* — 100% activity)

9-*cis* isomer
(22% activity)

The *trans* isomer is vitamin A, a compound necessary to our ability to see in dim light. The corresponding *cis* isomers can be similarly used by the retina cells, but these isomers cannot be used as efficiently as the all *trans* isomer shown, and they are much less effective in reducing night blindness.

Reactions of Alkenes

Because the carbon-to-carbon double bond is more reactive than the corresponding single bond, alkenes find widespread commercial use in a variety of reactions. They are especially important in the chemical industry, where they are used in the production of many other compounds. Three important types of reactions for which alkenes are well suited are addition reactions, oxidation reactions, and polymerization reactions.

9.12 Addition Reactions

An **addition reaction** takes place when two atoms react with the two carbons connected by the double bond, causing the double bond to become a single bond.

Ethene   Bromine   1, 2-Dibromoethane

This bromine reaction is often used to test for the presence of unsaturated bonds (that is, double or triple bonds) in a molecule. Bromine in water forms a reddish brown solution, but the dibromides formed from the addition reaction are colorless. Thus, if the reaction of bromine water with a hydrocarbon results in a colorless solution, the hydrocarbon may have contained unsaturated bonds.

Compounds other than the halogens can also be used in addition reactions. A few examples follow.

Hydrogenation reaction	$\begin{array}{c} H \quad\quad H \\ \diagdown \quad\quad \diagup \\ C{=}C \\ \diagup \quad\quad \diagdown \\ H \quad\quad H \end{array}$	$+\ H_2$	$\longrightarrow$	$\begin{array}{c} H \ \ H \\ \mid\ \ \mid \\ H{-}C{-}C{-}H \\ \mid\ \ \mid \\ H \ \ H \end{array}$
	Ethene	Hydrogen		Ethane

Hydrohalogenation reaction	$\begin{array}{c} H \quad\quad H \\ \diagdown \quad\quad \diagup \\ C{=}C \\ \diagup \quad\quad \diagdown \\ H \quad\quad H \end{array}$	$+\ HCl$	$\longrightarrow$	$\begin{array}{c} H \ \ H \\ \mid\ \ \mid \\ H{-}C{-}C{-}Cl \\ \mid\ \ \mid \\ H \ \ H \end{array}$
	Ethene	Hydrogen chloride		Ethyl chloride

Hydration reaction	$\begin{array}{c} H \quad\quad H \\ \diagdown \quad\quad \diagup \\ C{=}C \\ \diagup \quad\quad \diagdown \\ H \quad\quad H \end{array}$	$+\ HOH$	$\longrightarrow$	$\begin{array}{c} H \ \ H \\ \mid\ \ \mid \\ H{-}C{-}C{-}OH \\ \mid\ \ \mid \\ H \ \ H \end{array}$
	Ethene	Water		Ethyl alcohol

9.13 Oxidation Reactions

Earlier in the chapter we discussed the oxidation of alkanes by describing the combustion reaction. But compounds do not have to be burned in order to be oxidized. There are actually a wide range of reactions that fall into the category of oxidation reactions or, more precisely, oxidation-reduction (redox) reactions. For the purposes of our study of organic chemistry, we will define an **oxidation reaction** to be a reaction in which one of the reactant molecules gains oxygen atoms or loses hydrogen atoms. Conversely, a **reduction reaction** will be defined as a chemical reaction in which a molecule gains hydrogen atoms or loses oxygen atoms.

As was the case with the alkanes, alkenes can be oxidized completely to produce carbon dioxide and water.

$$CH_2{=}CH_2 + 3O_2 \longrightarrow 2CO_2 + 2H_2O$$

$$\text{Ethene} \quad\quad \text{Oxygen} \quad\quad \text{Carbon dioxide} \quad\quad \text{Water}$$

Note that in this reaction the molecule of ethene undergoes oxidation. Under certain conditions the oxidation of ethene may be incomplete, resulting in the formation of glycol (a type of alcohol) instead of carbon dioxide.

$$CH_2{=}CH_2 \xrightarrow{\ KMnO_4\ } \begin{array}{c} OH\ OH \\ \mid\ \ \ \mid \\ H{-}C{-}C{-}H \\ \mid\ \ \mid \\ H\ \ H \end{array}$$

$$\text{Ethene} \quad\quad \underset{\text{permanganate)}}{\text{(Potassium}} \quad\quad \text{Glycol}$$

9.14 Polymerization

You are probably most familiar with the word ethylene from seeing it in the name of a type of plastic, polyethylene. Polyethylene is a compound composed of thousands of ethylene (ethene) units bonded together. This is how three ethylene units would bond together:

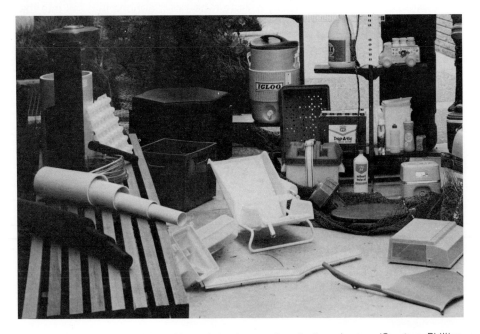

The process of joining many single units, called **monomers,** together to form very large molecules, called **polymers,** is **polymerization.** Thus, polyethylene is a polymer of the monomer ethylene. Synthetic (that is, man-made) polymers appear everywhere in our daily lives, from the plastic containers, bags, and wrappings we use, to the synthetic fibers such as orlon, rayon, and acrylics that we wear, to the synthetic rubber that we use for tires and for parts in appliances (Table 9.9 and Figure 9.16).

Figure 9.16 Examples of the wide variety of uses of synthetic polymers. (Courtesy Phillips Petroleum Company)

Table 9.9 Some Common Polymers

Monomer	Polymer	Uses
Propylene H_2C=$CHCH_3$	Polypropylene —$CH_2CHCH_2CHCH_2CH$— 　　　$\vert$　　$\vert$　　　$\vert$ 　　CH_3　CH_3　CH_3	Film and molded parts
Styrene H_2C=CHC_6H_5	Polystyrene —CH_2CH—CH_2CH—CH_2CH— 　　$\vert$　　　　$\vert$　　　　$\vert$ 　C_6H_5　　C_6H_5　　C_6H_5	Molded objects, insulation, and foam plastics
Methyl methacrylate 　CH_3 O 　$\vert$　$\vert\vert$ H_2C=C—$COCH_3$	Polymethyl methacrylate 　　CH_3　　CH_3　　CH_3 　　$\vert$　　　$\vert$　　　$\vert$ —CH_2C—CH_2C—CH_2C— 　　$\vert$　　　$\vert$　　　$\vert$ 　$COCH_3$ $COCH_3$ $COCH_3$ 　$\vert\vert$　　　$\vert\vert$　　　$\vert\vert$ 　O　　　O　　　O	Plexiglas and Lucite
Vinyl chloride H_2C=$CHCl$	Polyvinyl chloride (PVC) —$CH_2CHCH_2CHCH_2CH$— 　　$\vert$　　$\vert$　　$\vert$ 　　Cl　Cl　Cl	Plastic bottles and containers, plastic pipe, and insulation
Dichloroethene (Vinylidene chloride) H_2C=CCl_2	Polyvinylidene chloride —$CH_2CCl_2CH_2CCl_2CH_2CCl_2$—	Saran, wrapping film, fibers, and tubing
Acrylonitrile H_2C=$CHCN$	Polyacrilonitrile —$CH_2CHCH_2CHCH_2CH$— 　　$\vert$　　$\vert$　　$\vert$ 　　CN　CN　CN	Orlon and clothing fibers
Tetrafluoroethylene F_2C=CF_2	Polytetrafluoroethylene —$CF_2CF_2CF_2CF_2CF_2CF_2$—	Teflon and lubricating films

However, don't think that polymers are the result of man's ingenuity alone; the synthetic polymer industry was a result of man's attempts to imitate natural polymers. For example, tropical plants produce natural rubber in the form of a milky sap called latex, which is a large molecule composed of 4500 isoprene units (Figure 9.17). Isoprene is a five-carbon compound having two double bonds.

$$
\begin{array}{c}
CH_3 \\
\vert \\
H\text{—}C\text{=}C\text{—}C\text{=}C\text{—}H \\
\vert\qquad\vert\ \ \vert \\
H\qquad H\ \ H
\end{array}
$$

Isoprene (2-Methyl-1, 3-butadiene)

Figure 9.17 Collecting latex sap from rubber trees. (Carl Frank/ Photo Researchers)

A molecule of isoprene has double bonds between carbon atoms alternating with single bonds between carbon atoms. This alternating arrangement is given the name **conjugate double bonds**. Any compound having conjugate double bonds will be more stable than a similar compound having double bonds arranged in any other pattern.

Other polymers of isoprene form the large class of colored compounds that were first isolated from turpentine, and are known as the **terpenes.** One of these terpenes is carotene, so named because it was first isolated from carrots. Carotene is made up of eight isoprene units bonded in a fairly complex manner.

$$CH_3 \quad CH_3 \qquad\qquad CH_3 \qquad\qquad\qquad CH_3 \qquad CH_3\ CH_3$$
$$-(CH\!=\!CH\!-\!C\!=\!CH)_2CH\!=\!CH(CH\!=\!C\!-\!CH\!=\!CH)_2-$$
$$CH_3 \qquad\qquad\qquad\qquad\qquad\qquad\qquad\qquad CH_3$$

β-Carotene (Provitamin A)

As a solid, carotene is red in color; but when dissolved in oil or fat, the color ranges from orange to yellow depending upon its concentration. Carotene is the pigment responsible for the orange color of carrots and sweet potatoes, and the yellow color of butter, egg yolks, and chicken fat.

Table 9.10 Some Alkynes

Number of Carbons	Number of Triple Bonds	Common Name (IUPAC Name)	Structural Formula	Boiling Point °C
2	1	Acetylene (Ethyne)	$HC\equiv CH$	−84
3	1	Methylacetylene (Propyne)	$CH_3C\equiv CH$	−23
4	1	Ethylacetylene (1-Butyne)	$CH_3CH_2C\equiv CH$	8
		Dimethylacetylene (2-Butyne)	$CH_3C\equiv CCH_3$	27
	2	Biacetylene (1, 3-Butadiyne)	$HC\equiv C-C\equiv CH$	10
5	1	n-Propyl acetylene (1-Pentyne)	$CH_3CH_2CH_2C\equiv CH$	40
		Ethyl-methyl acetylene (2-Pentyne)	$CH_3CH_2C\equiv CCH_3$	56

People of Oriental ancestry have a small amount of carotene dissolved in the fatty layer under their skin, giving their skin its yellowish cast. Carotene is an important component in our diets because cells in our body can break up one of the isomers of carotene to form vitamin A. This fact gives some foundation to the old wives' tale that eating carrots will help you see better in the dark.

9.15 Alkynes

Alkynes are the class of hydrocarbons that contain triple bonds between carbon atoms. These triple bonds, containing three pairs of electrons between the two carbon atoms, put quite a strain on the molecule, making such bonds very reactive. Nevertheless, it is possible for a molecule to have more than one triple bond, and there are some molecules that contain both double and triple bonds (Table 9.10).

The simplest alkyne is ethyne, or acetylene, C_2H_2 (Figure 9.18).

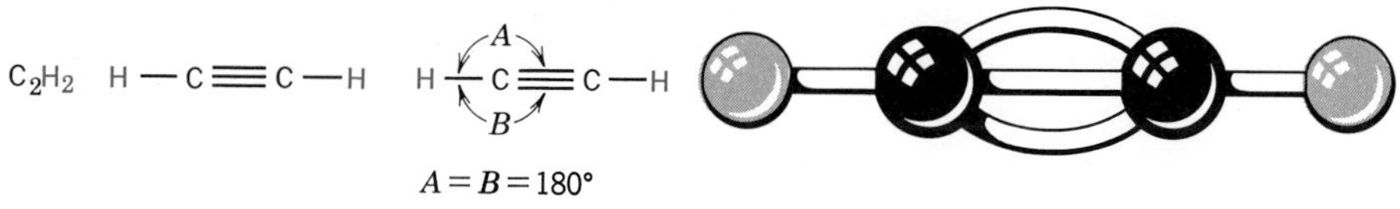

C_2H_2 $H-C\equiv C-H$ $H \overset{A}{\underset{B}{\equiv}} C\equiv C-H$

$A = B = 180°$

Figure 9.18 Ethyne (acetylene)

Acetylene is a flammable and explosive gas that burns with a bright flame, making it useful for lighting purposes. When acetylene is burned along with oxygen in an oxyacetylene welding torch, the resulting reaction is highly exothermic, yielding enough energy to cut and weld metals. Acetylene is so highly reactive that it forms a convenient starting material for the production of almost every simple organic compound.

9.16 Reactions of Alkynes

Alkynes undergo addition reactions in much the same manner as alkenes. Under specific chemical conditions, the hydrogenation reaction can be controlled in such a way that alkenes are produced.

Hydrogenation reactions

$$H-C\equiv C-H + H_2 \xrightarrow[\text{conditions}]{\text{Controlled}} \begin{array}{c} H \\ \diagdown \\ C=C \\ \diagup \\ H \end{array} \begin{array}{c} H \\ \diagup \\ \diagdown \\ H \end{array}$$

$$H-C\equiv C-H + 2H_2 \longrightarrow H-\overset{\overset{\displaystyle H}{|}}{\underset{\underset{\displaystyle H}{|}}{C}}-\overset{\overset{\displaystyle H}{|}}{\underset{\underset{\displaystyle H}{|}}{C}}-H$$

Halogenation reaction

$$H-C\equiv C-H + 2Br_2 \longrightarrow H-\overset{\overset{\displaystyle Br}{|}}{\underset{\underset{\displaystyle Br}{|}}{C}}-\overset{\overset{\displaystyle Br}{|}}{\underset{\underset{\displaystyle Br}{|}}{C}}-H$$

Hydrohalogenation reaction

$$H-C\equiv C-H + 2HCl \longrightarrow H-\overset{\overset{\displaystyle H}{|}}{\underset{\underset{\displaystyle H}{|}}{C}}-\overset{\overset{\displaystyle H}{|}}{\underset{\underset{\displaystyle Cl}{|}}{C}}-Cl$$

Alkynes also readily undergo combustion to form carbon dioxide and water.

Carbon in Rings

9.17 Cyclic Hydrocarbons

The great diversity and complexity of carbon chemistry again becomes evident when we realize that carbon atoms not only form straight chain molecules, but also form stable rings of various sizes, with the possibility of both single and double bonds. In fact, even a triple bond is possible in a very large ring (Table 9.11).

Table 9.11 Some Cyclic Hydrocarbons

Molecular Formula	Name	Structural Formula
C_3H_6	Cyclopropane	CH_2 / CH_2—CH_2
C_4H_8	Cyclobutane	CH_2—CH_2 / CH_2—CH_2
C_4H_6	Cyclobutene	CH_2—CH / CH_2—CH
C_5H_{10}	Cyclopentane	CH_2 / CH_2 CH_2 / CH_2—CH_2
C_5H_6	1, 3-Cyclopentadiene	CH / CH_2 CH / CH=CH
C_6H_{12}	Cyclohexane	CH_2 / CH_2 CH_2 / CH_2 CH_2 / CH_2
C_9H_{18}	1-Methyl-2-ethylcyclohexane	CH_2 / CH_2 $CHCH_3$ / CH_2 $CH CH_2CH_3$ / CH_2
$C_{12}H_{20}$	Cyclododecyne	CH_2—C≡C—CH_2 / $(CH_2)_8$

The most common cyclic hydrocarbons have rings composed of five or six carbon atoms, but both larger and smaller rings are possible.

The saturated cyclic hydrocarbon having the smallest number of carbon atoms is cyclopropane.

Cyclopropane

Cyclopropane is a highly potent anesthetic, but care must be taken in its use because it is both inflammable and explosive. Many hydrocarbons, in addition to cyclopropane, act as an anesthetic when inhaled. (Among the others are methane, acetylene, ethylene, and cyclobutane.) An anesthetic is a compound that, under the proper conditions, causes a person to lose the ability to feel pain, and in most cases also causes the person to lose consciousness. The anesthetic property of the hydrocarbons is due to the nonpolar characteristic of the molecules. The electrons in the covalent bond are shared fairly evenly between the carbon and the hydrogen atoms, and the bonds are symmetrically arranged about the carbon atoms.

An anesthetic compound acts on the nerves, preventing nerve impulses from moving along the nerve fibers. Each nerve is surrounded, in part, by a protective coating composed of molecules that are nonpolar in nature. When an anesthetic is inhaled, it enters the bloodstream from the lungs and travels to the tissues, where it dissolves in areas having large numbers of nonpolar molecules — and the protective coating about the nerves is just such a place. As the anesthetic accumulates in this coating, it causes a short circuit in the transmission of the nerve impulses. Since the nerve impulses no longer reach the brain, there is no longer any sensation of pain. The many deaths caused by inhaling methane, or sniffing the solvents in glue, have resulted from these hydrocarbon compounds dissolving in the protective coatings of the nerves to such an extent that they short-circuit the critical nerve impulses from the brain to the lungs and heart.

9.18 Cis and Trans Isomers

When carbon atoms are arranged in a ring, free rotation around even the carbon-to-carbon single bond is restricted. This leads to the existence of cis and trans isomers in ring compounds. A cis isomer of a ring compound will have specified atoms or groups of atoms lying on the same side of the carbon ring.

cis-1,2-Dibromocyclobutane

cis-1,2-Cyclobutanedicarboxylic acid

Conversely, a *trans* isomer will have these atoms or groups of atoms located on opposite sides of the carbon ring.

trans-1,2-Dibromocyclobutane

trans-1,2-Cyclobutanedicarboxylic acid

9.19 Benzene and Its Derivatives: the Aromatic Hydrocarbons

The most common and important of the cyclic hydrocarbons is benzene. Benzene and its derivatives comprise the class of compounds called the **aromatic hydrocarbons** (Table 9.12).

Table 9.12   Some Aromatic Hydrocarbons

Name	Strucual Formula	Name	Structural Formula
Benzene		Toluene (Methyl benzene)	CH_3
Bromobenzene	Br	Ethyl benzene	CH_2CH_3
Phenol	OH	Nitrobenzene	NO_2
Benzoic acid	COOH	Aniline	NH_2
Naphthalene		Anthracene	
Diphenyl			

In spite of their name, the aromatic hydrocarbons are no more "smelly" than other unsaturated hydrocarbons. However, the first natural compounds identified as part of this class had a definite and fairly pleasant odor – hence the name, aromatic. Benzene has the molecular formula C_6H_6, but its structure puzzled scientists for many years. The six carbon atoms in a molecule of benzene are arranged in a ring, and are often depicted with alternating single and double bonds.

Benzene

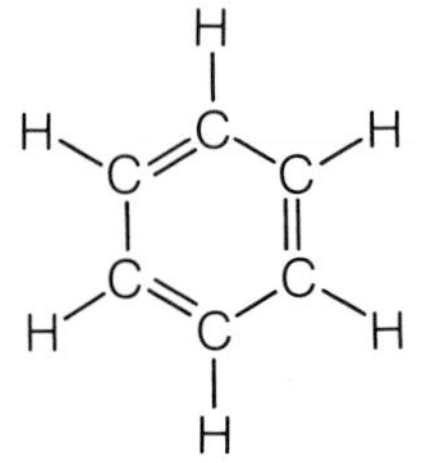

However, this structure does not accurately explain the properties of benzene. The actual benzene ring is much more stable than would be expected from this structure, and does not break open during reactions. Moreover, each carbon-to-carbon bond is the same strength, and the distance between each of the carbon atoms is equal. Modern theories of atomic behavior explain benzene's unusual characteristics in terms of partial or fractional bonds. One way to think about this is to picture the six carbon atoms held together by six equal $1\frac{1}{2}$ bonds that are less reactive than double bonds. Two-dimensional structural formulas do not adequately represent benzene's structure. It is

neither   nor

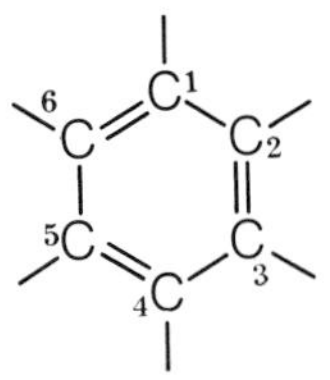

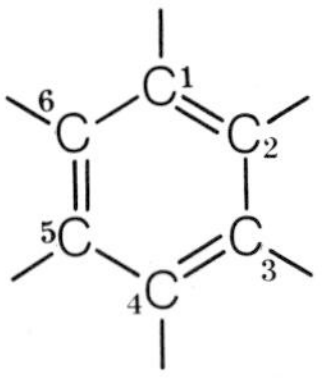

but halfway between these two structures. The name "resonance hybrid" is often used to describe this unusual structure. Symbolically we can represent this molecule by drawing a circle in the center of the benzene ring.

Benzene

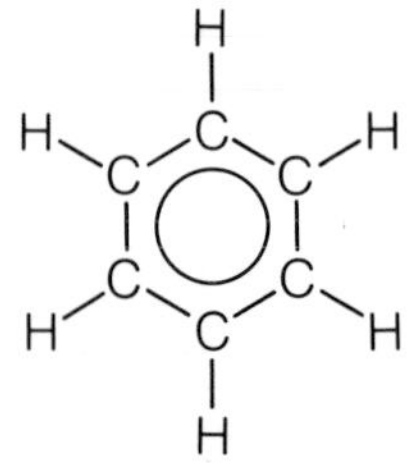

Its unusual structure makes the benzene molecule extremely stable, and this molecule forms part of the structure of many complex natural compounds. To simplify the writing of the structural formulas of such complex molecules, organic chemists have devised a schematic representation of a hydrocarbon or benzene ring.

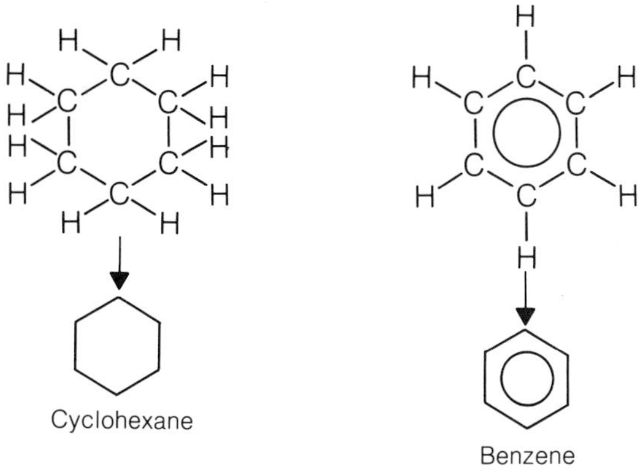

Cyclohexane

Benzene

To translate such a schematic diagram back to the full structural formula, place a carbon atom at each angle of the figure and draw a hydrogen atom on any spare bonds not involved in the ring. If any elements other than carbon or hydrogen are involved in the compound, they will be separately specified. Here are some examples:

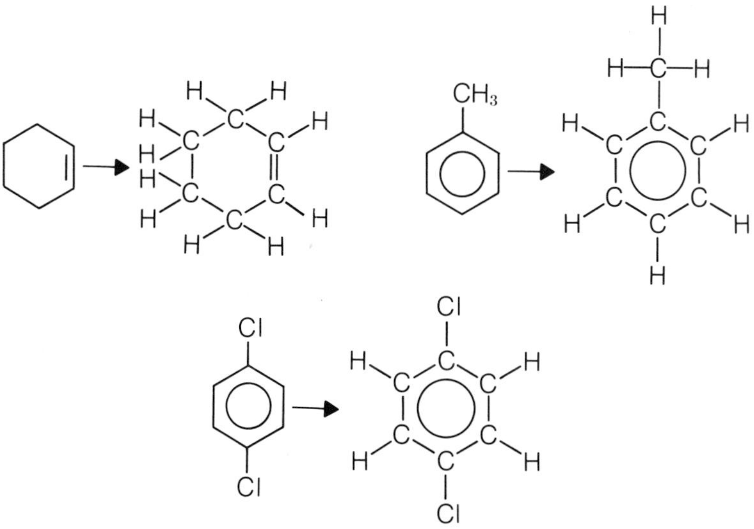

Benzene and other aromatic hydrocarbons can be recovered from coal tar, which is produced by heating bituminous (or soft) coal in the absence of air. Benzene is used by the chemical industry as a solvent, and as a starting material in the production of a variety of compounds. Its fumes are toxic and, when inhaled, can cause nausea and death from respiratory and heart failure. Again it should be stressed that the danger of these nonpolar compounds, many of which are used as solvents in such common products as paints, glues, and cleaners, is their ability to act on the central nervous system and to interfere with nerve action.

The benzene ring is found in many compounds that are critical to the life processes of the living organism. Phenylalanine, whose importance was illustrated in Chapter 1, is just one example.

Phenylalanine

Plants are able to synthesize benzene rings from carbon dioxide, water, and inorganic materials. However, animals cannot synthesize this compound, and their survival depends upon obtaining ample supplies of the essential aromatic compounds through their diets.

9.20 Reactions of Benzene

Extreme chemical conditions are required to cause a reaction involving the bonds between the carbon atoms on a benzene ring. As a result, the ring structure remains intact through most chemical reactions. Many aromatic compounds are produced both naturally and synthetically by substitution reactions — one or more hydrogens on the ring are replaced with other atoms or groups of atoms.

Substitution of a Halogen
We saw earlier that alkenes react readily with bromine, adding the bromine atoms across the double bond. However, only under special conditions will benzene react with bromine, and the resulting product is quite different from those found in reactions with alkenes. In reactions with benzene, the benzene ring remains intact and one bromine atom is substituted for a hydrogen atom.

Substitution reaction
(Benzene and bromine)

Addition reaction
(Ethene and bromine)

Reactions with Acids

Under certain conditions, benzene can be forced to react with concentrated acids in a substitution reaction.

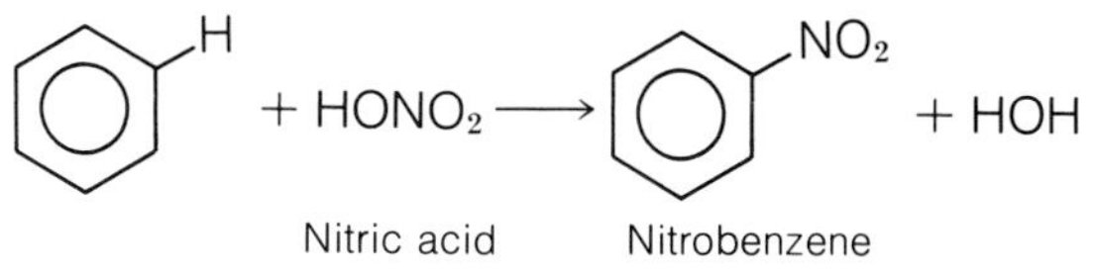
H
+ HONO₂ ⟶
NO₂
+ HOH
Nitric acid
Nitrobenzene

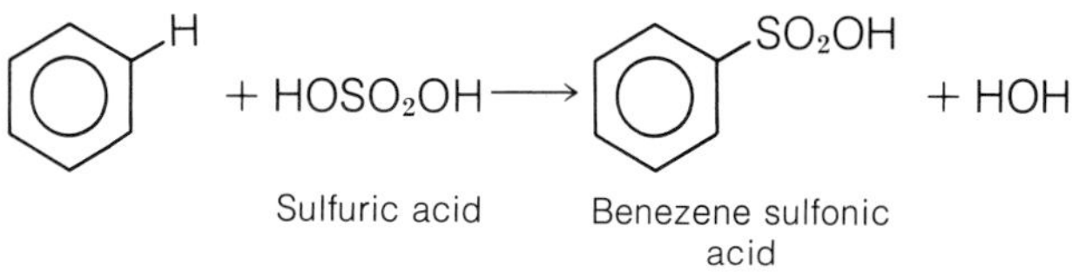
H
+ HOSO₂OH ⟶
SO₂OH
+ HOH
Sulfuric acid
Benezene sulfonic acid

9.21 Other Aromatic Compounds

Benzene rings often combine to form large molecules whose schematic structure resembles honeycombs. As a simple example, the compound naphthalene, which gives moth balls their characteristic odor, is composed of two benzene rings joined or fused together (see Table 9.12). Three benzene rings joined together form anthracene, an important starting material in the production of dyes (Figure 9.19).

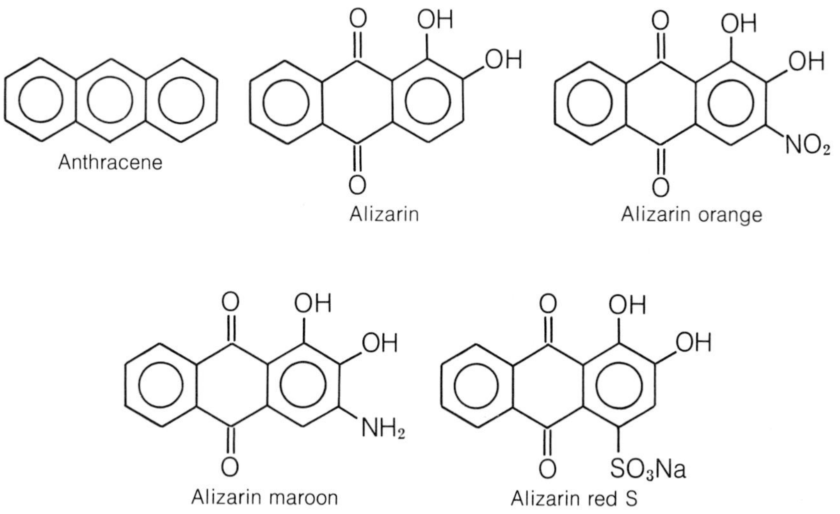
Anthracene
O
OH
OH
O
Alizarin
O
OH
OH
NO₂
O
Alizarin orange
O
OH
OH
NH₂
O
Alizarin maroon
O
OH
OH
O
SO₃Na
Alizarin red S

Figure 9.19 Some dyes that are produced from anthracene.

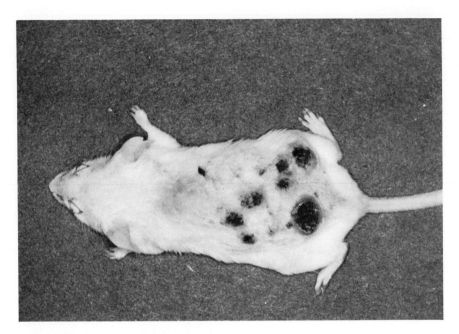

Figure 9.20 Skin tumors were produced by application of a solution containing 7,12-Dimethylbenzanthracene to the shaved skin on the back of this mouse. (Courtesy Kanematsu Suguira, Sloan Kettering Institute for Cancer Research)

Compounds containing multiple benzene rings are obtained in the production of coal tar. Their harmful effects to humans became evident when workers in European coal tar factories developed skin cancer. A

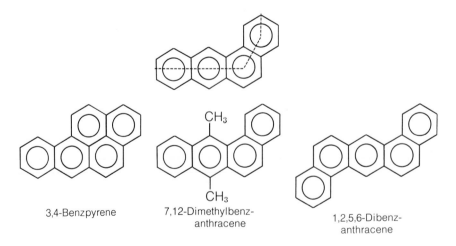

3,4-Benzpyrene   7,12-Dimethylbenz-anthracene   1,2,5,6-Dibenz-anthracene

Figure 9.21 These carcinogenic hydrocarbons found in coal tar all have a similar arrangement of fused benzene rings.

subsequent study of the chemical components of coal tar determined that several aromatic fused-ring compounds were capable of causing cancer in mice (Figure 9.20). Chemicals that cause cancer are known as **carcinogens.** It was found that the carcinogenic hydrocarbons in coal tar all had a similar arrangement of fused benzene rings (Figure 9.21). The compounds in Figure 9.21 are formed in the partial combustion of many large organic molecules. One of the most active carcinogens, 3,4-benzpyrene is discharged in vast quantities into the atmosphere of industrial nations each year, and is also one of the major carcinogens found in cigarette smoke. Only a few milligrams (0.001 grams) of 3,4-benzpyrene is sufficient to induce cancer in experimental animals.

The mechanism by which these rather inert compounds cause cancer in animals and man has puzzled scientists, but recent studies are beginning to shed light on this process. Our bodies contain a specific set of enzymes located primarily in the liver and kidneys, but also found in the lungs and other tissues, whose function it is to detoxify foreign chemicals that enter the body. One way in which this is accomplished is to make nonpolar compounds such as 3,4-benzpyrene more polar and, hence, more soluble in water and more readily excreted from the body.

3, 4-Benzpyrene   Metabolically activated derivative

Most of the products of the detoxifying reactions are less harmful to the cells than were the original reactants, but some of the products turn out to be very carcinogenic. Therefore, these enzymes, some of which are located in the lungs, can contribute to the production of carcinogenic hydrocarbons from cigarette smoke. The specific method by which these carcinogens turn a normal cell into a cancer cell is still an area of intensive research.

Hydrocarbon Derivatives

9.22 Halogenated Hydrocarbons

Hydrocarbon derivatives containing one or more atoms of the halogen family have a variety of uses in our society. Ethylene dibromide is a compound that is added to leaded gasoline to prevent the lead from settling out in the engine and ruining it.

$$\begin{array}{ccc} & Br & Br \\ & | & | \\ H-C-C-H \\ & | & | \\ & H & H \end{array}$$

Ethylene dibromide

The bromine in this compound combines with the lead atoms from the gasoline to form lead bromide, which is a vapor at engine temperatures and leaves the engine as part of the exhaust fumes.

Dichlorodifluoromethane, one of the compounds known as Freon®, consists of a methane molecule having two of its hydrogen atoms replaced by chlorine atoms, and two by fluorine atoms.

$$\begin{array}{c} F \\ | \\ Cl-C-F \\ | \\ Cl \end{array}$$

Dichlorodifluoromethane

Freon is odorless, nonpoisonous, noncorrosive, and nonflammable — making Freon far superior to other compounds for use as a refrigerant in refrigerators and other cold-storage equipment.

Fluorocarbons, compounds having carbon and fluorine atoms, are much more stable than hydrocarbons and are less affected by heat and other chemicals. Fluorocarbons find use as artificial rubber compounds, as lubricants, and as chemicals for fire extinguishers. An important polymer made from a fluorocarbon is teflon, which is constructed from the monomer tetrafluoroethene.

$$\begin{array}{ccc} F & & F \\ \backslash & & / \\ & C=C & \\ / & & \backslash \\ F & & F \end{array}$$

Tetrafluoroethene

Teflon is chemically inert in the presence of almost all compounds, and is widely used as a nonstick coating for pots and pans, and in electric insulation materials.

Extensive research is being conducted on the use of fluorocarbon compounds as substitutes for red blood cells, but there remains a question about possible harmful side effects of these compounds on the human body (Figure 9.22). Recent research indicates that the use of fluorocarbons as propellants in many aerosol products poses potential danger to the heart, and over time may have serious effects on human users of such products. Also, some scientists suspect that fluorocarbons released into the atmosphere may be damaging the protective layer of ozone gas in the

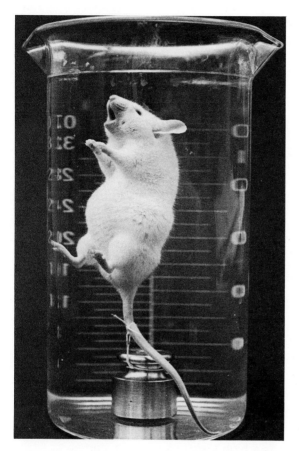

Figure 9.22 This mouse was submerged in the fluoro-carbon perfluorobutyl tetrahydrofuran for one hour. The oxygen needed by the mouse was dissolved in the fluoro-carbon, and when the mouse was removed, it was still alive and in healthy condition. (Courtesy Dr. Leland C. Clark, Jr., C. H. Research Foundation, Cincinnati)

upper atmosphere. If this is actually the case, a continuation of this process could have disastrous consequences, ranging from a significant increase in the occurrence of skin cancers to a possible change in the world's climates.

9.23 DDT

Widespread controversy surrounds the use of another halogenated hydrocarbon, dichlorodiphenyltrichloroethane—better known as DDT. DDT was first synthesized by a German chemist in 1874, but not until 60 years later was its effectiveness as an insecticide proven. Since that time there has been a succession of chlorinated hydrocarbons synthesized for use as pesticides (Table 9.13).

Table 9.13 Chlorinated Hydrocarbons Used in Pesticides

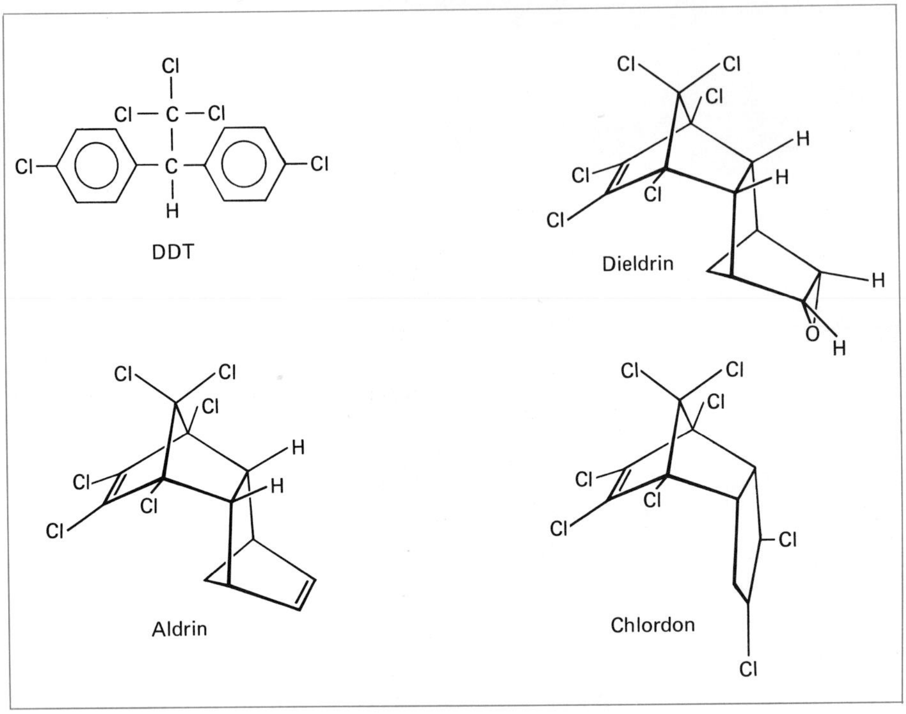

DDT

Dieldrin

Aldrin

Chlordon

DDT is a broad spectrum insecticide; that is, it indiscriminately kills both good and bad insects. It is still not known how DDT produces death in insects, but this compound has a high toxicity for insects and no apparent toxic effects to humans on brief exposure. It was not until Rachel Carson's book *Silent Spring* came out in 1962 that the harmful side effects of the chlorinated hydrocarbons were brought to the public's attention. DDT is a synthetic compound that breaks down slowly in the environment. Since it is a nonpolar compound it concentrates in the fatty tissues of living organisms, and this concentration increases as one moves up the food chain (Figure 9.23 and Table 9.14). Although the specific mechanism is not known, DDT is suspected of disrupting hormone production in animals, and this compound may have played a role in the extinction of

Table 9.14 Levels of DDT in Samples of Human Fat

	ppm DDT		ppm DDT
United States	12	Arctic regions	3
Israel	19	England	4
India	26	Canada	5

Figure 9.23 The higher up the food chain, the greater the concentration of DDT.

several species of birds (Figure 9.24). *Silent Spring* forcefully awakened the public to the delicate balance of nature, and to the ease with which that balance can be disrupted by synthetic chemicals released into the environment.

Figure 9.24 An egg found in the nest of a brown pelican off the coast of California. The weight of the adult bird sitting on the nest was enough to crush the thin shell resulting from high concentrations of DDT in the female bird. The concentration of DDT in the egg was measured at 2500 parts per million. None of the eggs laid by the 300-pair colony of birds hatched. (Courtesy Joseph R. Jehl, Jr.)

Additional Reading

Books
1. Isaac Asimov, *The New Intelligent Man's Guide to Science,* Basic Books, New York, 1965.

2. _____, *The World of Carbon,* Collier Books, New York, 1962.

3. Rachel Carson, *Silent Spring,* Houghton Mifflin, Boston, 1962.

4. Frank Graham, *Since Silent Spring,* Houghton Mifflin, Boston, 1970.

5. Alexander Oparin, *The Chemical Origin of Life,* Charles C. Thomas, Publisher, Springfield, Illinois, 1964.

Articles
1. L. D. Bodkin, "Carbon Monoxide and Smog," *Environment,* May 1974, page 34.

2. Earl Frieden, "The Chemical Elements of Life," *Scientific American,* July 1972, page 52.

3. P. Grover, "How Polycyclic Hydrocarbons Cause Cancer," *New Scientist,* June 14, 1973, page 685.

4. G. Haber, "The Crumbling Shield," *The Sciences,* December 1974, page 21.

5. Eugenia Keller, "The Origin of Life," *Chemistry,* December 1968, page 6; January 1969, page 12; April 1969, page 8.

6. _____, "The DDT Story," *Chemistry,* February 1970, page 8.

7. A. C. Lasaga, H. D. Holland, and M. J. Dwyer, "Primordial Oil Slick," *Science,* October 1, 1971, page 53.

8. S. E. Luria, "Genesis," *Natural History,* June–July 1973, page 10.

9. T. H. Maugh, II, "Chemical Carcinogenesis," *Science,* March 8, 1974, page 941.

10. Robert W. Medeiros, "Carbon Monoxide: The Invisible Enemy," *Chemistry,* January 1973, page 19.

11. G. Alex Mills, "Ubiquitous Hydrocarbons," *Chemistry*, February 1971, page 8; March 1971, page 13.

12. L. A. Purrett, "When Life Began on Earth," *Science News*, June 3, 1972, page 366.

Questions and Problems

1. Define the following terms:

   (a) Isomer
   (b) Organic chemistry
   (c) Hydrocarbon
   (d) Fractional distillation
   (e) Alkane
   (f) Alkene
   (g) Alkyne
   (h) Saturated hydrocarbon
   (i) Unsaturated hydrocarbon
   (j) Structural isomers
   (k) Geometric isomers
   (l) Substitution reaction
   (m) Addition reaction
   (n) Polymerization
   (o) Oxidation
   (p) Reduction
   (q) Aromatic hydrocarbon
   (r) Carcinogen

2. Carbon forms the backbone of life molecules. What are the properties of carbon that make it unique?

3. Define "organic chemistry."

4. What is the difference between the molecular and structural formula of a compound?

5. Identify each of the following compounds as either (a) an alkane, alkene, or alkyne, and (b) as saturated or unsaturated.

   (a) $CH_3CH_2CH_2CH_3$
   (b) 
   (c) 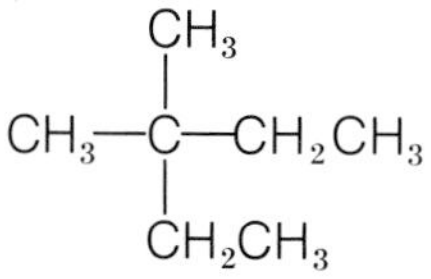
   (d) C_2H_2
   (e) 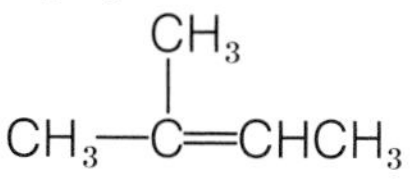
   (f) $CH_3(CH_2)_6CH_3$

6. Name the following compounds.

   (a) $CH_3CH_2CH_2CH_2CH_3$
   (b) $CH_3CH{=}CH_2$
   (c) $CH_3CH_2C{\equiv}CH$
   (d) 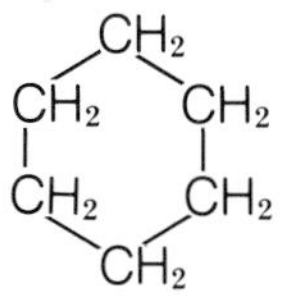

7. Write the structural formulas of the following compounds.

(a) Cyclopentane
(b) 1-Heptene

(c) 2-Hexyne
(d) 2,3-Dichlorobutane

8. Identify the structural isomers among the following compounds.

(a) $CH_3CH_2CH_2CH_2CH_3$

(b) $CH_3-\overset{\overset{\displaystyle CH_3}{|}}{\underset{\underset{\displaystyle CH_3}{|}}{C}}-CH_3$

(c) $H_2C\overset{\overset{\displaystyle CH_2}{\diagup}}{\underset{\underset{\displaystyle CH_2-CH_2}{|}}{\diagdown}}CH_2$

(d) $CH_3-\overset{\overset{\displaystyle CH_3}{|}}{\underset{\underset{\displaystyle H}{|}}{C}}-CH_3$

(e) $CH_3\overset{\overset{\displaystyle CH_3}{|}}{C}HCH_2CH_3$

(f) $CH_3\overset{\overset{\displaystyle CH_3}{|}}{C}H\overset{\underset{\displaystyle CH_3}{|}}{C}HCH_3$

9. Draw the geometric isomers for the following compounds.

(a) 2-Pentene
(b) 1,2-Dichloroethylene
(c) 1,2-Dichlorocyclopentane

10. Can dibromoacetylene exist as *cis* and *trans* isomers? Why or why not?

11. Propylene (propene) can undergo a reaction to form the polymer called polypropylene. Show the polymerization of three propylene molecules.

12. For each of the following pairs of molecules, identify the more reactive compound. Give the reason for your choices.

(a) $CH_3CH_2CH_3$ or $CH_3CH{=}CH_2$

(b) $CH_3C{\equiv}CCH_3$ or    $CH_3-\overset{\overset{\displaystyle CH_3}{|}}{\underset{\underset{\displaystyle H}{|}}{C}}-CH_3$

(c) 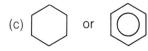   or

13. Identify each of the following as either a cyclic or aromatic hydrocarbon.

(a) 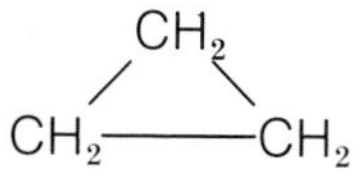
$$CH_2$$
$$CH_2 \!-\!\!-\!\!-\! CH_2$$

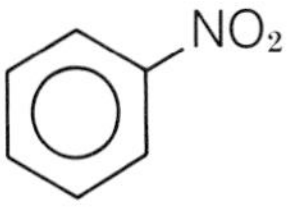

(c) 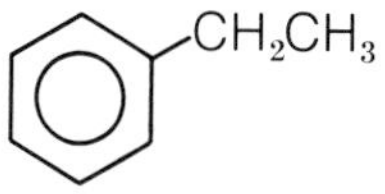
$$CH_2CH_3$$

(b)
$$NO_2$$

(d) 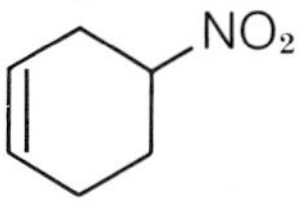
$$NO_2$$

14. Compounds found in several commercially available products have been found in laboratory experiments to be carcinogenic in mice. What does it mean for a compound to be carcinogenic? Should such products continue to be used by human consumers?

15. Ozone (O_3) is a compound which, when found in the atmosphere, is a mixed blessing. In the lower atmosphere, ozone forms brown smog which, in high concentrations, can cause respiratory problems in humans and can kill plant life. However, in the earth's upper atmosphere, ozone acts as a protective layer that screens out the harmful ultraviolet rays of the sun.
   1. Hydrocarbon pollutants released into the atmosphere increase the ozone at lower levels.
   2. Fluorocarbons that are released can reach the upper atmosphere unreacted, and may be removing the protective layer of ozone.
      (a) Why are fluorocarbons able to pass through the earth's lower atmosphere unreacted?
      (b) Discuss the implication for future generations of continued production and release of hydrocarbons and fluorocarbons into the atmosphere.

chapter 10

Oxygen

Learning Objectives

By the time you have finished this chapter, you should be able to:

1. Classify the organic molecules having oxygen-containing functional groups.

2. Discuss the reason for the comparatively high melting and boiling points of alcohols.

3. Explain why short carbon chain alcohols are soluble in water, while longer carbon chain alcohols are insoluble.

4. Identify the difference between a primary, secondary, and tertiary alcohol.

5. State two methods by which alcohols can be prepared in the laboratory.

6. Write the dehydration and oxidation reactions of ethanol.

7. Compare the polarity and water solubility of ethers and alcohols.

8. Identify the difference in structure between aldehydes and ketones.

9. Describe the effects of ethanol and methanol on the human body.

10. Describe one method of preparing aldehydes and ketones.

11. State the differences in the ease with which aldehydes and ketones can be oxidized and reduced.

12. Define an "acid" and contrast the strength of carboxylic acids and inorganic acids.

13. Describe one method of preparing organic acids.

14. Define a "condensation" reaction and give an example.

Joyce was thoroughly enjoying the wedding reception for her friend's son. She was feeling relaxed and jovial, and was having a great time. However, the next morning she awoke with a splitting headache and a general feeling of nausea. The substance responsible for these effects on her body was ethanol, the liquid commonly known as alcohol. Ethanol, or ethyl alcohol, is actually just one of many alcohols, all of which have molecules containing a special group of atoms. This group is called the hydroxyl group, and contains an oxygen atom and a hydrogen atom connected by a single bond, —O—H. When this hydroxyl group replaces one of the hydrogen atoms on a hydrocarbon, an alcohol is formed.

$$
\begin{array}{ccc}
& H \quad H & \\
& | \quad\ | & \\
H - & C - C & -H \\
& | \quad\ | & \\
& H \quad H &
\end{array}
\qquad
\begin{array}{ccc}
& H \quad H & \\
& | \quad\ | & \\
H - & C - C & -OH \\
& | \quad\ | & \\
& H \quad H &
\end{array}
$$

Ethane   Ethanol

You are probably quite familiar with the many effects that ethanol can have on the human body. It can create either a relaxed, easy feeling or a nauseating, aching one. As mentioned above, it is often responsible for a "great" evening, followed by a totally miserable morning. Ethanol, in the form of alcoholic beverages, is one of the most widely used of all drugs. It can be used to provide food energy, to either stimulate the appetite or to inhibit it, to aid digestion, to bring relief from tension, anxiety, or pain, and to function as a sedative.

As ethanol is swallowed, it is absorbed very rapidly by the mucous membranes found in the nose and throat, and its vapors are absorbed through the lungs. Absorption of ethanol into the body begins immediately after it is swallowed, with about 20% of the alcohol absorbed through the stomach, and the remainder through the small intestines. It can take up to six hours for the alcohol in a single drink to be absorbed and distributed throughout the body when the stomach is full, but only about one hour when the stomach is empty. Therefore, you certainly would feel the effects of a drink faster on an empty stomach than on a full one.

Ethanol dissolves completely in water. Once in the bloodstream it moves rapidly into the tissues, especially into organs having large blood supplies, such as the brain. This movement into the tissues continues until the concentration of ethanol in the tissues equals the concentration of ethanol in the blood. Because the ethanol will be equally distributed throughout the tissues of the body, the concentration of ethanol in one's breath or urine can be used as an accurate indicator of the level of ethanol in the blood. Police make use of this fact when they give a breath test to determine the state of intoxication of a driver.

A small amount of ethanol in the blood will act as a stimulant to most organs and body systems. But as the level increases, the ethanol becomes a depressant. This is especially true in the brain, where ethanol has a disruptive effect on the nerve cell membranes, making the brain less responsive to stimuli. When the blood alcohol level lies below 0.10%, a person is generally in a mood of pleasant relaxation; tensions and anxieties are eased. As the alcohol level rises, inhibitions become decreased as the control center in the brain becomes less active. At higher levels, the increased disruption of the nervous system results in a lack of muscular coordination, slurred speech, and difficulty in comprehending what is seen and heard. A concentration of alcohol in the blood above 0.36% can result in delirium, anesthesia, coma, and even death (Table 10.1).

Death from an overdose of ethanol occurs when the activity of the respiratory center in the brain becomes so disrupted that breathing stops. There have been reported instances of people making a wager that they could drink a pint of whiskey without removing their lips from the bottle, doing just that, collecting their bets, and falling dead while walking out the door.

Compounds Containing Oxygen

You are probably most familiar with oxygen as the component of air that we breathe in order to sustain life; without oxygen, we would die. But oxygen plays a further role in living systems, not just as free oxygen (O_2) or as part

Table 10.1   Blood Alcohol Levels and Stages of Intoxication

Percent Alcohol in Blood*	Stage of Intoxication	Clinical Effects
0.05 or less	Subclinical	Normal by ordinary observation.
0.05 to 0.09	Subclinical	Normal by observation; slight change by special tests.
0.09 to 0.21	Stimulation	Decreased inhibition; emotional instability; slight uncoordination; slowing of response to stimuli.
0.18 to 0.30	Confusion	Disturbance of sensation; decreased sense of pain; staggering gait; slurred speech.
0.27 to 0.39	Stupor	Marked stimuli response decrease; approaching paralysis.
0.36 to 0.48	Coma	Complete unconsciousness; depressed reflexes; subnormal temperature; anesthesia; impairment of circulation; heavy snoring; possible death.

*Overlapping stages of intoxication have been shown above; this was done to take into account the variation observed in blood alcohol levels and the corresponding physical and mental impairment in different individuals.

From E. C. Hoff, *Aspects of Alcoholism*, J. B. Lippincott Co., Philadelphia, 1963, p. 54. Used by permission.

of water (H_2O), but as an element in the structure of many molecules that are critically important in life processes. We have seen that alcohols, for example, are a class of compounds characterized by the presence of a hydroxyl group, —OH. Such special groups of atoms, only some of which contain oxygen, are called **functional groups.** Functional groups create a reactive spot on an organic molecule, giving the molecule specific chemical characteristics. In Chapter 9 we mentioned that the study of organic chemistry is made manageable by being able to categorize compounds having similar properties into classes, and then studying one or two examples of each class. In this chapter we will examine classes of organic compounds that have functional groups containing oxygen (Table 10.2).

10.1 Alcohols

To repeat, alcohols are organic compounds whose molecules contain the hydroxyl functional group, —OH. This hydroxyl group forms a reactive spot on the molecule, making alcohols an important class of organic compounds since they are easily formed, are quite reactive, and are good starting

Table 10.2 Functional Groups Containing Oxygen

Functional Group	Class of Compound	Typical Compound	
R—OH	Alcohol	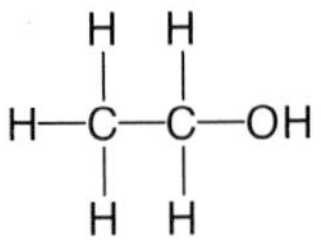 H—C—C—OH (H H / H H)	Ethanol
R—O—R	Ether	H—C—O—C—H (H / H)	Dimethyl ether
R—C—H (O)	Aldehyde	H—C—H (O)	Formaldehyde
R—C—R (O)	Ketone	H—C—C—C—H (H O H / H H)	Acetone
R—C—OH (O)	Acid	H—C—C—OH (H O / H)	Acetic acid
R—C—O—R (O)	Ester	H—C—C—O—C—C—H (H O / H H H)	Ethyl acetate

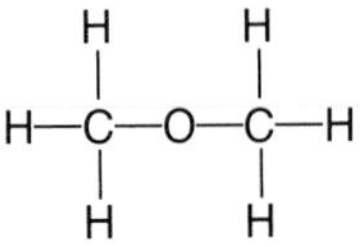

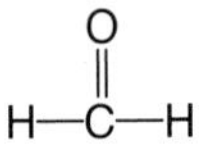

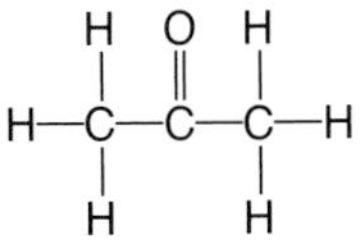

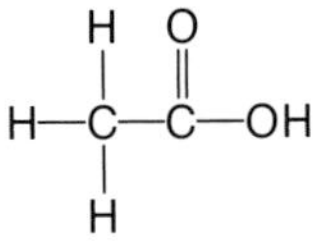

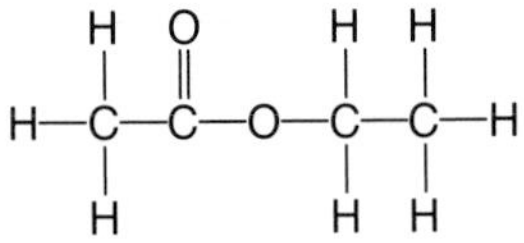

materials for the synthesis of many other compounds.

Alcohols can be divided into three categories according to the placement of the hydroxyl group on the molecule. **Primary alcohols** have the hydroxyl group attached to a carbon atom that is bonded to at most one other carbon atom.

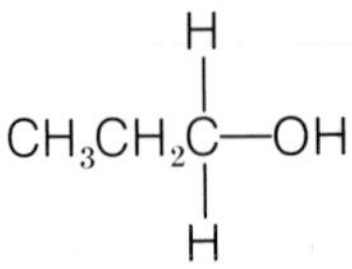

$$CH_3CH_2C—OH$$ (H / H)

1-Propanol
(n-Propyl alcohol)

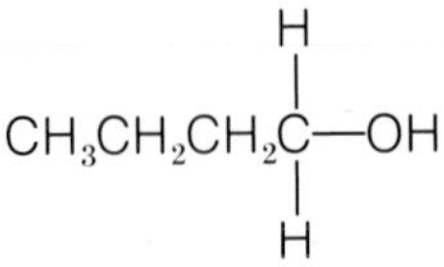

$$CH_3CH_2CH_2C—OH$$ (H / H)

1-Butanol
(n-Butyl alcohol)

$$\underset{\substack{\text{2-Methyl-1-propanol}\\ \text{(Isobutyl alcohol)}}}{CH_3-\overset{\displaystyle CH_3}{\underset{\displaystyle H}{C}}-\overset{\displaystyle H}{\underset{\displaystyle H}{C}}-OH}$$

2-Methyl-1-propanol
(Isobutyl alcohol)

Secondary alcohols have the hydroxyl group attached to a carbon atom that is bonded to two other carbon atoms.

2-Propanol
(Isopropyl alcohol)

2-Butanol
(sec-Butyl alcohol)

Tertiary alcohols have the hydroxyl group attached to a carbon atom that is bonded to three other carbon atoms.

2-Methyl-2-propanol
(tert-Butyl alcohol)

The hydroxyl group on an alcohol molecule forms a polar area on the otherwise nonpolar carbon chain. The oxygen atom has a stronger attraction for additional electrons than does either the hydrogen or carbon atoms, and will form an unequal, or polar, covalent bond with either hydrogen or carbon.

The presence of this polar hydroxyl group allows alcohols with short carbon chains to be soluble in polar solvents such as water. As the length of the carbon chain increases, however, the nonpolar character of the carbon chain overrides the attraction of the hydroxyl group for the water. This makes the larger molecules less and less soluble in such polar solvents, and more soluble in nonpolar solvents such as fats, benzene, and carbon tetrachloride (Table 10.3). Increasing the number of hydroxyl groups on an alcohol molecule increases the number of polar sites, making such molecules again more soluble in polar solvents.

Table 10.3 Solubility of Alcohols and Ethers

Compound	Formula	Solubility (Grams per 100 ml H_2O)
Methanol	CH_3OH	∞
Ethanol	CH_3CH_2OH	∞
1-Pentanol	$CH_3CH_2CH_2CH_2CH_2OH$	2.7
1,2-Pentanediol	$CH_3CH_2CH_2CHOHCH_2OH$	∞
1-Hexanol	$CH_3(CH_2)_4CH_2OH$	0.59
2-Hexanol	$CH_3(CH_2)_3CHOHCH_3$	Very slightly soluble
2,3-Hexanediol	$CH_3(CH_2)_2CHOHCHOHCH_3$	Soluble
1-Decanol	$CH_3(CH_2)_8CH_2OH$	Insoluble
Ethylene glycol	$HOCH_2CH_2OH$	∞
Glycerol	$CH_2OHCHOHCH_2OH$	∞
Diethyl ether	$CH_3CH_2OCH_2CH_3$	7.5
Dipropyl ether	$CH_3CH_2CH_2OCH_2CH_2CH_3$	Slightly soluble
Dibutyl ether	$CH_3(CH_2)_3O(CH_2)_3CH_3$	Insoluble

The polar hydroxyl group also affects the melting point and the boiling point of the alcohol molecule. The positive hydrogen atom on the hydroxyl group will enter into hydrogen bonding with the electronegative oxygen atom on a nearby alcohol molecule (Figure 10.1). The resulting attraction between alcohol molecules increases the amount of energy necessary to pull the molecules apart, thereby increasing the melting and boiling points. Alcohols have much higher boiling points than alkanes having comparable molecular weights (Table 10.4).

10.2 Methanol

The alcohol that is derived from the simplest hydrocarbon, methane, is methanol, CH_3OH. Its common name, wood alcohol, comes from the fact that it was first obtained by heating wood in the absence of air. Methanol is very poisonous; less than 10 cc (two teaspoons) can cause blindness, and 30 cc (two tablespoons) can cause death.

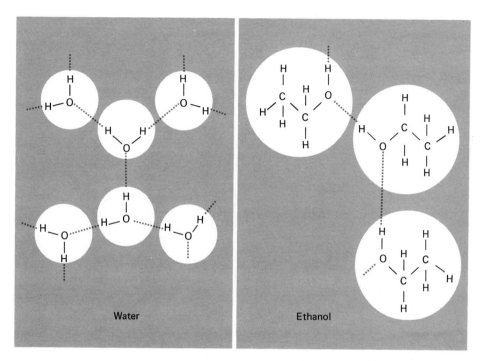

Figure 10.1 The hydrogen bonding in water and ethanol. The polar hydroxyl group on alcohols can enter into hydrogen bonding in much the same way as the −OH group in water. The hydrogen bond is indicated by the colored dotted line.

10.3 Ethanol

Ethanol has been known to mankind for thousands of years. Ethanol was first produced by mixing honey, fruits, berries, cereals, or other plant materials with water, and leaving them in the sun to produce a drink prized as food, as a ceremonial or religious potion, and as medicine. The production of ethanol through this process, known as fermentation, became quite an art long before the actual chemistry was understood (Figure 10.2). In fermentation, yeast cells in the fruit or cereal mixture use the nutrients found in the mixture to supply them with energy. The waste products of this energy-producing reaction are ethanol and carbon dioxide.

$$C_6H_{12}O_6 \xrightarrow{\text{Yeast}} 2CH_3CH_2OH + 2CO_2 + \text{Energy}$$

Glucose   Ethanol
(a sugar)

The ethanol produced by the yeast cells dissolves in the surrounding liquid. When the concentration of ethanol rises above 12 to 18% (depending upon the type of yeast) it becomes toxic to the cells, which die and settle to the bottom of the container, leaving behind an alcoholic beverage.

Table 10.4 Boiling Point Comparison

Hydrocarbons			Ethers			Alcohols		
Compound	Molecular Weight	Boiling Point °C	Compound	Molecular Weight	Boiling Point °C	Compound	Molecular Weight	Boiling Point °C
Methane CH_4	16	−161				Water H_2O	18	100
Ethane C_2H_6	30	−89				Methanol CH_3OH	32	65
Propane C_3H_8	44	−45	Dimethyl ether CH_3OCH_3	46	−23	Ethanol C_2H_5OH	46	78
n-Butane C_4H_{10}	58	−0.5	Methyl ethyl ether $CH_3OCH_2CH_3$	60	11	1-Propanol C_3H_7OH	60	97
n-Pentane C_5H_{12}	72	36	Diethyl ether $CH_3CH_2OCH_2CH_3$	74	34	1-Butanol C_4H_9OH	74	117

Figure 10.2 Wine making—old and new. (Top, New York Public Library Picture Collection; bottom, Courtesy The Taylor Wine Company)

Beverages having an alcoholic content higher than 18% can be produced by first distilling the ethanol to obtain it in pure form, and then mixing this alcohol in various proportions with other ingredients.

10.4 Polyhydric Alcohols

Alcohols can have more than one hydroxyl group on their molecules. We have already mentioned that the addition of more polar hydroxyl groups will make the molecule more water soluble, and will also raise the boiling point. (Remember, this is due to more hydrogen bonding.) Multiple hydroxyl groups, for a reason still unknown, make a compound sweet-tasting.

Ethylene glycol, a dihydric alcohol (that is, an alcohol containing two hydroxyl groups) is a colorless, sweet liquid that is water soluble. It is just as toxic as methanol when taken internally. Ethylene glycol has great commercial value as the basic ingredient in permanent antifreeze for automobiles, and in the production of synthetic fabrics.

$$
\begin{array}{cc}
\text{OH OH} & \text{OH OH OH} \\
| \; | & | \; | \; | \\
\text{H--C--C--H} & \text{H--C--C--C--H} \\
| \; | & | \; | \; | \\
\text{H H} & \text{H H H} \\
\text{Ethylene glycol} & \text{Glycerol}
\end{array}
$$

Glycerol or glycerin, which has three hydroxyl groups, is a thick, sweet-tasting liquid that is not toxic and is a component of all natural fats and oils. These properties make it suitable for wide industrial use. It protects the skin and, therefore, is used in hand lotions and cosmetics. Glycerol is also used in inks, tobaccos, cream-filled candies, and plastic clays to prevent loss of water (or dehydration). It is a sweetening agent and a solvent for medicines, a lubricant used in chemical laboratories, and a constituent of plastics, surface coatings, and synthetic fabrics.

One use of glycerol is in the production of the explosive called nitroglycerin.

$$
\begin{array}{ccc}
\text{ONO}_2 & \text{ONO}_2 & \text{ONO}_2 \\
| & | & | \\
\text{H--C} & \text{C} & \text{C--H} \\
| & | & | \\
\text{H} & \text{H} & \text{H}
\end{array}
$$

Nitroglycerin

Nitroglycerin was first produced in 1846 when an Italian chemist inadvertently mixed glycerol and nitric acid. The subsequent violent explosion almost killed him. Fifteen years later, after an explosion of nitroglycerin killed his brother, the Swedish scientist Alfred Nobel discovered a method for producing and transporting nitroglycerin safely. When mixed with a type of earth, nitroglycerin can be made insensitive to shock, and will require a percussion cap to be exploded. This mixture is known as dynamite. Nobel's invention earned him a fortune, but the wartime uses of dynamite upset him. This prompted him to leave a large part of his fortune for annual awards recognizing contributions to peace, literature, and science (Figure 10.3).

Figure 10.3 The Nobel peace prize medallion. This medallion is awarded each year along with a cash prize to persons who have contributed the greatest benefit to mankind in the fields of physics, chemistry, physiology or medicine, literature, economics, and peace. (Courtesy Nobel Committee of Norway)

10.5 Cyclic Alcohols

Hydroxyl groups can be found on all types of carbon chains. For example, menthol, which occurs in peppermint oil, is a cyclic alcohol containing a cyclohexane ring.

$$OH$$
$$-CHCH_3$$
$$H_3C \qquad CH_3$$

Menthol

The properties of menthol have been known for thousands of years. It causes an unusual cooling, refreshing sensation when rubbed on the skin, leading to its use in aftershave lotions and cosmetics. Menthol is a stimulant to inflamed mucous tissues, as may be found in the mucous membranes of the nose and throat, prompting its use in nose and throat sprays, in cough drops, and in cigarettes.

10.6 Aromatic Alcohols

Aromatic alcohols are produced when one of the hydrogen atoms on a benzene ring is replaced by a hydroxyl group (Table 10.5).

Table 10.5 Some Aromatic Alcohols

Name	Structural Formula	Name	Structural Formula
Phenol (Hydroxybenzene)	OH	Hydroquinone	OH ... OH
Thymol	CH_3 ... OH ... CH_3 ... H ... CH_3	Phloroglucinol (1, 3, 5-Trihydroxy-benzene)	OH ... HO ... OH
Catechol (1, 2-Dihydroxy-benzene)	OH ... OH	Picric acid (2, 4, 6-Trinitro-phenol)	OH ... O_2N ... NO_2 ... NO_2
Resorcinol	OH ... OH		

The simplest aromatic alcohol, phenol, is a powerful germicide, as are most of its derivatives. In 1865, the English surgeon Lister was the first to apply chemicals to a wound to prevent infection (Figure 10.4). For this purpose he used a dilute solution of phenol. As other antiseptic chemicals were found, the use of phenol was discontinued because of its damaging effects on the tissues to which it was applied, its absorption through the skin, and its extreme toxicity. However, it is still used by the drug industry as a standard for measuring the germicidal activity of other antiseptics.

Substituted phenols (phenols having other groups on the benzene ring) are also used as antiseptics. Hexachlorophene, a potent antiseptic compound used in soaps and deodorants, was withdrawn from the commercial market when it was shown to be absorbed through the skin and was found to cause brain damage in monkeys. Another complicated phenol that everyone would like to avoid is urushiol, which is one of the irritants in poison ivy.

Figure 10.4 Joseph Lister (1827–1912) first proved the value of using phenol as an antiseptic, and of heat sterilization of instruments to reduce infections after surgery. (The Bettman Archive)

Hexachlorophene

Urushiol

10.7 Preparation of Alcohols

Alcohols can be prepared through the addition of water to the double bond of an alkene in the presence of an acid (hydration reaction).

Ethylene   Water   Ethanol

About half of the ethanol used in the United States is produced in this way.

Alcohols can also be produced by the addition of hydrogen atoms to aldehydes or ketones (which we shall discuss shortly) in the presence of a catalyst. Remember, in organic chemistry any reaction in which a molecule gains hydrogen atoms or loses oxygen atoms is called a **reduction reaction.** The reduction of aldehydes or ketones produces alcohols.

$$
\underset{\text{Acetaldehyde}}{CH_3\overset{O}{\underset{\|}{C}}-H} + H-H \longrightarrow \underset{\text{Ethanol}}{CH_3-\overset{OH}{\underset{\underset{H}{|}}{C}}-H}
$$

$$
\underset{\text{Acetone}}{CH_3\overset{O}{\underset{\|}{C}}CH_3} + H-H \longrightarrow \underset{\substack{\text{Isopropyl alcohol} \\ \text{(2-Propanol)}}}{CH_3-\overset{OH}{\underset{\underset{H}{|}}{C}}-CH_3}
$$

Reactions of Alcohols

10.8 Dehydration

The dehydration reaction of an alcohol is the removal of a water molecule from an alcohol, producing an unsaturated hydrocarbon. Some alcohols undergo dehydration more easily than others; tertiary alcohols are the easiest to dehydrate and primary alcohols are the hardest. The dehydration process can be accomplished by heating, but in most cases a dehydrating agent such as sulfuric acid is required.

$$
\underset{\text{Ethanol}}{H-\overset{H}{\underset{\underset{H}{|}}{C}}-\overset{OH}{\underset{\underset{H}{|}}{C}}-H} \xrightarrow{H_2SO_4} HOH + \underset{\text{Ethene}}{\overset{H}{\diagdown}\underset{H}{\diagup}C=C\overset{H}{\diagup}\underset{H}{\diagdown}}
$$

$$
\underset{\substack{\text{n-Propyl alcohol} \\ \text{(1-Propanol)}}}{H-\overset{H}{\underset{\underset{H}{|}}{C}}-\overset{H}{\underset{\underset{H}{|}}{C}}-\overset{OH}{\underset{\underset{H}{|}}{C}}-H} \xrightarrow{H_2SO_4} HOH + \underset{\substack{\text{Propylene} \\ \text{(Propene)}}}{H-\overset{H}{\underset{\underset{H}{|}}{C}}-\overset{H}{\underset{\underset{H}{|}}{C}}-C\overset{H}{\diagup}\underset{H}{\diagdown}}
$$

10.9 Oxidation

We have defined an oxidation reaction as one in which a molecule gains oxygen atoms or loses hydrogen atoms. The oxidation of an alcohol involves the loss of two hydrogen atoms, one from the hydroxyl group and one from the carbon to which the hydroxyl group is attached. These hydrogen fragments form water by reaction with the oxygen from the oxidizing agent. Potassium permanganate, $KMnO_4$, or potassium dichromate, $K_2Cr_2O_7$, are commonly used as oxidizing agents for alcohols.

Primary Alcohols. The oxidation of primary alcohols produces members of the class of organic compounds called aldehydes.

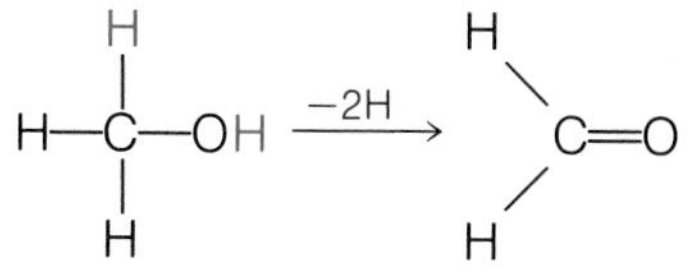

Methanol → Formaldehyde

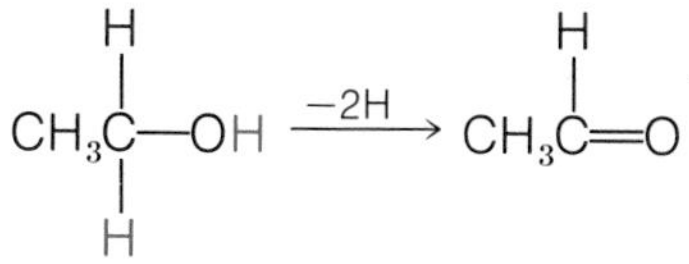

Ethanol → Acetaldehyde

Secondary Alcohols. The oxidation of secondary alcohols produces members of the class of organic compounds called ketones.

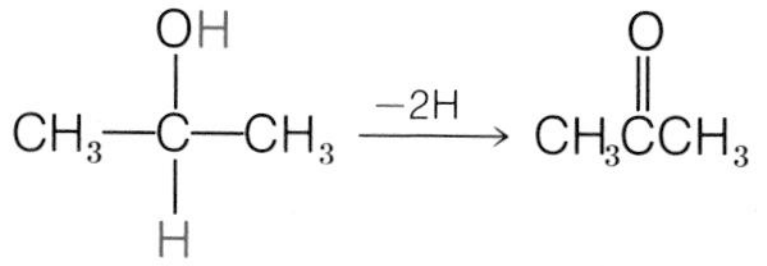

Isopropyl alcohol
(2-Propanol) → Acetone
(2-Propanone)

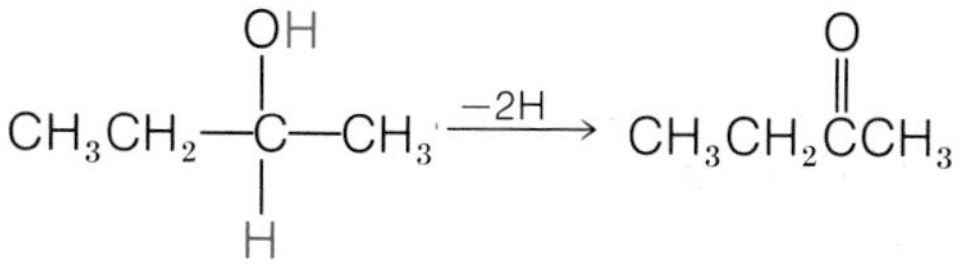

sec-Butyl alcohol
(2-Butanol) → Methyl ethyl ketone
(2-Butanone)

Tertiary Alcohols. Tertiary alcohols cannot be easily oxidized because the carbon attached to the hydroxyl group does not have a hydrogen atom attached to it.

10.10 Ethers

An organic molecule having an oxygen atom bonded to two carbon atoms belongs to the class of compounds called **ethers,** —C—O—C—. Since the placement of the oxygen atom between the carbon atoms eliminates most of the polarity found in alcohols, ethers are only slightly more soluble in water than are alkanes, and are generally soluble in nonpolar solvents. Because there is no possibility of hydrogen bonding, ethers boil at much lower temperatures than do alcohols with comparable molecular weights (Table 10.4). Ethers are extremely flammable, and great care must be taken in their use. Most ethers have some anesthetic properties (Table 10.6).

Table 10.6 Some Ethers

Name	Structural Formula
Dimethyl ether	CH_3OCH_3
Methyl ethyl ether	$CH_3OCH_2CH_3$
Diethyl ether	$CH_3CH_2OCH_2CH_3$
Diisopropyl ether	$CH_3CH-OCHCH_3$ $\quad\quad\ \ \ CH_3\quad\ CH_3$
Divinyl ether	$CH_2{=}CHOCH{=}CH_2$
Diphenyl ether	
Methyl phenyl ether (Anisole)	
Eugenol	

Diethyl ether is the compound used as the anesthetic that is most commonly referred to as ether (Figure 10.5). Although it takes a long time for diethyl ether to have full effect on a patient (10 to 15 minutes), it is quite useful in long operations because there is a very large margin of safety in its use. That is, there is a large difference between the concentration that causes anesthesia and the concentration that will kill the patient. Often a different chemical having a more rapid anesthetic effect is used first, and then the patient is switched to diethyl ether. Divinyl ether is a fast-acting anesthetic that can be used first, but the double bonds in its structure make it highly reactive, and care must be taken in handling this compound to prevent decomposition. Eugenol, which is obtained from cloves, is a mild local anesthetic that is used by dentists to lessen pain when filling tooth cavities.

10.11 Preparation of Ethers

Ethers can be produced by the removal of water from two molecules of alcohol in the presence of a dehydrating agent such as sulfuric acid. This method can be used to produce ethers with molecules having the same

Figure 10.5 The first public demonstration of the use of
ether as a surgical anesthetic took place at Massachusetts
General Hospital in Boston on October 16, 1846. (Courtesy
Massachusetts General Hospital)

alkyl groups on each side of the oxygen atom.

$$CH_3OH + HOCH_3 \xrightarrow{\text{H}_2\text{SO}_4} CH_3OCH_3 + HOH$$

Methanol   Methanol   Dimethyl ether

10.12 Reactions of Ethers

Ethers are relatively unreactive compounds. They undergo combustion, and
under certain conditions can be broken apart — a reaction important in the
digestion of sugars.

10.13 Aldehydes and Ketones

Two classes of compounds, **aldehydes** and **ketones,** result when an
oxygen atom is double bonded to a carbon atom, creating the **carbonyl**

functional group $-\overset{\overset{\textstyle O}{\|}}{C}-$. The difference in properties between the aldehydes
and the ketones is due to the position of the carbonyl group. Aldehydes

have the carbonyl group attached to a terminal carbon—that is, a carbon

atom that is bonded to at most one other carbon atom, $-\overset{\displaystyle O}{\overset{\displaystyle \|}{C}}-\overset{}{C}-H$. The

carbonyl group in a ketone will appear somewhere in the middle of the molecule, and will be attached to a carbon atom that is bonded to two

other carbon atoms, $-C-\overset{\displaystyle O}{\overset{\displaystyle \|}{C}}-C-$. Aldehydes and ketones are both highly

reactive compounds, with aldehydes being even more reactive than ketones. The large class of foods called carbohydrates contains aldehyde and ketone functional groups, and will be the subject of a later chapter.

10.14 Formaldehyde

Formaldehyde is the simplest of the aldehydes.

Formaldehyde

You may be aware of the use of formaldehyde as a tissue preservative. It combines readily with proteins, killing microorganisms and hardening tissues. If you have ever come in contact with formaldehyde, you know the irritating effect it can have on the membranes of your eyes, nose, and throat. In our earlier discussion of alcohols it was mentioned that very small amounts of methanol taken in the body are sufficient to cause blindness. This occurs because the liver, in attempting to rid the body of methanol, changes this compound into formaldehyde. This formaldehyde in the blood-stream then concentrates in the retina of the eye, destroying the cells and causing blindness. When formaldehyde is polymerized with the alcohol phenol, substances called resins, which are the gummy saps of evergreens, are formed. The stone called amber is actually tree resin that has hardened over millions of years.

10.15 Other Aldehydes

Many aldehydes have pleasant odors or tastes, and are used in making perfumes and artificial flavorings. Some familiar flavorings are shown in Table 10.7. The aldehydes used in perfumes and flavorings are only slightly soluble in water, but will readily dissolve in ethanol. As a result, many perfumes and flavorings use ethanol as their solvent.

Table 10.7   Some Simple Aldehydes and Their Uses

Name	Formula	Uses
Formaldehyde	H—C—H (with O double bonded to C)	Tissue preservative; used in the syn-thesis of plastics
Acetaldehyde	CH_3—C—H (with O double bonded to C)	Metabolic intermediate
Propionaldehyde	CH_3CH_2—C—H (with O double bonded to C)	Disinfectant; used in synthesis of plastics
Citral	CH_3 C=CHCH$_2$CH$_2$C=CHC—H with CH$_3$ groups and CH$_3$, O	Lemon flavoring
Benzaldehyde	C$_6$H$_5$—C—H (with O double bonded to C)	Almond flavoring
Cinnamaldehyde	C$_6$H$_5$—CH=CHC—H (with O double bonded to C)	Cinnamon flavoring
Vanillin	HO— benzene ring —C—H, with CH$_3$O substituent	Vanilla flavoring

10.16 Ketones

Ketones are widely used in the chemical industry as solvents. One of the more common ketone solvents is acetone.

$$CH_3CCH_3$$
Acetone

Acetone can be produced in the body as the result of a metabolic side reaction, not normally occurring except when there is some metabolic disorder. In the disease diabetes, some normal metabolic reactions do not occur and acetone is produced in large quantities. This compound accumulates in the tissues and appears in the urine and breath of untreated diabetics (Table 10.8).

Table 10.8 Some Ketones and Their Uses

Name	Formula	Uses
Acetone	$$CH_3\overset{O}{\overset{\|}{C}}CH_3$$	Solvent; found in urine of diabetics
Methyl ethyl ketone	$$CH_3\overset{O}{\overset{\|}{C}}CH_2CH_3$$	Solvent; fingernail polish remover
Progesterone	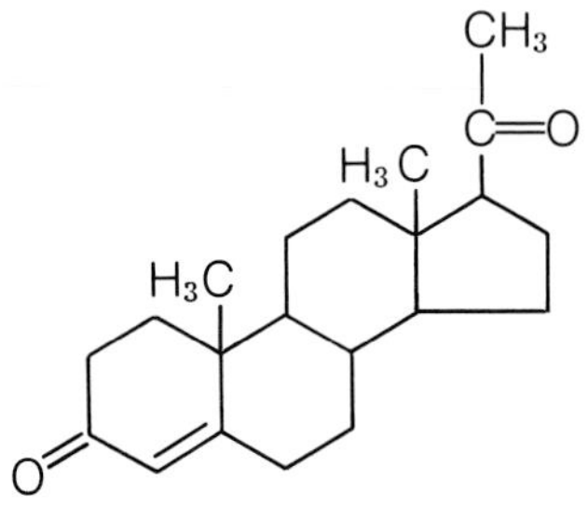	Pregnancy hormone
Carvone	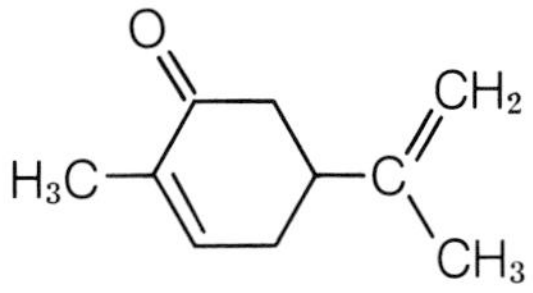	Oil of spearmint

Some unusual cyclic ketones, isolated from animals, have recently become very popular in the cosmetic industry. Are you familiar with musk oil? Civetone, isolated from the African civet cat, and muscone, isolated from the male musk deer, have very nauseating odors in high concentrations. However, when used in minute quantities in perfumes, they intensify and give long-lasting qualities to the scent.

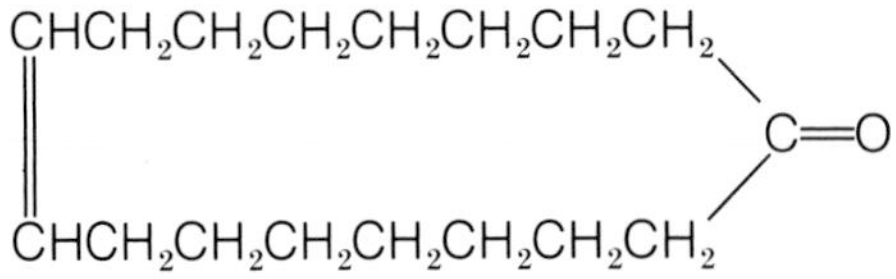

Civetone

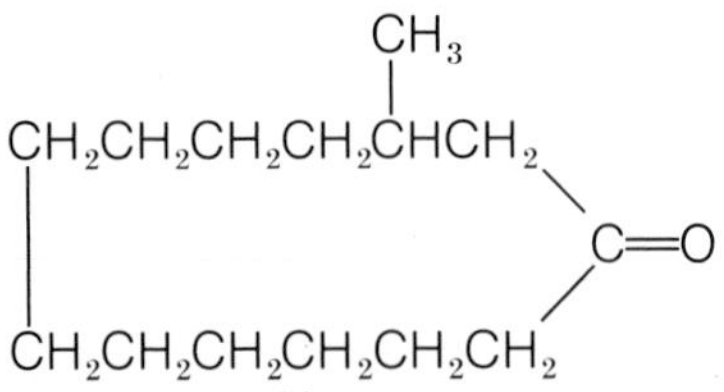

Muscone

10.17 Preparation of Aldehydes and Ketones

Aldehydes and ketones are produced by the oxidation of primary and secondary alcohols in the presence of oxidizing agents such as potassium permanganate or potassium dichromate.

Isobutyl alcohol → Isobutyraldehyde (KMnO₄)

Benzyl alcohol → Benzaldehyde (KMnO₄)

$$CH_3CH_2\underset{\underset{H}{|}}{\overset{\overset{OH}{|}}{C}}CH_2CH_3 \xrightarrow{KMnO_4} CH_3CH_2\overset{\overset{O}{\|}}{C}CH_2CH_3$$

3-Pentanol → 3-Pentanone (Diethyl ketone)

1-Phenylethanol → Acetophenone (Methyl phenyl ketone) (K₂Cr₂O₇)

Reactions of Aldehydes and Ketones

10.18 Reduction

Primary alcohols can be produced from aldehydes, and secondary alcohols can be produced from ketones by the reduction of these compounds in the presence of a proper catalyst such as platinum (symbol, Pt).

Formaldehyde + H₂ →(Pt) Methanol

Acetone + H₂ →(Pt) 2-Propanol

10.19 Oxidation

Aldehydes are very easy to oxidize, and such oxidation produces organic acids, which we will discuss shortly. Ketones, however, can be oxidized only under extreme conditions. This, therefore, gives us a convenient method of differentiating between aldehydes and ketones in the laboratory.

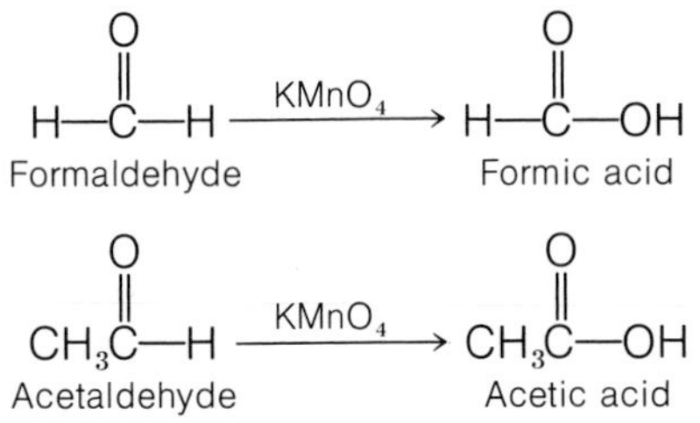

$$H-\overset{\overset{O}{\|}}{C}-H \xrightarrow{KMnO_4} H-\overset{\overset{O}{\|}}{C}-OH$$

Formaldehyde   Formic acid

$$CH_3\overset{\overset{O}{\|}}{C}-H \xrightarrow{KMnO_4} CH_3\overset{\overset{O}{\|}}{C}-OH$$

Acetaldehyde   Acetic acid

10.20 Organic Acids

One definition of an acid is any substance that gives off hydrogen ions. Organic acids are molecules that contain the **carboxyl** functional group, $-\overset{\overset{O}{\|}}{C}-OH$. The bond between the oxygen atom and the hydrogen atom in the carboxyl group is extremely polar, so much so that when placed in water some of the molecules ionize as follows:

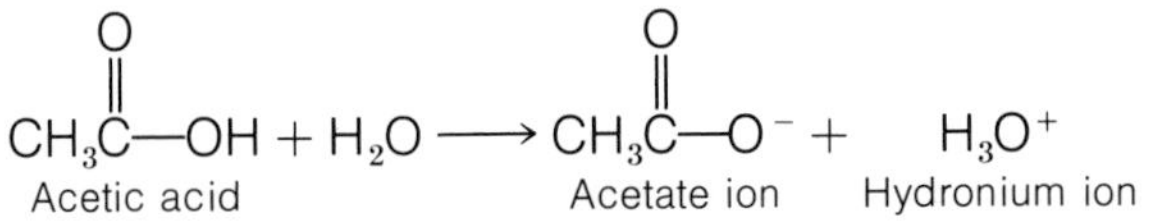

$$CH_3\overset{\overset{O}{\|}}{C}-OH + H_2O \longrightarrow CH_3\overset{\overset{O}{\|}}{C}-O^- + H_3O^+$$

Acetic acid   Acetate ion   Hydronium ion

The stronger the acid, the more readily it gives up its hydrogen to form the hydronium ion (Figure 10.6). Organic acids are very weak compared to inorganic acids such as nitric or sulfuric acids. These organic, or carboxylic, acids are found in fats. For this reason they are referred to as fatty acids, and we will discuss these compounds in the chapter on lipids. Many hydroxy acids (which are acids having a hydroxyl (—OH) group in addition to the carboxyl group—lactic acid, for example) are produced and utilized in important metabolic reactions. These hydroxy acids will also be discussed in later chapters (Table 10.9).

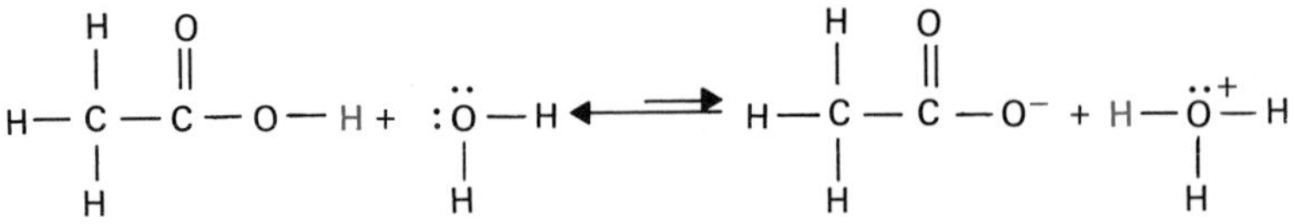

$$H-\overset{\overset{\displaystyle H}{|}}{\underset{\underset{\displaystyle H}{|}}{C}}-\overset{\overset{\displaystyle O}{\|}}{C}-O-H + \overset{\cdot\cdot}{\underset{\underset{\displaystyle H}{|}}{O}}-H \rightleftharpoons H-\overset{\overset{\displaystyle H}{|}}{\underset{\underset{\displaystyle H}{|}}{C}}-\overset{\overset{\displaystyle O}{\|}}{C}-O^- + H-\overset{\cdot\cdot+}{\underset{\underset{\displaystyle H}{|}}{O}}-H$$

Figure 10.6 When placed in water, some of the acetic acid molecules break apart to form the negative acetate ion and the positive hydronium ion.

Table 10.9   Some Common Carboxylic Acids

Name	Formula	Source
Formic acid	$H-\overset{\displaystyle O}{\overset{\|}{C}}-OH$	Stings of ants and bees
Acetic acid	$CH_3-\overset{\displaystyle O}{\overset{\|}{C}}-OH$	Vinegar
Propanoic acid (Propionic acid)	$CH_3CH_2-\overset{\displaystyle O}{\overset{\|}{C}}-OH$	Does not occur in nature; its calcium salt is used to retard spoilage in bread
Palmitic acid	$CH_3(CH_2)_{14}-\overset{\displaystyle O}{\overset{\|}{C}}-OH$	Palm oil; found in all fats; used to manufacture soaps
Stearic acid	$CH_3(CH_2)_{16}-\overset{\displaystyle O}{\overset{\|}{C}}-OH$	Animal fats; used to make soaps
Oleic acid	$CH_3(CH_2)_7CH=CH(CH_2)_7-\overset{\displaystyle O}{\overset{\|}{C}}-OH$	Most oils and fats
Benzoic acid	$\langle\bigcirc\rangle-\overset{\displaystyle O}{\overset{\|}{C}}-OH$	Balsams
Lactic acid	$CH_3\underset{\displaystyle OH}{CH}-\overset{\displaystyle O}{\overset{\|}{C}}-OH$	Sour milk
Succinic acid	$HO-\overset{\displaystyle O}{\overset{\|}{C}}-CH_2CH_2-\overset{\displaystyle O}{\overset{\|}{C}}-OH$	Sugar cane and sugar beets
Adipic acid	$HO-\overset{\displaystyle O}{\overset{\|}{C}}-CH_2CH_2CH_2CH_2-\overset{\displaystyle O}{\overset{\|}{C}}-OH$	Beet juice; used in manufacture of nylon
Citric acid	$HO-\overset{\displaystyle O}{\overset{\|}{C}}-CH_2-\underset{\displaystyle OH}{\overset{\displaystyle \overset{O}{\overset{\|}{C}}-OH}{C}}-CH_2-\overset{\displaystyle O}{\overset{\|}{C}}-OH$	Tart taste of citrus fruits
Tartaric acid	$HO-\overset{\displaystyle O}{\overset{\|}{C}}-\underset{\displaystyle H}{\overset{\displaystyle OH}{C}}-\underset{\displaystyle H}{\overset{\displaystyle OH}{C}}-\overset{\displaystyle O}{\overset{\|}{C}}-OH$	Grapes; its monopotassium salt is cream of tartar
Salicylic acid	$\langle\bigcirc\rangle\underset{\displaystyle OH}{}-\overset{\displaystyle O}{\overset{\|}{C}}-OH$	Bark of willows

Formic acid is very irritating to tissues. It is found in the stings of bees and ants, and is the compound that causes inflammation. Formic acid is another product that is formed by the liver in the oxidation of methanol. When released into the bloodstream it causes severe acidosis, a condition that disrupts the oxygen-carrying capacity of the blood.

Acetic acid has been known to man as long as alcohol, for it is the substance that is formed when wine turns sour. This souring process occurs when special bacteria oxidize ethanol to acetic acid, producing the characteristic taste and smell of vinegar. Acetic acid is the compound that is used by the living cell to produce the longer-chain fatty acids, all of which have an unmistakably strong smell. Butyric acid is the unpleasant odor of rancid butter, and is one of the compounds that causes "body odor." Capric acid gives limburger cheese its smell.

$$CH_3CH_2CH_2\overset{\displaystyle O}{\overset{\displaystyle \|}{C}}-OH \qquad CH_3CH_2CH_2CH_2CH_2CH_2CH_2CH_2CH_2\overset{\displaystyle O}{\overset{\displaystyle \|}{C}}-OH$$

Butyric acid   Capric acid

Benzoic acid and its sodium salt, sodium benzoate, are used in the making of bread and cheese to prevent spoilage from the growth of mold or bacteria (Figure 10.7). Salicylic acid, whose structure is similar to that of benzoic acid, cannot be used in foods because it is so irritating to tissues.

Figure 10.7 These are some of the many foods containing sodium benzoate as a preservative. (Ron Nelson)

It is capable of destroying horny growths such as corns and warts, and is used in products treating such conditions.

Oxalic acid, a dicarboxylic acid, will lose its two hydrogen atoms to form the oxalate ion, which is a normal product of the body's metabolism.

Oxalic acid	Oxalate ion $(C_2O_4^{2-})$	Hydronium ion

Oxalate ions bond strongly with calcium ions to form an insoluble solid. Although this process occurs in the urine of humans, normally the crystals of calcium oxalate are kept from forming. In some cases, however, they can accumulate to form small stones known as kidney stones.

$$Ca^{2+} \quad + \quad C_2O_4^{2-} \quad \longrightarrow \quad CaC_2O_4$$

Calcium ion	Oxalate ion	Calcium oxalate

10.21 Preparation of Acids

Acids are prepared by the oxidation of primary alcohols or aldehydes in the presence of a strong oxidizing agent.

$$CH_3CH_2CH_2CH_2OH \xrightarrow{KMnO_4} CH_3CH_2CH_2\overset{O}{\overset{\|}{C}}-OH$$

1-Butanol → Butyric acid

$$CH_3\overset{O}{\overset{\|}{C}}-H \xrightarrow{KMnO_4} CH_3\overset{O}{\overset{\|}{C}}-OH$$

Acetaldehyde → Acetic acid

Aromatic acids can be prepared by oxidizing a carbon side chain on a benzene compound with a strong oxidizing agent. The carbon atom attached to the ring becomes the carbon atom in the carboxylic group, and the rest of the carbons on the side chain go to form carbon dioxide.

Ethyl benzene $\xrightarrow{K_2Cr_2O_7}$ Benzoic acid $+ CO_2$

10.22 Esters

An important class of compounds derived from carboxylic acids is the **esters,** whose molecules contain the ester functional group.

$$\underset{\text{Ester functional group}}{-\overset{\overset{\text{O}}{\|}}{\text{C}}-\text{O}-\text{C}-} \quad \text{Ester linkage}$$

Esters can be produced by means of a reaction in which a molecule of water is removed from a molecule of a carboxylic acid and a molecule of an alcohol.

$$\underset{\substack{\text{Carboxylic}\\\text{acid}}}{\overset{\overset{\text{O}}{\|}}{\text{RCOH}}} + \underset{\text{Alcohol}}{\text{HOR}} \longrightarrow \underset{\text{Ester}}{\overset{\overset{\text{O}}{\|}}{\text{RCOR}}} + \underset{\text{Water}}{\text{HOH}}$$

Such a reaction, in which a molecule of water is removed from two reactant molecules with the resulting formation of one product molecule, is known

Table 10.10   Some Common Esters

Name	Formula	Source or Flavor
Ethyl formate	$CH_3CH_2-O-\overset{\overset{\text{O}}{\|}}{C}-H$	Rum
Isobutyl formate	$CH_3\overset{\overset{\text{CH}_3}{\|}}{C}HCH_2-O-\overset{\overset{\text{O}}{\|}}{C}-H$	Raspberries
Amyl acetate	$CH_3(CH_2)_4-O-\overset{\overset{\text{O}}{\|}}{C}-CH_3$	Banana
Isopentyl acetate	$CH_3\overset{\overset{\text{CH}_3}{\|}}{C}HCH_2CH_2-O-\overset{\overset{\text{O}}{\|}}{C}-CH_3$	Pear
Octyl acetate	$CH_3(CH_2)_7-O-\overset{\overset{\text{O}}{\|}}{C}-CH_3$	Orange
Ethyl butyrate	$CH_3CH_2-O-\overset{\overset{\text{O}}{\|}}{C}-CH_2CH_2CH_3$	Pineapple
n-Pentyl butyrate	$CH_3(CH_2)_4-O-\overset{\overset{\text{O}}{\|}}{C}-CH_2CH_2CH_3$	Apricots

as a **condensation reaction.** This type of reaction accounts for many of the chemical reactions occurring in biological systems. Esters are noted for their pleasant aromas, and are the source of the distinctive smells of flowers and the tastes of fruits (Table 10.10).

Nitroglycerin, which was mentioned previously as the active substance in dynamite, is an ester formed from an inorganic acid (nitric acid) and an alcohol (glycerol). In addition to its use in dynamite, nitroglycerin is used to relieve pain in certain heart disorders. This compound acts to enlarge smaller blood vessels and to relax smooth muscles found in the arteries, thereby reducing high blood pressure and the accompanying pain experienced in certain heart disorders.

$$\begin{array}{cc}
\text{Glycerol} & \text{Nitric acid}
\end{array}$$

$$\text{Nitroglycerin}$$

10.23 Salicylic Acid Esters

Salicylic acid, which is found in the bark of the willow tree, has a molecule containing an acid group and an alcohol group, both of which can undergo condensation reactions to form esters. We stated previously that salicylic acid is a very irritating substance with a disagreeable taste, and is not taken internally. But the products of condensation reactions with this compound are not as irritating and have a wide range of uses. For example, if the acid group on salicylic acid is condensed with methanol, methyl salicylate is formed. This compound is more commonly known as oil of wintergreen, and is used in perfumes, candies, and in ointments such as Ben Gay®, which cause a mild burning sensation on your skin in an attempt to take your mind off your sore muscles.

Salicylic acid   Methanol   Methyl salicylate

If acetic acid reacts with the alcohol group on salicylic acid, acetylsalicylic acid, one of the most widely used drugs in the world, is formed.

Salicylic acid　　　Acetic acid　　　Acetylsalicylic acid
(Aspirin)

Aspirin, as acetylsalicylic acid is more commonly known, has the properties of an analgesic (pain reliever) and antipyretic (fever reducer). Despite the claims of various aspirin producers, extensive testing has shown that there is no difference between competing brands of aspirin except for their price. Adult aspirin tablets contain only 5 grains (0.33 g) of acetylsalicylic acid. The bulk of the tablet consists of a binder that the manufacturer uses to put the aspirin in tablet form. The apparent difference in size between different brands of aspirin tablets is due to the binder used, and not the aspirin itself.

Aspirin is by far the most commonly taken medicine in the United States, accounting for an annual production of 30 million pounds a year. (This works out to approximately 200 tablets per year for every man, woman, and child!) Although the exact mechanism underlying aspirin's therapeutic effects is not known, its use is considered relatively safe. However, the habitual use of aspirin can produce several unpleasant side effects. Aspirin can irritate the lining of the stomach, and has been shown to produce stomach ulcers in rats. As molecules of aspirin pass through the stomach wall, they injure the cells of this tissue, causing small hemorrhages. In most people, the common dosage of two five-grain aspirin tablets will cause only small blood loss in the stomach—from 0.2 to 5 ml. Some people, however, are much more sensitive to this drug, and experience stomach bleeding at potentially dangerous levels.

10.24 Polyesters

Esters have become extremely important in the production of synthetic fibers. Ester polymers called **polyesters** impart many desirable characteristics to synthetic fabrics formed from them. These advantages include the ability of the fabric to be set into permanent creases and pleats, and its tendency to resist turning gray or yellow with long use. The familiar material Dacron is a polyester made of esters of ethylene glycol and terephthalic acid. Notice that both ethylene glycol and terephthalic acid have two sites where ester bonds can be formed.

HOCH₂CH₂OH   Terephthalic acid structures

Ethylene glycol   Terephthalic acid

$$\text{—COH} + \text{HOCH}_2\text{CH}_2\text{OH} + \text{HOC—COH} + \text{HOCH}_2$$

↓

Dacron   +   Water

Supplies of ethylene glycol for use in permanent antifreeze in automobiles are becoming increasingly scarce as this raw material is diverted to the manufacture of synthetic fabrics (Figures 10.8 and 10.9).

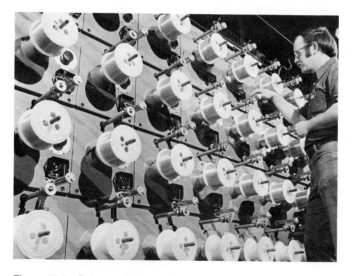

Figure 10.8 Polyester fiber is being wound on these spools to be used in the production of such products as zippers, threads, and paper machine screen belts. (Courtesy The Goodyear Tire and Rubber Company)

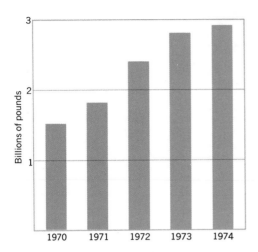

Figure 10.9 Polyester fiber production (billions of pounds) in the United States from 1970 to 1974.

10.25 Preparation of Esters

As has been mentioned, esters can be produced by means of a condensation reaction between a carboxylic acid and an alcohol. This reaction can be made to occur in the laboratory in the presence of a strong acid.

$$CH_3OH + HOCCH_3 \xrightarrow{H^+} CH_3OCCH_3 + H_2O$$

Methanol Acetic acid   Methyl acetate Water

10.26 Reactions of Esters

Esters may be broken apart by an addition reaction with water to form the carboxylic acid and the alcohol. This reaction, called **hydrolysis,** is the exact reverse of the condensation reaction by which the ester may have been formed, and can be made to occur by heating the ester in a water solution of a strong inorganic acid.

Additional Reading

Books

1. Isaac Asimov, *The World of Carbon,* Collier Books, New York, 1962.

2. H. Wallgren and H. Barry, *Actions of Alcohol,* Elsevier, Amsterdam, 1970.

3. For further detail on the classes of compounds with oxygen-containing functional groups, see any organic chemistry textbook.

Articles

1. "Aspirin and its Competitors," *Consumer Reports,* August 1972, page 540.

2. C. S. Lieber, "The Metabolism of Alcohol," *Scientific American,* March 1976, page 25.

3. "The Skin-Disinfectant Debate," *Chemistry,* May 1972, page 3.

4. Valerie Webb, "Hydrogen Bond 'Special Agent,'" *Chemistry,* June 1968, page 16.

Questions and Problems

---

1. Define the following terms.

    (a) Functional group
    (b) Alcohol
    (c) Primary, secondary, and tertiary alcohol
    (d) Polyhydric alcohol
    (e) Dehydration reaction
    (f) Oxidation reaction
    (g) Reduction reaction

    (h) Ether
    (i) Aldehyde
    (j) Ketone
    (k) Carbonyl group
    (l) Carboxyl group
    (m) Ester
    (n) Condensation reaction
    (o) Hydrolysis reaction

2. Identify each of the following as either an alcohol, ether, carboxylic acid, ester, aldehyde, or ketone.

(a)
$$CH_3CH\overset{\overset{\displaystyle CH_3}{|}}{\underset{\underset{\displaystyle H}{|}}{C}}\overset{\overset{\displaystyle O}{\|}}{C}-OH$$

(b) $CH_3CH_2OCH_3$

(c)
$$CH_3CH_2O\overset{\overset{\displaystyle O}{\|}}{C}CH_3$$

(d)
$$CH_3CH_2\overset{\overset{\displaystyle OH}{|}}{\underset{\underset{\displaystyle H}{|}}{C}}\overset{\overset{\displaystyle O}{\|}}{C}-OH$$

(e)

(f)

(g)

(h)

(i)
$$CH_3\overset{\overset{\displaystyle CH_3}{|}}{\underset{\underset{\displaystyle H}{|}}{C}}\overset{\overset{\displaystyle O}{\|}}{C}-H$$

(j)
$$CH_3CH_2\overset{\overset{\displaystyle O}{\|}}{C}CH_3$$

3. Why is it that propane boils at −45°C and methyl ether at −23°C, but ethanol boils at 78°C?

4. Explain why 2-hexanol is only very slightly soluble in water, while 2,3-hexanediol is soluble in water.

5. Identify each of the following compounds as either a primary, secondary, or tertiary alcohol.

(a)
$$CH_3\overset{\overset{\displaystyle OH}{|}}{\underset{\underset{\displaystyle H}{|}}{C}}CH_2CH_3$$

(b)
—CH_2OH

(c)
$$CH_3CH_2\overset{\overset{\displaystyle CH_3}{|}}{\underset{\underset{\displaystyle CH_3}{|}}{C}}-OH$$

(d)
$$CH_3CH_2\overset{\overset{\displaystyle OH}{|}}{\underset{\underset{\displaystyle H}{|}}{C}}-H$$

(e)
$$CH_3-\overset{\overset{\displaystyle CH_3}{|}}{\underset{\underset{\displaystyle CH_3}{|}}{C}}\overset{\overset{\displaystyle OH}{|}}{\underset{\underset{\displaystyle H}{|}}{C}}-CH_3$$

(f)
$\overset{\overset{\displaystyle OH}{|}}{\underset{\underset{\displaystyle CH_3}{|}}{C}}-CH_3$

6. Assume you are making chocolate-covered cherries as a fund-raising project. Why might you want to add some glycerol to the cherry filling?

7. Write an equation for the formation of:

   (a) 2-Propanol from propene
   (b) 3-Pentanol from 3-pentanone (diethyl ketone)

8. Write an equation for the

   (a) Dehydration of 1-butanol
   (b) Oxidation of 3-pentanol

9. Write an equation for the formation of diethyl ether.

10. Why might a drink of methanol cause blindness and death?

11. Describe the products of the oxidation of the following compounds.

   (a) A primary alcohol   (c) A tertiary alcohol
   (b) A secondary alcohol

12. How would you go about distinguishing between an aldehyde and a ketone in the laboratory?

13. Where might you find aldehydes and ketones in products used in your home?

14. What is one method you might use to produce propanoic acid in the laboratory?

15. Write an equation for the condensation reaction between

   (a) Methanol and propanoic acid
   (b) 1-Propanol and salicylic acid
   (c) Propanoic acid and salicylic acid

16. How would you go about making the flavoring for wintergreen candies?

chapter 11

Nitrogen

Learning Objectives

By the time you have finished this chapter, you should be able to:

1. Identify the nitrogen-containing functional groups.

2. Given the structural formula of an amine, identify it as a primary, secondary, or tertiary amine.

3. Show how a nitrogen atom is able to form a coordinate covalent bond.

4. Define a "base" and write an equation showing the reaction of an amine and water.

5. Given the structural formula of an amide, identify the molecule as a primary, secondary, or tertiary amide.

6. Describe two methods by which amides can be produced in the laboratory.

7. Write an equation for the hydrolysis of a primary and a secondary amide.

8. Define "heterocyclic rings."

9. Describe three specific physiological effects of alkaloids and their use in drug therapy.

"Well, this is it—they've just called for the hundred meter freestyle. Now my team's hopes for the state championship rest on me, Ed Scott, the kid who barely made the team his senior year! Two minutes to go—I must get control of myself. I was fine until they called the race, now my heart is pounding (can't everyone hear it?). My hands are sweating and I have this sick feeling in the pit of my stomach. Shake it off, Ed—try to relax those muscles! You'll need everything you've got to beat that guy from Crescent Tech. Slow down that breathing, man, you're panting like the race was over! TAKE YOUR MARK! Concentrate, Ed. GET SET! Well, it's now or never—body, swim like you've never swum before. GO!!"

Faced with a stressful situation, we've all had reactions much like Ed's. A sudden, loud squeal of brakes close by, or a near catastrophe on a ski slope can leave you with your heart pounding, your body sweating, and your stomach feeling as though it were climbing into your throat. These diverse body reactions are all caused by one chemical compound that is produced in a pair of hat-shaped organs called the adrenals, located on top of the kidneys. The compound is one of many chemical messengers, called hormones, that travel through the bloodstream and cause changes in body cells. This particular compound is called epinephrine, but you are probably more familiar with its common name, adrenalin. Under normal conditions, small amounts of epinephrine are released into the blood to help control blood pressure and to maintain the level of sugar in the blood. But this compound is also the body's means of meeting emergencies and dealing with stress situations such as emotional excitement, exercise, extreme temperature changes, severe hemorrhaging, and the administration of certain anesthetics.

Epinephrine marshals many parts of the body in attempting to meet the stress situation. It increases the rate and strength of the heart beat, increasing the output of the heart. It raises the blood pressure by causing constriction of blood vessels in all parts of the body, except for the vessels in such vital organs as the skeletal muscles, heart, brain, and liver. It relaxes the smooth muscles in the lungs, making epinephrine a very effective drug in the treatment of acute asthma attacks. It also increases the rate and depth of breathing, enabling more oxygen to get into the lungs. Epinephrine causes an increase in blood sugar, and a general increase in the metabolic activity of the cells. It slows down the action of the digestive tract, and accelerates blood clotting. It delays the fatigue of skeletal muscles, and increases the strength of contraction of those muscles. This last property has allowed people to show amazing strength in stressful situations, as in the case of the man who lifted up a car after an auto accident to free his son who was trapped underneath. Each of these changes in normal body function caused by epinephrine enables the body to meet the initial challenge of the stress. The effects are fairly short-lived, since the epinephrine released into the blood stream is inactivated by the liver in about three minutes.

Epinephrine is a fairly complex molecule containing a benzene ring, hydroxyl groups, and a functional group that contains nitrogen.

$$HO-\underset{}{\bigcirc}-\underset{\underset{OH}{|}}{CH}-CH_2-\underset{\underset{H}{|}}{N}-CH_3 \qquad (OH)$$

Epinephrine (Adrenalin)

In the preceding chapters we have discussed important biological compounds composed of carbon, hydrogen, and oxygen. Nitrogen is the fourth most abundant element in the human body, making up 1.4% of the total number of atoms. Nitrogen can form strong bonds with carbon, oxygen, and hydrogen, allowing this element to be found in many different arrangements in organic molecules. As in the previous chapters, we can categorize these nitrogen-containing molecules by similarities in chemical properties resulting from the presence of important functional groups (Table 11.1).

Table 11.1   Some Nitrogen-Containing Functional Groups

Functional Group	Class of Compound	Typical Compound	
R—NH₂	Amine	CH_3NH_2	Methylamine
R—C(=O)—NH₂	Amide	$CH_3C(=O)—NH_2$	Acetamide
R—C(R)=NH	Imine	$H_2N—C(NH_2)=NH$	Guanidine
R—C≡N	Nitrile	$H_2C=CHC≡N$	Acrylonitrile

Nitrogen is the first member of group V on the periodic table. An atom of nitrogen has five valence electrons, and will form three covalent bonds to become stable.

Nitrogen

Such bonding can occur in several ways—by the formation of three single bonds, a double and a single bond, or one triple bond (Figure 11.1). Nitrogen is present in the atmosphere as the molecule N_2, two nitrogen atoms connected by a triple bond. This molecule accounts for 80% of the gases in the atmosphere; it is stable and relatively unreactive. Nitrogen in the form of N_2 is useless to most forms of life. We inhale it with every breath we take, but exhale it again since our bodies cannot force this molecule to react with other atoms. Only one type of organism is able to force atmospheric nitrogen into chemical combinations with other atoms, thereby making it usable by other forms of life. These organisms are bacteria living in the soil or in the roots of leguminous plants such as peas and alfalfa.

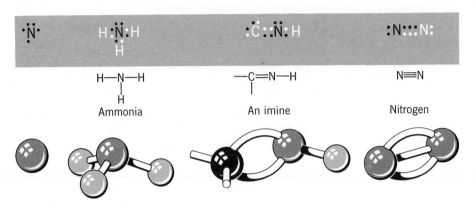

Figure 11.1 Nitrogen can form single, double, and triple covalent bonds.

The bacteria convert gaseous nitrogen into compounds that can be used by the plants. Animals then get the nitrogen they need by eating such plants, or by eating other animals that have eaten such plants.

Compounds Containing Nitrogen

11.1 Amines

Amines are organic compounds containing the functional group $-NH_2$, and are known for their very strong, pungent odors. For example, methylamine, CH_3NH_2, smells like spoiled fish. Amines are produced in the natural decay processes of living organisms, and such amines are called **ptomaines.**

$$H_2NCH_2CH_2CH_2CH_2NH_2$$
Putrescine
(1,4-Diaminobutane)

$$H_2NCH_2CH_2CH_2CH_2CH_2NH_2$$
Cadaverine
(1,5-Diaminopentane)

The names of these two ptomaines might give you a hint as to their smells. It was once thought that such amines were responsible for the vomiting and diarrhea that comes from eating spoiled food, so these symptoms were referred to as ptomaine poisoning. However, it is now known that such bodily reactions result from more complicated causes.

Amines can be classified as primary, secondary, or tertiary amines depending upon the number of carbon atoms bonded directly to the nitrogen atom. In a primary amine, the nitrogen atom will be bonded to one carbon atom and to two hydrogen atoms.

$$H_3C-\underset{\underset{H}{|}}{\overset{\overset{H}{|}}{N}}-H$$
Methylamine, a primary amine

In a secondary amine, the nitrogen will be bonded to two carbon atoms and one hydrogen atom.

$$H_3C-\underset{\underset{\displaystyle H}{|}}{N}-CH_3$$

Dimethylamine, a secondary amine

In a tertiary amine, the nitrogen will be bonded to three carbon atoms (Figure 11.2 and Table 11.2).

$$H_3C-\underset{\underset{\displaystyle CH_3}{|}}{N}-CH_3$$

Trimethylamine, a tertiary amine

Table 11.2   Some Amines

Name	Formula	Type
Methylamine	CH_3NH_2	Primary
Ethylamine	$CH_3CH_2NH_2$	Primary
Isopropylamine (2-Aminopropane)	$CH_3-CH-NH_2$ $\quad\quad\;\;\vert$ $\quad\quad\;\;CH_3$	Primary
Methylethylamine	CH_3CH_2-N-H $\quad\quad\quad\;\;\vert$ $\quad\quad\quad\;\;CH_3$	Secondary
Dimethylethylamine	$CH_3-N-CH_2CH_3$ $\quad\quad\;\;\vert$ $\quad\quad\;\;CH_3$	Tertiary
Phenylamine (Aniline)	$C_6H_5-NH_2$	Primary
N-Methylaniline	$C_6H_5-N-CH_3$ $\quad\quad\quad\;\vert$ $\quad\quad\quad H$	Secondary
Diphenylamine	$C_6H_5-N-C_6H_5$ $\quad\quad\quad\vert$ $\quad\quad\quad H$	Secondary
N, N-Dimethylaniline	$C_6H_5-N-CH_3$ $\quad\quad\quad\;\vert$ $\quad\quad\quad CH_3$	Tertiary

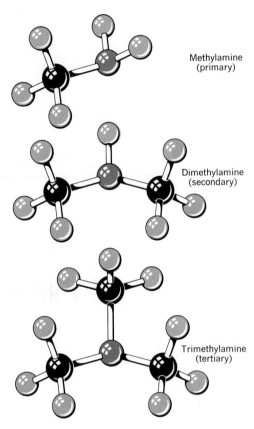

Figure 11.2 Examples of primary, secondary, and tertiary amines.

Methylamine
(primary)

Dimethylamine
(secondary)

Trimethylamine
(tertiary)

In certain instances, nitrogen can form a strong fourth bond. After forming three covalent bonds, an atom of nitrogen has a pair of electrons that it can share with another atom, forming a bond known as a **coordinate covalent,** or **donor-acceptor, bond.** If a nitrogen atom in an amine forms a fourth bond, or coordinate bond, an organic ammonium ion is formed.

$$\underset{\text{Trimethylamine}}{\overset{\overset{\displaystyle CH_3}{|}}{H_3C-N-CH_3}}\qquad\qquad \underset{\text{Tetramethylammonium ion}}{\overset{\overset{\displaystyle CH_3}{|}}{\underset{\underset{\displaystyle CH_3}{|}}{H_3C-N-CH_3}}}$$

Choline is a compound containing an organic ammonium ion, and derivatives of the choline molecule are substances critical to the structure of the cell and to the transmission of nerve impulses.

$$\underset{\text{Choline}}{\left[\overset{\overset{\displaystyle CH_3}{|}}{\underset{\underset{\displaystyle CH_3}{|}}{HOCH_2CH_2-N-CH_3}}\right]^{+} OH^-}\qquad \underset{\substack{\text{Acetylcholine}\\ \text{(neurochemical transmitter)}}}{\left[\overset{\overset{\displaystyle O}{\|}}{CH_3COCH_2CH_2}-\overset{\overset{\displaystyle CH_3}{|}}{\underset{\underset{\displaystyle CH_3}{|}}{N-CH_3}}\right]^{+} OH^-}$$

11.2 Basic Properties of Amines

The ability to form this fourth, or coordinate, bond gives the amines their basic properties. As you will remember, we defined an acid to be a substance that donates or gives off hydrogen ions, H^+, and a base to be a substance that accepts or picks up hydrogen ions. The stronger the base, the greater its attraction for hydrogen ions. Amines are all derivatives of the weak inorganic base ammonia, NH_3. When placed in water, ammonia picks up hydrogen ions to form the ammonium ion.

$$NH_3 + H^+ \longrightarrow NH_4^+$$

Ammonia   Ammonium ion

Organic amines are weak bases. In water, methylamine will attract some of the hydrogen ions present to form the methylammonium ion.

$$\underset{\text{Methylamine}}{CH_3\ddot{N}:} + H^+ \rightleftharpoons \underset{\text{Methylammonium ion}}{CH_3-\overset{+}{N}-H}$$

The strength of an amine as a base depends upon the organic groups attached to the nitrogen atom. Carbon chains attached to the nitrogen atom increase the basic properties of the molecule, while benzene rings attached to the nitrogen atom greatly decrease the basic properties of the molecule.

Diethylamine	>	Methylamine	>	Ammonia	>	Aniline
100× stronger base than ammonia		10× stronger base than ammonia		weak inorganic base		100,000× weaker base than ammonia

11.3 Reactions of Amines

Amines form salts when they react with acids.

$$\underset{\text{Methylamine}}{CH_3NH_2} + \underset{\substack{\text{Hydrochloric}\\\text{acid}}}{HCl} \longrightarrow \underset{\substack{\text{Methylammonium}\\\text{chloride}}}{CH_3NH_3^+Cl^-}$$

$$\underset{\text{Trimethylamine}}{CH_3-\overset{\overset{\displaystyle CH_3}{|}}{\underset{\underset{\displaystyle CH_3}{|}}{N}}} + \underset{\substack{\text{Hydrobromic}\\\text{acid}}}{HBr} \longrightarrow \underset{\substack{\text{Trimethylammonium}\\\text{bromide}}}{CH_3-\overset{\overset{\displaystyle CH_3}{|}}{\underset{\underset{\displaystyle CH_3}{|}}{N}}-H^+Br^-}$$

When heated with alkyl halides, amines form compounds known as quaternary ammonium salts.

$$(CH_3)_3N \quad + \quad CH_3I \quad \longrightarrow \quad (CH_3)_4N^+I^-$$

Trimethylamine   Methyl iodide   Tetramethylammonium iodide

The reaction of aromatic primary amines with nitrous acid is one of the most useful in synthetic organic chemistry. The products of this reaction are diazonium salts, unstable compounds that are intermediates in the production of azo dyes, a large class of synthetic dyes.

Aniline   Nitrous   Hydro-   Benzenediazonium
                acid   chloric   chloride
                      acid   (a diazonium salt)

11.4 Amides

Amides may be thought of as derivatives of organic acids, in which the —OH group has been replaced by the —NH$_2$ group (Table 11.3).

Table 11.3 Some Amides

The bond between the carbon and the nitrogen atoms in an amide molecule is known as the **amide linkage.** It is a very stable bond and is found as a repeating linkage in large molecules such as proteins, nylon, and other industrial polymers.

$$CH_3\overset{\overset{\displaystyle O}{\|}}{C}-OH$$

Acetic acid

$$CH_3\overset{\overset{\displaystyle O}{\|}}{C}-NH_2$$

Acetamide

Amide linkage

Amides having short carbon chains are soluble in water because of the presence of the polar amide group. This solubility decreases as the molecular weight of the amide increases. Amides, like alcohols, have melting and boiling points that are higher than alkanes with comparable molecular weights. These high melting and boiling points result from hydrogen bonding between the polar amide groups on neighboring amide molecules (Figure 11.3).

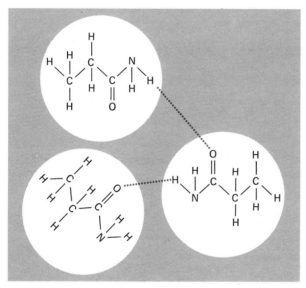

Figure 11.3 Hydrogen bonding between molecules of propionamide.

The field of organic chemistry was first established in 1828 with the synthesis of the amide called urea.

$$H_2N-\overset{\overset{\displaystyle O}{\|}}{C}-NH_2$$

Urea

For several hundred years scientists had been trying unsuccessfully to synthesize in the laboratory various compounds that are found in living organisms. Their lack of success was thought to be caused by the absence of a "vital force" that had to be present for these compounds to be formed.

Figure 11.4 Friedrich Wohler (1800–1882) is known not only for his synthesis of an organic compound from inorganic reagents, but also for many other contributions to the field of chemistry. He discovered the element beryllium, and was the first to study human metabolism. (Culver Pictures)

In 1828 the German scientist Friedrich Wöhler (Fig. 11.4) mixed together two common inorganic compounds, ammonium chloride and sodium cyanate, in an attempt to form ammonium cyanate.

$$NH_4Cl + NaOCN \longrightarrow H_2NCONH_2 + NaCl$$

| Ammonium chloride | Sodium cyanate | Urea | Sodium chloride |

Much to his amazement he obtained the organic compound urea. However, the idea of vital force had many firm believers, and it took many other experiments over the course of 20 years for this theory to be discarded. The synthesis of organic compounds in the laboratory then became an established field of research. Today, hundreds of new organic compounds are produced yearly, most of which are not actually found in living substances. For this reason, the term "organic chemistry" is now used in a more general sense to denote studies of the chemistry of carbon compounds.

Urea is a waste product of the body's metabolism. We have previously seen that compounds containing carbon, hydrogen, and oxygen are broken down by the body to produce waste products of carbon dioxide (CO_2) and water (H_2O). The carbon dioxide is then eliminated through the lungs, and water is eliminated through the breath, sweat, and urine. When nitrogen-containing compounds are broken down by the body, the nitrogen is converted into the compound ammonia, NH_3. Ammonia, however, is extremely toxic to living tissues. Aquatic animals are able to get rid of ammonia quickly into the surrounding water, but land animals must convert this ammonia to a less toxic substance that can be transported through the body. Such animals convert ammonia into urea, a compound that can accumulate in the blood without harm and is eliminated by the kidneys.

Primary, secondary, and tertiary amides contain nitrogen bonded to one, two, or three carbon atoms, respectively, just as in the case of the corresponding amines. An example of a secondary amide having medical importance is xylocaine, a local anesthetic that has largely replaced novocaine for use by dentists.

Xylocaine

A tertiary amide of interest is the hallucinogen LSD, lysergic acid diethylamide. LSD appears to disrupt the transmission of nerve impulses in the brain. The structure of LSD has certain features resembling those of serotonin, a chemical produced by the body to control the transmission of nerve impulses. Nerve cells in the brain seem to confuse LSD with serotonin, preventing the serotonin from carrying out its inhibitory function (Figure 11.5). This results in the generation of uncontrolled nerve impulses in the brain, creating hallucinations and other behavioral abnormalities.

LSD

Serotonin

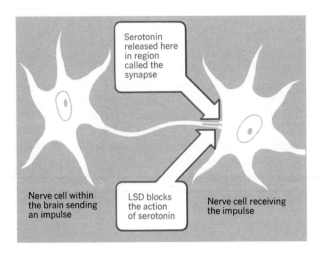

Figure 11.5 The hallucinogen LSD affects brain cells by blocking the action of serotonin.

Tetracyclines are complicated molecules containing an amide group along with other functional groups (you might see how many you can identify). The tetracyclines are formed by certain molds, and are widely used for their antibiotic properties. You may be familiar with the names of one or more of these compounds in connection with prescription drugs.

Tetracycline

Aureomycin

Terramycin

11.5 Preparation of Amides

Amides can be produced through the reaction of organic acid derivatives, such as acid chlorides or esters, and ammonia.

$$CH_3CH_2\overset{\overset{\displaystyle O}{\|}}{C}-Cl + 2NH_3 \longrightarrow CH_3CH_2\overset{\overset{\displaystyle O}{\|}}{C}-NH_2 + NH_4Cl$$

Propionyl chloride   Propionamide

ACID CHLORIDE

$$CH_3\overset{\overset{\displaystyle O}{\|}}{C}-OCH_3 + NH_3 \longrightarrow CH_3\overset{\overset{\displaystyle O}{\|}}{C}-NH_2 + HOCH_3$$

Methylacetate   Acetamide

ESTER

Reactions of Amides

11.6 Hydrolysis

Compounds that contain amide groups are among the most important in our bodies. For example, proteins are complex compounds containing many amide linkages. Primary amides hydrolyze (that is, are split apart by water) very slowly, forming acids and ammonia. The slowness of this reaction explains the stability of the proteins in our bodies. Secondary and tertiary amides will hydrolyze to form the corresponding acid and an amine. Under certain conditions, and in the presence of specific enzymes such as those found in the digestive tract, the hydrolysis of amides will occur in a relatively short time.

$$CH_3\overset{\overset{\displaystyle O}{\|}}{C}-NH_2 + H_2O \xrightarrow{\text{Catalyst}} CH_3\overset{\overset{\displaystyle O}{\|}}{C}-OH + NH_3$$

Acetamide   Acetic acid   Ammonia
(primary amide)

$$CH_3CH_2\overset{\overset{\displaystyle O}{\|}}{C}-\underset{\underset{\displaystyle H}{|}}{N}CH_2CH_3 + H_2O \xrightarrow{\text{Catalyst}} CH_3CH_2\overset{\overset{\displaystyle O}{\|}}{C}-OH + H_2NCH_2CH_3$$

N-Ethylpropionamide   Propionic acid   Ethylamine
(secondary amide—the
N indicates the ethyl group is
attached to the nitrogen)

$$CH_3\overset{\overset{\displaystyle O}{\|}}{C}-\underset{\underset{\displaystyle CH_3}{|}}{N}-C_6H_5 + H_2O \xrightarrow{\text{Catalyst}} CH_3\overset{\overset{\displaystyle O}{\|}}{C}-OH + H-\underset{\underset{\displaystyle CH_3}{|}}{N}-C_6H_5$$

N-Phenyl-N-methyl-   Acetic acid   N-Methylaniline
acetamide (tertiary
amide)

11.7 Dehydration

When amides are heated in the presence of a strong dehydrating agent, a molecule of water is removed and a member of the class of compounds called nitriles is formed.

$$\underset{\text{Acetamide}}{CH_3\overset{\displaystyle O}{\overset{\|}{C}}-NH_2} \xrightarrow{P_4O_{10}} \underset{\text{Acetonitrile}}{CH_3C\equiv N} + H_2O$$

11.8 Nitriles

Carbon is triply bonded to nitrogen in the extremely toxic compound hydrogen cyanide, $H-C\equiv N$. When inhaled or absorbed through the skin, this compound irreversibly blocks the mechanism by which cells get their energy, thereby killing all cells. Hydrogen cyanide is the gas used in penitentiary gas chambers. However, if the carbon atom triply bonded to the nitrogen is bonded to another carbon atom (instead of to a hydrogen atom), a group of much less toxic compounds called **nitriles** is formed.

Industrially, nitriles are polymerized to form many synthetic fabrics such as orlon, acrylan, and dynel, which are used in knitwear and carpets.

Acrylonitrile → Polyacrylonitrile (Orlon and acrylan)

11.9 Imines

Carbon-to-nitrogen double bonds are characteristic of the **imine** functional group, $\diagdown C{=}N{-}H$. This functional group is found in a molecule of the strong base guanidine. Guanidine forms part of the compound arginine, which is found in very high concentrations in sperm cells. Guanidine groups are also found in molecules of creatine, which, when bonded to a phosphate group, are the molecules that store energy for our muscles.

Guanidine   Arginine   Creatine

Two guanidine groups are part of the complicated structure of the antibiotic streptomycin, which was one of the first antibiotics to be isolated from the soil. Streptomycin resembles natural compounds found in infectious organisms, and is absorbed by these cells. But the structure of this compound is different enough to block the synthesis of protein in the cell, thereby killing the organism. Streptomycin is used in the treatment of tularemia, known as rabbit fever, and in the treatment of tuberculosis.

Streptomycin

11.10 Nitrogen in Rings

Our discussion of compounds containing rings has thus far been limited to molecules with rings composed of carbon atoms only. However, the elements nitrogen, oxygen, and sulfur can join with carbon atoms to form a ring. The resulting compounds are called **heterocyclic** compounds, and contain two or more types of atoms in the ring. Heterocyclic rings are found in an enormous number of naturally occurring compounds, and form an entire field of study by themselves. Table 11.4 shows some nitrogen-containing heterocyclic rings that are commonly found in biological systems, and we will study compounds containing such rings in later chapters.

Table 11.4   Some Heterocyclic Rings Containing Nitrogen

Name	Structural Formula	Found in the Structure of—
Pyrrolidine		Amino acids
Pyrrole		Chlorophyll, hemoglobin and vitamin B_{12}
Imidazole		An amino acid
Indole		An amino acid
Pyridine		The vitamin niacin
Pyrimidine		Nucleic acids
Purine		Nucleic acids

Before leaving our study of organic compounds containing nitrogen, however, we will take a brief look at some complex compounds that have profound physiological effects.

11.11 Alkaloids

Alkaloids comprise a large class of nitrogen-containing compounds having complex structures, often including nitrogen in heterocyclic rings. The alkaloids can be further categorized in terms of the specific ring structures found in the molecule. Most of the complicated alkaloids are found in a specific species of plant, and are often part of the plant's defense mechanism. These alkaloids, in various plant mixtures, have been used as drugs for centuries. Some of the naturally occurring alkaloids have also been synthesized in the laboratory. Moreover, in trying to duplicate the structure of an alkaloid, chemists have created compounds having properties far superior to the natural compound when used as a drug. By studying and replacing certain functional groups on these alkaloids, chemists have produced compounds having desirable properties, such as relieving pain or inducing sedation, without the accompanying side effects of dependence (addiction) or tolerance (requiring increased dosages to produce the same results in extended use) that often result from the use of the natural alkaloid. A complete discussion of the alkaloids and their many effects is not within the scope of this book, but we will briefly examine some interesting examples.

11.12 Epinephrine and Norepinephrine

Epinephrine belongs to a class of relatively simple alkaloids containing one benzene ring. We have already discussed the many effects that epinephrine can have on the body. Looking again at the structure of epinephrine we can recognize that it is a secondary amine.

Epinephrine

Another compound having a structure very closely related to epinephrine is norepinephrine, also produced by the adrenals. This compound, often called noradrenalin, is a primary amine whose structure differs only slightly from that of epinephrine—it lacks a methyl group on the nitrogen.

Norepinephrine

Just this small change makes a difference in the action of the two compounds in the body. Norepinephrine is also produced by certain nerve endings, and acts as a chemical messenger between nerve cells. The primary function of norepinephrine in the body is to maintain muscle tone in the blood vessels, thereby controlling blood pressure.

11.13 Amphetamines

Benzedrine (also called amphetamine) is a synthetic compound having a structure similar to that of epinephrine. Benzedrine stimulates the cortex of the brain, producing a decreased sense of fatigue and an increased alertness. Along with ephedrine, a natural substance extracted from a tree in the pine family, benzedrine has been prescribed to reduce fatigue, overcome sleepiness, and suppress the appetite. Benzedrine has also been used to treat bronchial congestion. With continued use of benzedrine for any of these purposes, dependence and tolerance can result. Over the last decade the abuse of this drug has increased to such an extent that it is now under strict regulation and control. Benzedrex, a derivative of benzedrine having the same decongestant effects, is now used in inhalers to relieve nasal congestion. This compound does not bring about dependence.

Benzedrine   Ephedrine   Benzedrex

11.14 Nicotine

Some alkaloids contain a nitrogen atom as part of the ring. Nicotine, which is found in the leaves of tobacco, has two nitrogen-containing rings.

Nicotine

Nicotine in pure form is a rapid-acting, extremely toxic drug. Cases have been reported of gardeners who have died from handling nicotine as an insecticide; death from respiratory failure occurred within a few minutes. The action of nicotine in the human body is complex. It stimulates the central nervous system, causing irregular heartbeat and blood pressure, induces vomiting and diarrhea, and first stimulates and then inhibits glandular secretions. Nonsmokers can absorb only about 4 mg of nicotine before symptoms of nausea, vomiting, diarrhea, and weakness begin. But a smoker builds up a tolerance to nicotine, and may absorb twice as much without apparent ill effects. The smoke from one cigarette may contain as much as 6 mg of nicotine, but only about 0.2 mg is absorbed into the body.

11.15 Caffeine

Caffeine, another alkaloid whose structure has two nitrogen-containing rings, occurs naturally in coffee beans, tea leaves, and the seeds of the chocolate tree. In addition to being found in coffee, tea, and chocolate, caffeine is also added to some carbonated beverages and nonprescription medicines (such as some aspirins).

Caffeine

Caffeine is a stimulant to the central nervous system, causing restlessness and mental alertness. For some people, taking caffeine can become a habit—they may become extremely reliant upon their morning coffee—but there is no evidence that caffeine is addictive.

11.16 Atropine and Curare

Some other alkaloids worthy of mention have more complex nitrogen-containing structures (Table 11.5). Atropine causes dilation of the pupils, and is used in eye surgery. Taken internally it relieves abdominal pain from severe muscle contractions.

Table 11.5 Alkaloids with Nitrogen-Containing Rings

It is used as premedication for gas anesthesia because it dries up secretions in the nose and throat. Curare has long been known for its chemical potency. South American Indians applied it to the tips of their arrows to paralyze their prey. It works by blocking the action of the nerve-transmission chemical acetylcholine, resulting in the paralysis of muscles. Curare is used in low doses to treat muscle spasms, and at higher doses to paralyze patients in certain types of surgery.

11.17 Cocaine

Cocaine is a local anesthetic and a stimulant to the central nervous system. It has effects similar to, but stronger than, the amphetamines. Because it does not produce tolerance, cocaine has been increasingly popular with drug users. In trying to synthesize compounds having the local anesthetic properties of cocaine, without the side effects, chemists have synthesized many useful anesthetics. One such anesthetic is procaine, which has been widely used in dentistry in the form of its derivative Novocaine.

Procaine

11.18 Opiates

Morphine and codeine are alkaloids that can be isolated from the opium poppy. These complicated molecules have a narcotic effect on humans, and have benefited mankind as well as created great sociological problems through their use and abuse. Morphine reduces pain, causes drowsiness and changes in mood, and produces mental fogginess. It has been used to provide pain relief in radical surgery and in cases of wartime injury. However, repeated use of morphine results in drug addiction and tolerance. Morphine also causes constriction of the smooth muscles such as those found in the intestines and can, therefore, be used in the treatment of diarrhea. Paregoric, commonly given to children suffering from diarrhea, is a mixture with an opium base. Codeine, although less effective than morphine, is also less likely to cause drug addiction. It is mainly used to relieve coughing.

Heroin is a synthetic alkaloid derived from morphine. It is the most powerful pain reliever of the morphine-type alkaloids, but also has increased narcotic action and is very addictive.

11.19 Barbiturates

Barbiturates are derivatives of barbituric acid (Table 11.6). These compounds act by depressing the central nervous system, and are thought to inhibit nerve response centers. The properties of specific barbiturate compounds depend upon the nature of the substituted groups on their molecules. Barbiturates can be used as hypnotics, sedatives, anticonvulsants, and anesthetics.

Table 11.6 Some Barbiturates

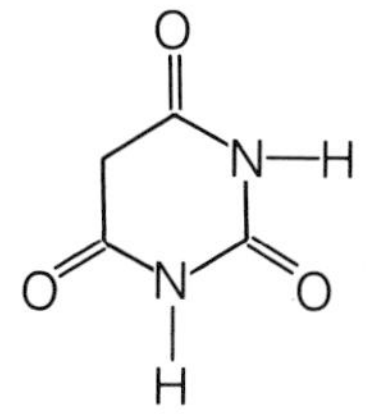

Barbituric acid

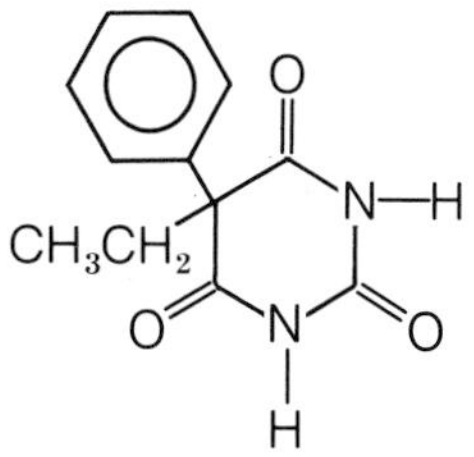

Phenobarbital
(a tranquilizer)

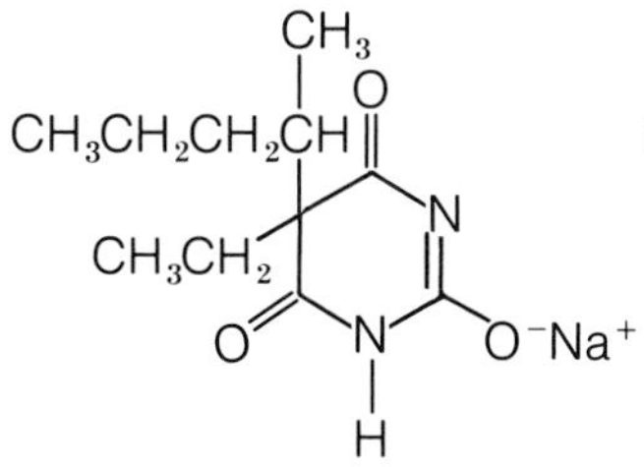

Pentobarbital sodium
(Nembutal-sleeping pills)

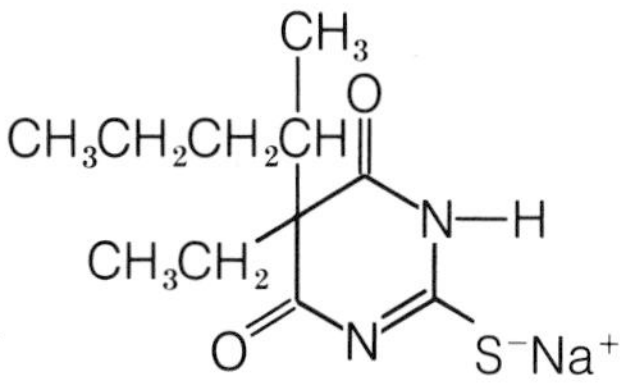

Sodium pentothal
(an anesthetic)

Their most common use (and abuse) is as a sleep-producing drug. Excessive use of these compounds can produce physical dependence. One class of barbituric acid derivatives, the thiobarbiturates, contains a sulfur atom in the molecule. The most important member of this class is sodium pentothal, a fast-acting intravenous anesthetic from which recovery is fairly rapid and often without side effects.

In this chapter we have discussed only a few of the nitrogen-containing compounds found in living systems. Other large classes of nitrogen-containing compounds, such as proteins and nucleic acids, will be the subjects of later chapters.

Additional Reading

Books
1. Isaac Asimov, *The World of Nitrogen*, Abelard-Schuman, Ltd., London, 1958.

2. R. P. Mariella and R. A. Blau, *Chemistry of Life Processes,* Harcourt, Brace and World, New York, 1968.

3. G. A. Swan, *An Introduction to the Alkaloids,* Wiley, New York, 1967.

4. See the chapters on nitrogen-containing compounds in any general organic chemistry textbook.

Articles
1. Joan Z. Majtenyi, "Antibiotics — Drugs from the Soil," *Chemistry,* January 1975, page 6.

2. R. G. Naves and B. Strickland, "Barbiturates," *Chemistry,* March 1974, page 15.

Questions and Problems

1. Define the following terms:

   (a) Amine
   (b) Ptomaines
   (c) Coordinate covalent bond
   (d) Primary, secondary, and tertiary amine
   (e) Amide

   (f) Amide linkage
   (g) Imine
   (h) Nitrile
   (i) Heterocyclic ring
   (j) Alkaloid
   (k) Barbiturate

2. Identify each of the following compounds as an amine, amide, imine, or nitrile.

   (a) 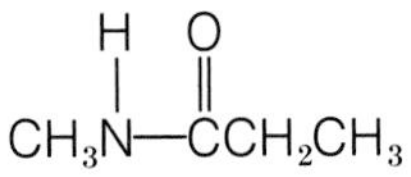
   $$\underset{\underset{CH_3N-CCH_2CH_3}{}}{\overset{H\ \ O}{}}$$

   (b) $CH_3C{\equiv}N$

   (c) 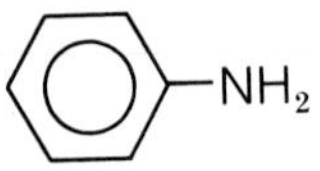 $\bigcirc\!\!-NH_2$

   (d) 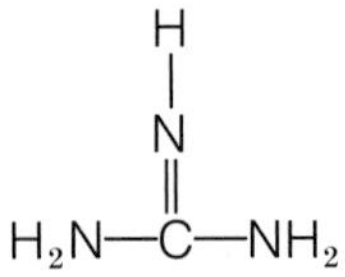 $H_2N-\overset{\overset{H}{|}}{\underset{\underset{NH_2}{\|}}{C}}$ H₂N—C—NH₂ (with N—H above)

   (e) 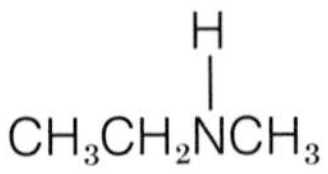
   $$CH_3CH_2\overset{\overset{H}{|}}{N}CH_3$$

   (f) 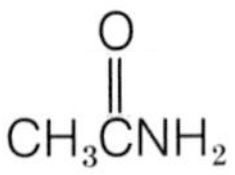
   $$CH_3\overset{\overset{O}{\|}}{C}NH_2$$

3. Identify each of the following compounds as a primary, secondary, or tertiary amine.

   (a) 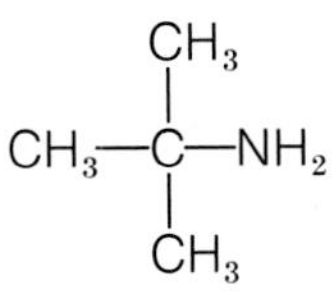
   $$CH_3-\overset{\overset{CH_3}{|}}{\underset{\underset{CH_3}{|}}{C}}-NH_2$$

   (d) $CH_3CH_2NHCH_3$

(b) $\langle\bigcirc\rangle$—N—CH$_3$ (e) CH$_3$CH$_2$NCH$_3$

                CH$_3$ CH$_3$

(c) CH$_3$CH$_2$NH$_2$ (f)

                                     N
                                     H

4. What substances can cause the foul smell of decaying organic material?

5. Explain why an amine can act as a base in a chemical reaction.

6. Identify each of the following as a primary, secondary, or tertiary amide.

          O   O

          ‖   ‖

(a) CH$_3$NCCH$_3$ (c) CH$_3$CH$_2$N—CCH$_3$

         CH$_3$ H

              O   O

              ‖   ‖

(b) $\langle\bigcirc\rangle$—CNH$_2$ (d) $\langle\bigcirc\rangle$—N—CCH$_2$CH$_3$

                                              CH$_3$

7. Look at the structural formula of aureomycin on page 322. Identify all of the functional groups in the molecule.

8. Write an equation for the formation of:

   (a) Acetamide from acetyl chloride.
   (b) Benzamide from ethylbenzoate.

9. Write an equation for the hydrolysis of:

   (a) N-methylacetamide   (b) Propionamide

10. Draw the structure of two heterocyclic rings that contain nitrogen.

11. Morphine is a very effective pain reliever. Why must its continued use in a patient be avoided?

12. What changes in functional groups must take place on the morphine molecule to produce heroin?

13. Describe three specific effects of nicotine on the human body. Propose a theory to explain why smokers can tolerate larger concentrations of nicotine than nonsmokers before feeling ill effects.

14. Describe three different uses of alkaloids in drug therapy.

chapter 12

The Remaining Twenty Elements

Learning Objectives

By the time you have finished this chapter, you should be able to:

1. Describe three important functions of calcium in the body.

2. Write the reaction for the removal of calcium ions from the blood by the citrate ion.

3. Indicate two ways in which the normal metabolism of calcium can be disrupted.

4. Write the formula of the phosphate ion and describe three important functions of this ion in our bodies.

5. Describe two functions of magnesium in the living organism.

6. Describe five important functions of the sodium, potassium, and chloride ions in the human body.

7. Write the formula for the sulfhydryl and thioether functional groups.

8. Define "trace element."

9. Describe the functions of four trace elements.

10. Using specific examples, explain how trace elements can function in both a complementary and antagonistic fashion.

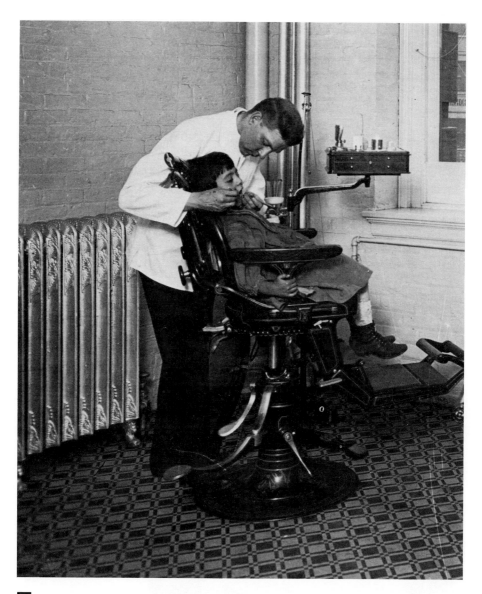

Fred McKay arrived in Colorado Springs, Colorado, in 1901 to begin a practice in dentistry. The young dentist quickly became alarmed when he discovered the badly stained teeth characteristic of the natives of that area. He was unable to find any mention of such a condition in the dental literature, so he named this condition "Colorado Brown Stain." The severity of this staining ranged from small opaque, paper-white areas visible on the surface of normally translucent teeth (a condition called mottling), to a very noticeable dark brown stain on the teeth, the enamel surfaces of which were often pitted. McKay launched an effort to record the occurrence and severity of this condition in various parts of the country, and tried to interest other members of the dental profession in joining in a study of Colorado Brown Stain. By 1916, after a study encompassing 26 communities, McKay had concluded that there was some factor in the drinking water of these communities that was responsible for the mottled teeth. Moreover, he determined that this factor had a noticeable effect only during the formative stages of permanent tooth development in children.

In areas of the country experiencing this widespread occurrence of mottled tooth enamel, parents were understandably distressed by the appearance of their children's teeth. McKay's discovery that this condition was related to the composition of the drinking water prompted communities such as Oakley, Idaho, to switch their source of drinking water. This change completely eliminated the occurrence of mottled teeth in the children of that town. McKay's investigations continued, and in 1929 he was able to make the important observation that patients suffering from mottled teeth were less susceptible to dental caries, commonly known as cavities.

In 1931, independent investigations by three different laboratories resulted in the identification of the fluoride ion, F^-, as the factor in drinking water causing the mottling of teeth during their formative stages. Subsequent measurements of the fluoride concentration in various water supplies revealed that the residents of Colorado Springs had for half a century been drinking water containing 2 parts per million (ppm) of fluoride ion, and that the previous water supply of Oakley, Idaho, contained 6 ppm fluoride (the new water supply for this community contained less than 0.5 ppm fluoride ion). By 1942, it had been established that a level of 1 ppm fluoride in the drinking water would significantly lower the incidence of dental caries without producing mottled teeth (Table 12.1 and Figure 12.1).

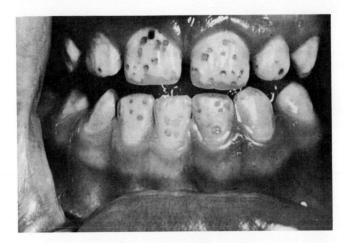

Figure 12.1 These teeth show severe mottling, or fluorosis, caused by the long-term drinking of water containing fluoride in a concentration greater than 10 parts per million. (Courtesy National Institutes of Health)

Table 12.1 Fluoride Levels in Drinking Water

Parts per Million of Fluoride	Effects
Less than 0.5	High incidence of dental caries; no mottling
1	Low incidence of dental caries; little or no mottling
Greater than 2.5	Disfiguring mottling
10 to 20	Severe mottling or fluorosis

In 1950, the United States Public Health Service officially endorsed the fluoridation of public water supplies to a level of 1 ppm fluoride. Many cities and towns quickly acted to fluoridate their water supplies following studies which showed that the rate of decayed, missing, and filled (DMF) teeth in children exposed to such water was reduced by as much as 60% (Table 12.2).

The actual mechanism by which fluoride prevents dental caries is not understood. Dental caries are created when acids, produced by bacteria in

Table 12.2 Improvement in Dental Decay Observed in Some Fluoridation Studies

Location of Study	Years of Observation	Age Groups	Percent Reduction in Dental Decay
Brantford, Ontario	7	6	59
		7	70
		8	52
		9	46
		13	33
Evanston, Illinois	4	6	74
		7	56
		8	35
Grand Rapids, Michigan	8	6	71
		7	53
		8	49
		9	48
		13	40

C. M. Taylor and O. F. Pye, *Foundations of Nutrition,* The Macmillan Company, New York. Copyright © 1966, The Macmillan Company, page 313. Used by permission.

the mouth, eat into the enamel coating of the teeth. When a supply of fluoride is available to children, fluoride ions are deposited as inorganic material in the structure of the tooth enamel while the teeth are forming. It is thought that this fluoride compound may make the tooth enamel more resistant to the action of the acids secreted by mouth bacteria. Scientists also have discovered that fluoride ions in a test tube are very effective in stopping the action of enzymes. This fact has led to the theory that a high concentration of fluoride in the tooth enamel may inhibit the action of bacterial enzymes responsible for converting sugar into the decay-causing acids.

The fluoridation of public water supplies is not without controversy. Although low levels of fluoride have not been shown to have any adverse effects on the human body, high levels of fluoride in the body can cause debilitating disease and even death. The body not only deposits fluorides as part of tooth enamel, but also deposits these ions in the bones. High levels of fluoride intake can result in a condition called skeletal fluorosis, marked by increased bone density and abnormal bone growth. In some areas of the world the natural level of fluorides is sufficient to produce severe skeletal fluorosis. The best documented cases of this condition have occurred in India, but this disorder has also arisen in Tanzania, Japan, and in the Southwestern United States. Skeletal fluorosis can cause abnormalities in the spine, with the development of pointed protrusions growing out of the vertebrae. A person suffering this condition has trouble breathing and cannot bend his back, giving the disease the slang name "poker back." Such bone growths can continue to a point where the spinal cord is constricted, leading to uncontrolled muscle contractions in the legs, and eventual paralysis. It should be noted that in regions in India where this condition is prevalent, the level of fluoride intake is greater than 10 ppm.

If very small concentrations of fluoride can have such wide-ranging effects on our bodies, it is reasonable to ask if there are other elements that are equally effective in very small concentrations. In the last few chapters we discussed carbon, hydrogen, oxygen, and nitrogen, and saw how these four elements, accounting for 99.3% of all the atoms in the body, form an enormous variety of compounds. There are 20 other elements that have so far been shown in laboratory experiments to be necessary for life. Although they make up only 0.7% of the atoms in the human body, they perform many diverse functions critical to life. Depriving a living organism of any of these elements will result in disease and death (Table 12.3).

These 20 elements can be divided into two groups. One group consists of seven elements that are found in substantially larger concentrations than the remainder; this group of elements contains potassium, magnesium, sodium, calcium, phosphorus, sulfur, and chlorine. The remaining elements are found in such very small amounts that they are referred to as the **trace**

Table 12.3   Elements Necessary for Life

Element	Percent of Total Number of Atoms in the Human Body	Number of Grams in a 70-kg Man
Hydrogen	63	6580
Oxygen	25.5	43,550
Carbon	9.5	12,590
Nitrogen	1.4	18 15
Calcium	0.31	1700
Phosphorus	0.22	680
Potassium	0.06	250
Sulfur	0.05	100
Chlorine	0.03	115
Sodium	0.03	70
Magnesium	0.01	42
Iron	<0.01	7
Manganese, cobalt, copper, zinc, molybdenum, vanadium, chromium, tin, fluorine, silicon, selenium, iodine	<0.01	<1

elements. All 20 elements are found in the living system either as ions, or covalently bonded in organic molecules. Their functions are quite varied, and may depend upon their chemical forms or their locations in the body's tissues and fluids. They maintain the electrical balance in the cells and the body fluid. They must be present in particular ratios to permit the contraction of muscles and the transmission of nerve impulses. Their concentrations control the movement of body fluids in the tissues and cells. They are part of the secretions essential to digestive processes. They activate or form integral parts of enzymes, and are contained in vitamins, hormones, and in proteins important in transporting materials throughout the body (Figure 12.2).

Figure 12.2 These elements must be present in your body for a normal and healthy life.

The Intermediate Seven

12.1 Calcium

Calcium is the fifth most abundant element in the body, accounting for 0.31% of the atoms in the body, and making up about 2% of an adult's body weight. You've no doubt heard that calcium is important for the formation of strong bones and teeth. Indeed, 99% of the calcium in your body is found in your bones and teeth, imbedded in the form of several inorganic salts in a framework of proteins (Figure 12.3). Calcium is a group II element; it has two valence electrons and is found in the body in the form of the stable calcium ion, Ca^{2+}. The level of calcium ion in the blood is determined by a complicated relationship involving two hormones, parathormone and calcitonin, and vitamin D. The level of these substances in the blood controls the amount of calcium absorbed in the intestines, the level of calcium ion in the blood, and the amount of calcium deposited in the bones and teeth.

The calcium ions not found in the skeleton play several other critical roles in the body. Calcium ions must be present in the correct concentration to allow the contraction of muscles, and are especially important in maintaining the rhythmic contraction of the heart muscle. Calcium ions affect nerve transmission by having a stabilizing effect on the nerve

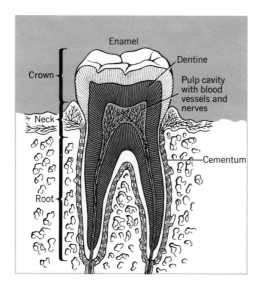

Figure 12.3 The structure of a tooth. Enamel, dentine, and cementum are composed of the following inorganic materials imbedded in a protein framework:

Main Inorganic Material

Apatite	$Ca(OH)_2 \cdot 3[Ca_3(PO_4)_2]$

Trace Inorganic Material

Fluorapatite	$CaF_2 \cdot 3[Ca_3(PO_4)_2]$
Chlorapatite	$CaCl_2 \cdot 3[Ca_3(PO_4)_2]$
Dahllite	$CaCO_3 \cdot 3[Ca_3(PO_4)_2]$

membrane. Too much calcium in the blood results in a deadening of nerve impulses and muscle response, making the individual unresponsive to any stimuli. Too little calcium in the blood can result in a high (hyper) irritability of the nerves and muscles. Under such hyperirritability the slightest stimulus, such as a loud noise, cough, or touch, can send a person into convulsive twitching. Such a state is extremely exhausting and will soon result in death.

A specific level of calcium ions in the fluid of the brain is critical in maintaining body temperature. Too high a concentration of calcium will result in the lowering of body temperature. Calcium ions must be present for the blood to clot. Any condition that will remove the calcium ions from the blood will prevent the clotting process from occurring. Leeches, fleas, and other blood-feeding creatures are able to secrete a substance that reacts with the calcium ions in the blood, preventing the blood from clotting while they feed on it and digest it. Citrate ions and oxalate ions will react with calcium ions in freshly collected blood, stopping the blood-clotting reactions. Sodium citrate is used as an anticoagulant in whole blood to be used for transfusions.

$$2C_6H_5O_7^{3-} + 3Ca^{2+} \longrightarrow Ca_3(C_6H_5O_7)_2$$

Citrate ion

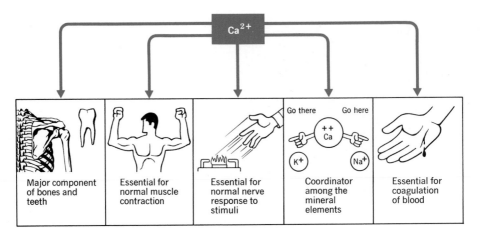

Figure 12.4 Functions of calcium in the body.

Calcium ions, in addition to activating part of the clotting process, also activate a variety of other enzymes. The calcium ion may be considered the "coordinator" among the other mineral ions, regulating the flow of these ions in and out of the cell (Figure 12.4).

Dairy products are the major source of calcium in our diets; an adult who drinks one pint of milk a day is meeting his minimum daily requirement for calcium. The intricate mechanism controlling the level of calcium in the blood can be upset by various factors including low levels of calcium in the diet, or abnormal levels of vitamin D, calcitonin, or parathormone in the blood. Too little vitamin D in the body produces rickets, a disease that causes bones to soften and bend out of shape (Figure 12.5). Although vitamin D is formed in the body when substances in the skin are exposed to sunlight, supplemental vitamin D has been added to milk supplies in the United States to prevent the occurrence of rickets. However, it has been shown that too much vitamin D can be as harmful as too little; it causes a thickening of bones and a calcification of soft bones.

Exposure to cadmium ions can lower the level of calcium in our bodies, causing a disruption of normal calcium metabolism. In the late 1950s, residents of Japan's Jinzu River basin were affected by a strange malady called the "Hai-Hai" or "Ouch-Ouch" disease. This disease caused severe and painful decalcification of bones, often resulting in multiple fractures. This malady was found to be caused by the presence of cadmium ions, Cd^{2+}, in rice irrigated by water discharged from upstream industries into the river.

12.2 Phosphorus

Although calcium plays a major role in the formation of bones and teeth,

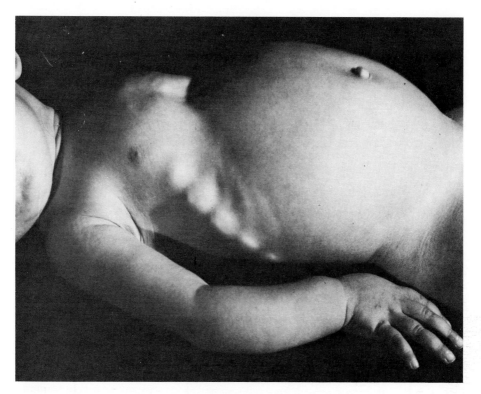

Figure 12.5 This child shows the characteristic deformities of rickets: bones that become soft from a lack of Vitamin D and bend out of shape. (Courtesy of the Children's Hospital Medical Center, Boston, Mass.)

phosphorus is an important element in the inorganic calcium salts that are found in these bones and teeth (Figure 12.3). Ninety percent of the phosphorus in the body is found in the bones and teeth in the form of the negative phosphate ion, PO_4^{3-}. Phosphorus, like nitrogen, is a member of group V; it can form three covalent bonds, and has two unshared electrons that can enter into a coordinate covalent bond, as occurs in the phosphate ion.

Phosphoric acid, H_3PO_4

Phosphate ion, PO_4^{3-}

The phosphate ion results when the inorganic acid called phosphoric acid loses three hydrogen ions.

$$H_3PO_4 \xrightarrow{-H^+} H_2PO_4^- \xrightarrow{-H^+} HPO_4^{2-} \xrightarrow{-H^+} PO_4^{3-}$$

| Phosphoric acid | Dihydrogen phosphate ion | Monohydrogen phosphate ion | Phosphate ion |

The three inorganic phosphate ions resulting from the removal of hydrogen ions from phosphoric acid help to maintain the neutrality of the fluids in the body. The phosphate group is capable of replacing any hydroxyl group (OH) on an organic compound to form an organic phosphate compound. Organic phosphates are found in phospholipids (which make up the cell membrane and nerve tissues), DNA and RNA (which control heredity and protein synthesis), and coenzymes (compounds that work with enzymes in the body). These compounds will be discussed in detail in later chapters. When certain organic phosphates undergo hydrolysis, considerable chemical energy is liberated. Such phosphates are known as high energy phosphates, and are the compounds that supply the immediate energy needs of the cell. These important functions make phosphorus essential to all body tissues (Figure 12.6). Fortunately, phosphorus is so widely distributed in our daily foods that there is little chance of becoming deficient in this element.

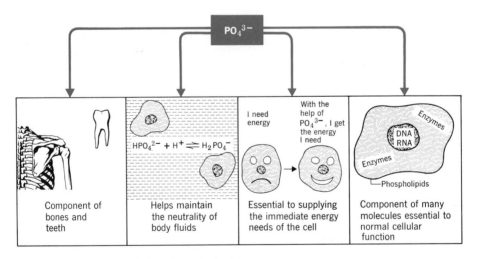

Figure 12.6 Functions of phosphorus in the body.

12.3 Magnesium

Magnesium ions, Mg^{2+}, make up 0.01% of the atoms in the body. These ions activate many of the enzymes important in the addition and removal of phosphate groups from compounds in the cell (Figure 12.7). Magnesium forms part of the chlorophyll molecule, which traps sunlight in the photosynthesis process and gives plants their green color (Figure 12.8).

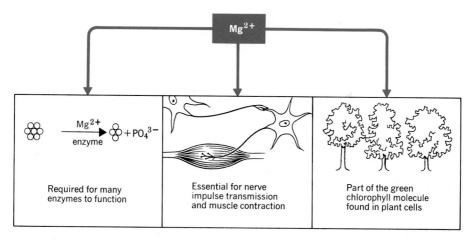

Figure 12.7 Functions of magnesium in living organisms.

Magnesium is found in a wide variety of foods such as green vegetables, nuts, cereals, and seafoods, so we are fairly well assured of enough magnesium in our diets.

If the magnesium level in the body is lowered, a person may suffer emotional irritability and aggressiveness, muscle spasms, and convulsions. Magnesium deficiencies have been observed in chronic alcoholics, in infants having a protein deficiency disease called Kwashiorkor, and in postsurgical patients on restricted diets. Too much magnesium decreases muscle and nerve response, and high levels can produce local or general anesthesia and paralysis.

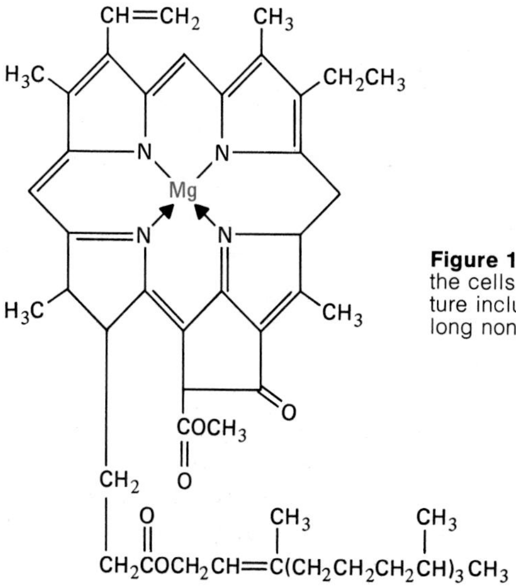

Figure 12.8 The structure of chlorophyll, found in the cells of photosynthetic plants. Its complex structure includes a magnesium ion in the center, and a long nonpolar side chain.

12.4 Potassium, Sodium, and Chlorine

The roles of potassium, sodium, and chlorine in the body are intricately interrelated. Sodium and potassium ions are often present in the form of their chloride salts. The potassium ion is usually found within the cell, while the sodium ion is generally found in the fluids surrounding the cell. The primary function of the potassium, sodium, and chloride ions is to control the electrical balance in the cells, tissue fluids, and blood. Such electrical balance is necessary to maintain the normal flow of fluids and to control the balance between acids and bases in the body. These three ions also play an important role in the transport of oxygen and carbon dioxide in the blood. In addition, sodium and potassium ions (along with calcium and magnesium ions) help maintain the proper level of nerve and muscle response. The effects of sodium and potassium are antagonistic to those caused by calcium and magnesium. Therefore, the relative magnitude of the concentrations of these four ions is critical to normal nerve and muscle performance.

Sodium chloride and potassium chloride act to keep large protein molecules in solution, and also to regulate the proper viscosity or thickness of the blood. The acid in the stomach that begins the digestion of certain foods is hydrochloric acid, HCl, which is derived from the sodium chloride in the blood. Other digestive compounds found in the gastric juices, pancreatic juices, and bile are likewise formed from the sodium and potassium salts in the blood. The response of the retina of the eye to light impulses is another body process that depends upon the correct concentrations of sodium, potassium, and chloride ions (Figure 12.9).

In light of the many important functions depending upon these three ions, it is easy to see that an imbalance in the level of any one of them can have serious effects on the body. Experimental animals have been subjected to diets deficient in these ions, and have experienced slow growth, slow heart rate, muscular atrophy, and sterility. Plant material is high in potassium ions, but high levels of potassium ions in the diet will result in excessive excretion of the sodium ion from the body. Herbivorous animals (animals that eat plant material) must, therefore, be supplied with a high level of salt or sodium chloride in their diets to maintain the proper balance between the sodium and potassium ions in their bodies. Such animals have been known to travel hundreds of miles and to risk their lives in order to reach a salt lick. Under normal dietary conditions, humans are not subject to a deficiency of these ions. One rare disorder, hyperkalemic paralysis, produces a high level of potassium in the body and results in general muscular weakness or paralysis.

It is no news to you that strenuous exercise, especially in hot weather, results in heavy perspiration. Perspiration is composed primarily of water, but there are also many ions dissolved in this fluid (among which are potassium, sodium, and chloride ions, giving sweat its salty taste). If the concentration of these ions in the body is significantly reduced by heavy

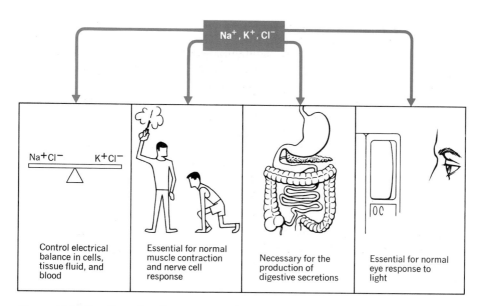

Figure 12.9 Functions of sodium, potassium, and chlorine in the body.

perspiration, an imbalance will occur that affects muscle and nerve response. Nausea, vomiting, exhaustion, and muscle cramps can result. For this reason, athletes often guard against such ion loss by taking salt tablets before heavy exercise, or by drinking specially formulated beverages, such as Gatorade©, to replenish their lost salts.

12.5 Sulfur

Sulfur is the last of the seven elements in this grouping. It is found in group VI on the periodic table; it appears just under oxygen and exhibits properties similar to oxygen. It will form two covalent bonds. One functional group in which sulfur is found is the thiol or sulfhydryl group, which closely resembles the hydroxyl group.

$$:\ddot{S}: H \qquad -S-H \qquad\qquad :\ddot{O}: H \qquad -O-H$$
Sulfhydryl group   Hydroxyl group

One class of simple compounds containing sulfhydryl groups is called the mercaptans.

$$CH_3CH_2SH \qquad\qquad CH_3CH_2OH$$
Ethyl mercaptan   Ethanol

These compounds smell terrible. You are probably familiar with a skunk's odor; the compound that produces that odor is butyl mercaptan.

$$CH_3CH_2CH_2CH_2SH$$
n-Butyl mercaptan

Because natural gas is odorless and toxic, gas companies add mercaptans to it before piping it to consumers. This allows leaks to be easily detected.

Other compounds containing sulfhydryl groups need not have any odor. Proteins, for example, are polymers of smaller units called amino acids (to be discussed in detail in Chapter 15). The important amino acid called cysteine has a sulfhydryl group, and is critical to the structure and function of protein molecules of which it is a part.

$$H-O-\overset{\overset{O}{\|}}{C}-\overset{\overset{H}{|}}{\underset{\underset{NH_2}{|}}{C}}-CH_2-S-H$$

Cysteine

Sulfur can replace the oxygen atom in an ether linkage to produce a thioether, or organic sulfide, —C—S—C—. Among the compounds containing this functional group are those giving onions and garlic their distinctive flavors.

$$H-\overset{\overset{H}{|}}{C}=\overset{\overset{H}{|}}{C}-S-\overset{\overset{H}{|}}{C}=\overset{\overset{H}{|}}{C}-H$$

Divinyl sulfide
(onion flavor)

$$H-\overset{\overset{H}{|}}{C}=\overset{\overset{H}{|}}{C}-CH_2-S-CH_2-\overset{\overset{H}{|}}{C}=\overset{\overset{H}{|}}{C}-H$$

Diallyl sulfide
(garlic flavor)

Methionine is an amino acid that contains a thioether group. All body proteins, as well as special protein molecules such as enzymes and hormones, contain this amino acid.

$$HO-\overset{\overset{O}{\|}}{C}-\overset{\overset{H}{|}}{\underset{\underset{NH_2}{|}}{C}}-CH_2CH_2-S-CH_3$$

Methionine

The body is able to produce cysteine from methionine, making this molecule extremely important in sulfur metabolism (Table 12.4).

Table 12.4 The Role of Sulfur in the Body

1. Found in proteins and other biological compounds.	
2. Found in several functional groups:	
Thiol or sulfhydryl	—S—H
Disulfide	—S—S—
Thioethers or organic sulfides	—C—S—C—

We have seen that the roles of these seven elements in our bodies are varied, complex, and often intricately related. The maintenance of proper concentrations of these elements through good nutrition is vitally important to assure a healthy body.

The Remaining Thirteen Elements

12.6 The Trace Elements

The remaining 13 elements that have been shown to be essential to life are often called the **trace elements.** Since these elements are found only in trace (extremely small) amounts in the living system, it has been quite difficult to establish the necessity of these elements and their functions in the body. The list of trace elements contains four nonmetals: silicon, fluorine, selenium, and iodine. The others are metals. One is tin, and the remaining eight are transition metals: vanadium, chromium, manganese, iron, cobalt, copper, zinc, and molybdenum (Table 12.5).

Although these elements are required in only minute amounts, they are critical to the proper functioning of the living system. A total absence of any one of these trace elements means certain death to the organism. The chemical action of many of these elements is quite complex, and the exact functions of several of them are still largely unknown. Many of these trace elements form important components of enzymes. Since enzymes can be used over and over again, they can be effective even when found in only very low concentrations in the cells of the body. The transition metals have unique chemical binding properties that make them especially important in enzyme molecules.

12.7 Iodine

One of the nonmetal trace elements, iodine, has long been known to be essential to life. Seventy to eighty percent of the iodine in the body is concentrated in the thyroid gland, a small gland in the neck. Iodine is part of a hormone produced by the thyroid gland. This hormone regulates the body's chemical activity (that is, its metabolism), and is an indispensable factor in normal growth (Figure 12.10). An insufficient amount of iodine in the diet causes enlargement of the thyroid, a medical condition known as simple goiter (Figure 12.11). This increase in the size of the gland is a compensatory reaction on the part of the body in response to the low level of iodine. The body attempts to increase the production of thyroid hormone by increasing the number of cells in the thyroid, but the attempt is futile as long as the concentration of iodine remains low. Salt-water fish are a rich source of iodine, and many cases of goiter were formerly observed in the Midwest where supplies of such fish were uncommon. The incidence

Table 12.5 The Trace Elements

Element	Atomic Number	Function
Fluorine, F	9	Constituent of bones and teeth
Silicon, Si	14	Essential for chicks, function unknown
Vanadium, V	23	Essential in lower plants, marine animals and rats
Chromium, Cr	24	Increases the effectiveness of insulin
Manganese, Mn	25	Essential for activity of several enzymes
Iron, Fe	26	Essential for hemoglobin and many enzymes
Cobalt, Co	27	Constituent of vitamin B_{12}, required for activity of several enzymes
Copper, Cu	29	Constituent of many enzymes, essential for utilization of iron in hemoglobin
Zinc, Zn	30	Essential for activity of many enzymes
Selenium, Se	34	Necessary for normal liver function
Molybdenum, Mo	42	Required for the activity of several enzymes
Tin, Sn	50	Essential in diets of rats, function unknown
Iodine, I	53	Constituent of thyroid hormones

Adapted from "The Chemical Elements of Life" by Earl Frieden, Copyright © 1972 by Scientific American, Inc. All rights reserved.

of goiter in such regions has been greatly reduced by the use of table salt containing the iodide ion, I^-. Although the amount of iodine in your body is only $\frac{1}{2,500,000}$ of your entire body weight (an amount of iodine about the size of a pinhead), the absence of this trace amount of iodine would be fatal.

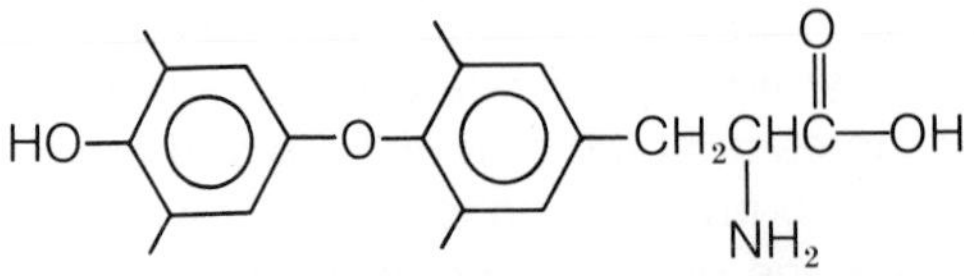

Figure 12.10 Thyroxin is the iodine-containing hormone produced by the thyroid.

Figure 12.11 These villagers from Paraguay are suffering from endemic goiter. The swollen thyroid gland in their necks results from a lack of the trace element iodine in their diets. (Courtesy World Health Organization. Photo by Paul Almasy)

12.8 Iron

Television advertising has made each of us aware that we need iron, and has informed us that a tired feeling might indicate a lack of this element in our blood. Four atoms of iron are found in every molecule of hemoglobin, the molecule in the red cells of our blood that carries oxygen from the

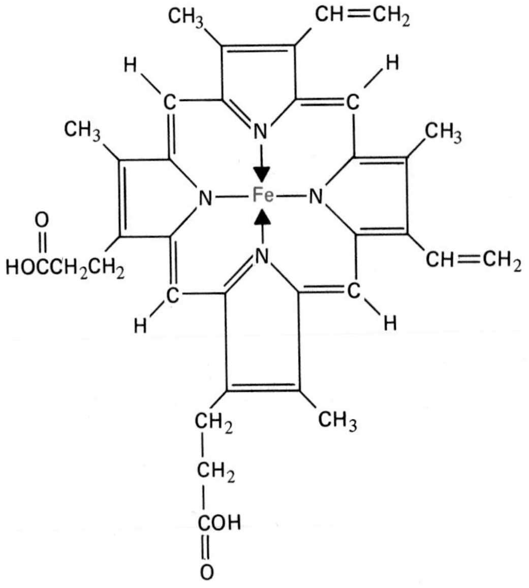

Figure 12.12 Four of these iron-containing heme groups are found in each hemoglobin molecule. Each heme is capable of carrying one oxygen molecule, so one hemoglobin molecule can carry four oxygen molecules.

lungs to the tissues, and that causes the blood to appear red (Figure 12.12). If you were able to extract all of the iron in a healthy body, you would end up with enough to make only two small nails (about 5 to 7 grams). Yet this amount of iron is critical, for only a small reduction in the level of iron in the blood will result in a condition known as anemia, which causes general body weariness, fatigue, and apathy. Anemias from iron deficiencies result in low hemoglobin levels, and often occur in infants at six months and in women from 30 to 50 years of age.

Iron for the diet is found in large amounts in organ meats such as liver, kidney, and heart, in egg yolks, dried vegetables of the pea family, and shellfish. Iron is absorbed in the small intestine in the form of the ferrous ion, Fe^{2+}. Iron absorption is increased by the presence of vitamin C, which reduces the ferric ion, Fe^{3+}, in the intestines to the ferrous ion. Under normal conditions, only 5 to 15% of the iron in the foods we eat is actually absorbed into our bodies. In previous centuries, some of the iron required by man was supplied from cast-iron cookware as the foods cooked in iron pots leached small amounts of this metal from the surface of the pot. Now that cast-iron pots have largely been replaced by aluminum, this regular source of iron has been removed from our diet, and more cases of low-level anemias are being observed in the population. A deficiency of iron in the diet results in listlessness and fatigue, reduced resistance to disease, and an increase in heart and respiratory rate.

Children with iron deficiencies show a decreased growth rate and abnormal red cell growth. High levels of iron in the body are also detrimental to health, causing cirrhosis of the liver, fibrosis of the pancreas resulting in diabetes, and congestive heart failure.

The functioning of iron in the body clearly illustrates how the concentration of one trace element can be closely dependent upon the concentration of some other trace element. In this instance, an enzyme containing copper must be present for a hemoglobin molecule to be formed. Therefore, the concentration of hemoglobin in the body depends not only upon the level of iron, but also upon the concentration of copper. A high concentration of copper will result in a high usage of iron. Such relationships between trace elements can also work in an antagonistic fashion. For example, a high concentration of the trace element molybdenum in the diet will cause a decrease in the absorption of copper by the body, resulting in a decrease in the formation of hemoglobin.

12.9 Copper

Since 1818, copper has been known to be a component of living tissue. The concentration of copper in the human body is very small, less than one-thirtieth of the amount of copper in a penny, yet this concentration is critical. Copper is necessary for the normal functioning of all living cells; too much or too little copper will result in malfunctions of the cell.

Copper performs a variety of functions in the body. It is a component of several important enzymes, one of which helps in the formation of blood vessels, tendons, and bones. Copper-deficient animals will exhibit weakness and fragility of blood vessels and bones (Figure 12.13). The formation of the protective sheath about the nerves is dependent upon copper, and a deficiency of copper plays a role in a degeneration of the nervous system in which nerve impulses are no longer properly transmitted. Copper also helps protect us from the harmful ultraviolet rays of the sun; this element is part of an enzyme that assists in the formation of melanin, the dark pigment of the skin that is our natural protection from ultraviolet radiation. Our cells would not be able to extract energy from foods without another copper-containing compound. Earlier we mentioned that copper must be present for the formation of hemoglobin. And, finally, the ability to taste foods may also be dependent upon the presence of copper.

Because copper is a part of critical enzymes, low levels of this element can cause serious illness. But high levels of copper can be equally toxic. Since copper is abundant in all foods and in drinking water, the body has developed intricate mechanisms to regulate the absorption and excretion of copper. The copper level in the body depends upon a balance between copper, molybdenum, and sulfate in the diet. A disorder called Wilson's disease results from a genetic abnormality in which the body's ability to eliminate copper is impaired. The liver, kidneys, and brain of an individual suffering from this disease will show abnormally high levels of copper,

Figure 12.13 These three dogs are from the same litter. The dogs at the right and center have been fed diets deficient in copper, and show rough hair and deformed legs not found in the normal dog on the left (who was fed a diet containing copper). (Courtesy Baxter and Van Wyk, *Bulletin of the Johns Hopkins Hospital,* 93:1, 1953)

possibly leading to mental illness and death. Wilson's disease can be treated with a chemical that binds with copper atoms and thus detoxifies them.

12.10 Cobalt

A lack of cobalt in the diet results in a disease called pernicious anemia, which produces symptoms of fatigue and general weakness (Figure 12.14). This disease is not caused by a lack of hemoglobin, but rather results from a lack of erythrocytes, red blood cells that carry the hemoglobin molecules (and thus the oxygen) to the cells. Cobalt is a part of vitamin B_{12}, which is essential for the formation of erythrocytes (Figure 12.15). However, too much vitamin B_{12} in the diet will stimulate the production of too many erythrocytes, producing a condition called polycythemia.

12.11 Zinc

Zinc is essential for plant growth and is found in all naturally growing materials. It is part of several liver enzymes, one of which is important in the oxidation of alcohol to less toxic substances. A high level of alcohol in the body may cause the enzyme to break down, thus producing toxic conditions in the liver. Alcoholics having cirrhosis of the liver show a high level of zinc in the urine.

Low levels of zinc have recently been associated with poor appetite and poor growth in children, and may cause a total loss of the sense of taste in some adults. Diets supplemented with zinc have been shown to lead to the return of normal taste response, and improvement in the appetite and growth rate in affected children.

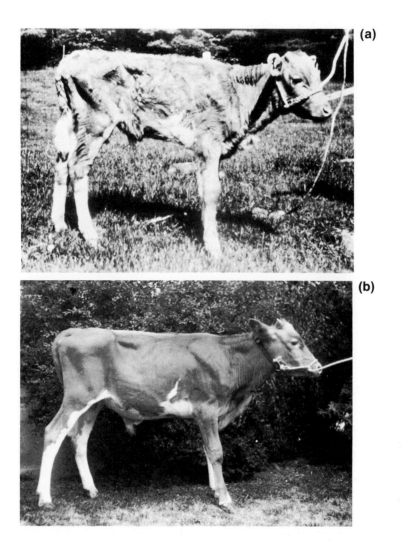

Figure 12.14 (a) This calf shows the symptoms of cobalt deficiency. (b) The same calf after it was fed cobalt for several weeks. (Courtesy Keener, *Univ. N.H. Agr. Expt. Sta., Bull. 411,* Durham)

12.12 Manganese

Manganese activates a number of enzymes, and is found in high concentrations in areas of the cell where cellular energy is produced. It is essential for normal thyroid function and cartilage and bone growth. The manganese ion, Mn^{2+}, is a powerful inhibitor of certain nerve impulse transmissions. Low levels of manganese in the diet can lead to decalcification of bones, bone fragility, and skeletal malformation. Low levels have also been shown to produce abnormal menstrual cycles in animals. Miners working to extract this metal have been observed to have high levels of

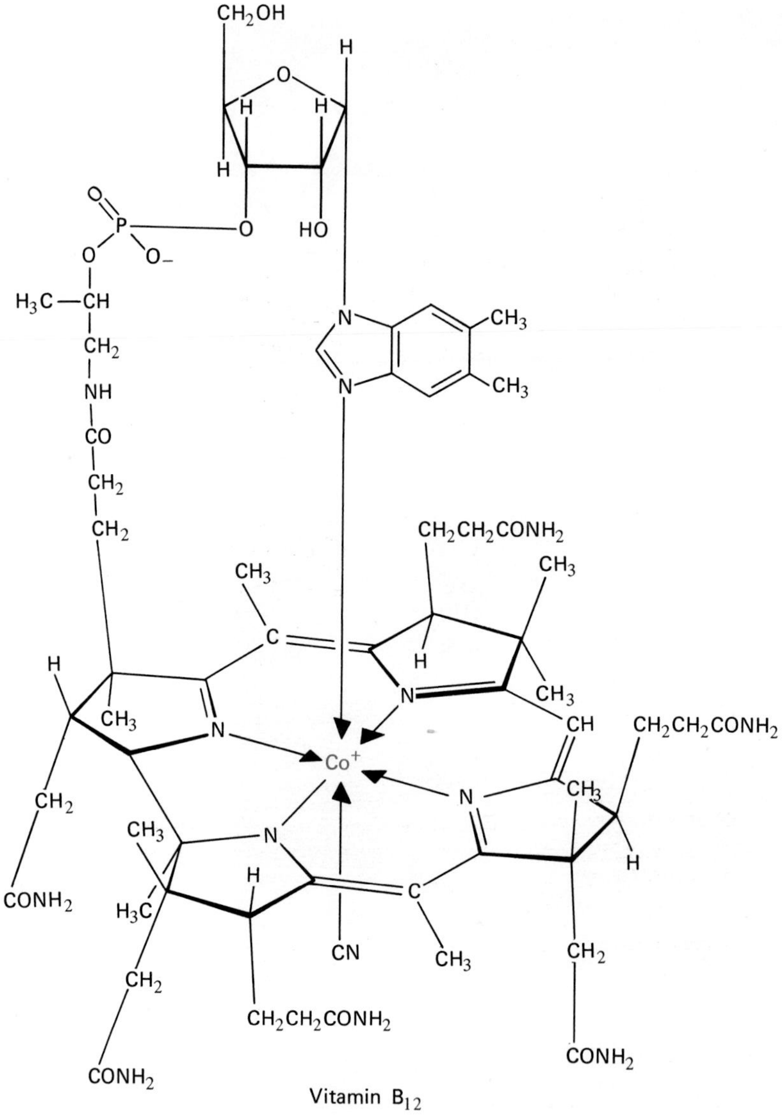

Figure 12.15 Vitamin B$_{12}$, or cobalamin, has a very complex structure containing a central cobalt ion.

manganese in their bodies. The resulting high brain concentrations of this ion produced headaches, psychotic behavior, and drowsiness. Blueberries and wheat bran are the richest dietary sources of manganese.

12.13 Selenium

Selenium, in very small amounts, is essential for the normal functioning of the liver of rats, and for the prevention of white muscle disease in cattle, sheep, hogs, and poultry. This element, however, is quite toxic in slightly larger amounts. In areas of Wyoming and South Dakota, the soil and foliage is rich in selenium. Livestock there suffer from the "blind staggers" — selenium poisoning that causes impairment of vision, muscle weakness, necrosis of the liver, and death from respiratory failure. Selenium is in the same chemical family as sulfur, and living organisms have difficulty distinguishing between the two elements. When more than trace amounts of selenium are present, it is substituted for sulfur in many cellular compounds. These selenium compounds have lower stability and are more reactive than the corresponding sulfur compounds, and will disrupt the normal functioning of the cell. High concentrations of selenium are reported to have caused cancer in laboratory animals.

Elevated concentrations of cobalt can magnify the toxic effects of selenium, producing enlargement of the heart and liver. This may have been the cause of the syndrome known as "beer drinkers' cardiomyopathe," which produced many deaths in several cities in 1965 and 1966. This syndrome resulted when small amounts of cobalt were added to beer to stabilize the foam, compounding the effect of the selenium that occurred naturally in fairly high amounts in the water supply of those cities.

12.14 Chromium

The level of sugar in the blood is critical to the functioning of body tissues, especially the brain. This blood sugar level remains remarkably constant under normal conditions. Many factors help to regulate the amount of sugar in the blood, among which is the compound called insulin. Insulin is secreted by the pancreas, and works as a controlling factor in lowering the blood sugar level. When the pancreas fails to secrete enough insulin, the disease called diabetes results. Chromium appears to play a role in lowering the blood sugar level in the body by increasing the effectiveness of insulin. Ample supplies of the trace element chromium are available to us since it is found in all plant and animal tissues.

12.15 Molybdenum

Molybdenum participates in the energy transfer reactions in the cell, is necessary for the function of certain intestinal enzymes, and is involved in the mechanism that controls the amount of copper absorbed by the body. Dietary sources of molybdenum include plants in the pea family, cereals, organ meats, and yeast.

12.16 Silicon, Vanadium, and Tin

The functions of silicon, vanadium, and tin are still under study. Very little is known about their exact roles, but they have been shown to be essential in the diets of various plants and animals.

Trace element research is bringing about a greater understanding of the ways in which these elements influence living organisms, and of the diseases that result from changes in their concentrations. Doctors are beginning to study trace elements as possible aids in treating disease, monitoring stress, and predicting heart disease. Although the interactions between trace elements are quite complex, a thorough understanding of these interactions is necessary if man is to remain healthy in the face of an industrialized society that is changing the natural levels of these and other trace elements in the soil, water, food, and air.

Additional Reading

Books

1. Frank J. McClure, *Fluoridation*, United States Department of Health, Education and Welfare, Bethesda, Maryland, 1970.

2. Henry Schroeder, *The Trace Elements and Man*, The Devin-Adair Company, Old Greenwich, Connecticut, 1973.

3. E. J. Underwood, *Trace Elements in Human and Animal Nutrition*, Academic Press, 1962.

Articles

1. G. W. Evans, "Biological Functions of Copper," *Chemistry*, June 1971, page 10.

2. Earl Frieden, "The Chemical Elements of Life," *Scientific American*, July 1972, page 52.

3. T. H. Maugh, II, "Trace Elements: A Growing Appreciation of Their Effects on Man," *Science*, July 20, 1973, page 253.

4. R. F. Miller and R. F. Dacheux, "Information Processing in the Retina: Importance of Chloride Ions," *Science*, July 20, 1973, page 266.

5. Michael Prival, "Fluorides in the Water," *Environment*, April 1973, January–February 1974, and June 1974.

6. Paul G. Seybold, "The Chemical Senses: Taste," *Chemistry*, March 1975, page 6.

7. Thressa Stadtman, "Selenium Biochemistry," *Science*, March 8, 1974, page 915.

8. D. R. Williams, "Life's Essential Elements," *Education in Chemistry*, March 1973, page 56.

Questions and Problems

1. For what body tissues is fluoride important?

2. What was the cause of "Colorado Brown Stain"?

3. What level of fluoride ion intake has been shown to prevent dental caries without producing mottling of teeth?

4. How can the fluoride ion be both beneficial and harmful to living tissue?

5. In considering mineral elements necessary for living tissues, calcium and phosphorus are often discussed together. Why? What happens to body growth if there is an inadequate supply of these elements in the diet?

6. What steps would you take to prevent the clotting of a recently drawn blood sample? Write the equation for the reaction that would take place.

7. What three factors control the amount of calcium in the blood and body fluids?

8. Discuss how a trace metal such as cadmium, which is not essential to the body's metabolism, can disrupt the metabolism of an essential element.

9. Write an equation for the formation of the phosphate ion from phosphoric acid.

10. What are three functions of the phosphate ion in the body?

11. What effect will high levels of magnesium and calcium have on nerve function? How does this effect compare with the effect of high levels of sodium and potassium ions?

12. Sodium, potassium, and chlorine are three essential elements whose varied functions in the human body are intricately related. Discuss five essential functions of these elements.

13. What harmful effects can excessive perspiration have on a person? How can one counteract the results of excessive perspiration?

14. Identify each of the following compounds as containing either a sulfhydryl, organic sulfide, or disulfide functional group.

(a) $CH_3CH_2\!-\!S\!-\!CH_3$ (c) $CH_3CH_2\!-\!S\!-\!S\!-\!CH_2CH_3$

(b) CH_3CHSH
$\quad\;\; |$
$\quad\; CH_3$

15. In what large class of compounds is sulfur found in the human body?

16. Why are large concentrations of selenium harmful to tissues? Support your conclusions using the periodic table.

17. What is meant by the term "trace element"?

18. What is the function of iodide in the human body? What condition results from too little iodide in the diet?

19. What three trace elements are involved in the synthesis of hemoglobin?

20. Describe the functions of the following trace elements:

(a) Copper (e) Selenium
(b) Cobalt (f) Chromium
(c) Zinc (g) Molybdenum
(d) Manganese

21. Using specific examples, explain how:

(a) Trace elements can play complementary roles in living tissues.
(b) Trace elements can play antagonistic roles in living tissues.

section IV
the compounds
of life

chapter 13

Carbohydrates

Learning Objectives

By the time you have finished this chapter, you should be able to:

1. Define monosaccharides, disaccharides, and polysaccharides.

2. Given the structure of a monosaccharide, identify the compound as an aldose or ketose.

3. Write the linear structure of glucose.

4. List the three hexoses and one pentose that play important roles in human metabolism.

5. Name the two polysaccharides constituting starch, and describe the difference in their structures.

6. Give the structural reason why we cannot digest cellulose.

7. Define "metabolism," "catabolism," and "anabolism."

8. Write the general reaction for the following processes:
(a) Photosynthesis   (d) Glycolysis
(b) Glycogenesis   (e) Fermentation
(c) Glycogenolysis

9. Describe in general the digestion and metabolic uses of the starch in a slice of toast.

10. Describe the meaning of "high-energy bond" and give an example of compounds in which they are found.

11. Describe the two stages of cellular oxidation of glucose, indicating where in the cell they occur and in which step the most energy is released.

12. State two ways in which the body can control carbohydrate metabolism.

13. Define the following terms:
(a) Diabetes mellitus   (e) Renal threshold
(b) Blood sugar level   (f) Glucose tolerance test
(c) Hypoglycemia   (g) Insulin shock
(d) Hyperglycemia

ARIZONA COLLEGE OF MEDICAL ●
DENTAL ● LEGAL CAREERS
4020 N. 19th AVENUE
PHOENIX, ARIZONA 85015

Carbohydrates

363

Cheryl was an active 10-year-old who enjoyed outdoor activities. She had especially been looking forward to playing little league baseball during the summer, and was terribly disappointed when a bladder infection kept her from practicing with her team. It was soon after Cheryl's infection that her mother began noticing a change in her daughter's behavior. Cheryl became increasingly irritable and tired. She complained that her skin itched, and she had developed several open sores from scratching. She drank large amounts of liquids practically continuously, requiring her to urinate frequently.

One sunny afternoon Cheryl's mother returned home to find her daughter lying on the couch and complaining of extreme thirst, blurred vision, and difficulty in breathing. Cheryl's face was flushed, but her skin was dry and cold. She was breathing deeply and rapidly, and her breath had a strange fruity odor. Cheryl's mother quickly called the doctor who, upon hearing the symptoms, said he would meet them at the hospital immediately. Meanwhile, Cheryl was becoming less and less responsive, and was moaning with pain. On the way to the hospital she lapsed into a coma.

At the hospital, the doctor found Cheryl's blood pressure to be low, and her temperature to be subnormal. Analysis of her urine revealed high concentrations of the sugar glucose, along with ketone compounds. Her blood sugar level was 660 mg/100 ml, and the level of acetone in her blood was 52 mg/100 ml. The doctor diagnosed Cheryl's condition as a diabetic coma, and began treatment by administering doses of rapid-acting insulin at hourly intervals. The hospital staff monitored Cheryl's blood and urine sugar levels closely. As her blood sugar began to reach a normal level, Cheryl was given doses of sugars to prevent the blood sugar level from falling below normal. She also was given saline intravenously to replenish the salts and fluids that she had lost.

As Cheryl was being treated for her diabetes, two floors above her in the hospital a 34-year-old woman named Diane was about to give birth to her first child. Diane's labor had begun early in the eighth month, so there was some concern over the condition of the infant. The baby was stillborn. To determine the cause of the baby's death, the doctor put Diane through a series of laboratory tests, one of which measured her body's ability to handle a large dose of glucose. From this test and others, Diane was found to have the adult form of the disease diabetes.

Diabetes results from a low level or a total lack of the hormone insulin, which is important in controlling the level of the blood sugar glucose. Both Cheryl's juvenile form of diabetes (which accounts for 10% of all diabetes patients) and Diane's adult type can be treated and controlled; however, neither can be cured. Cheryl was put on a strict low carbohydrate diet, and was taught to administer daily doses of insulin by injection. Doctors were able to control Diane's diabetes by restricting her diet and prescribing oral drugs that would stimulate her body to produce insulin.

The blood sugar glucose belongs to a large class of compounds known as carbohydrates. In this chapter we will study the structure, properties, and functions of the carbohydrates. Later, we will return to discuss in greater detail the disease diabetes.

Carbohydrates

Carbohydrates are a class of compounds that includes polyhydric aldehydes, polyhydric ketones, and large molecules that can be broken down to form polyhydric aldehydes and ketones. These compounds include sugars, starches, cellulose, dextrins, and gums. Carbohydrates are found mainly in plants, where they make up about 75% of the solid plant matter. They function both as supporting structures of the plant and as storehouses for the plant's energy supply. The carbohydrate cellulose is the most important component of the supporting tissue of plants (such as the wood in trees), and the carbohydrate starch is the energy storage molecule (Figures 13.1 and 13.2).

Figure 13.1 This giant sequoia tree is the largest living object in the world. It is as tall as a six-story building, wider at the base than an average city street, and is estimated to be between 2500 and 3000 years old. The supportive structure of the tree is composed of the carbohydrate cellulose. (Courtesy National Park Service)

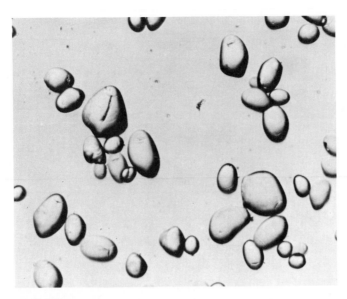

Figure 13.2 These are granules of starch from a potato. Starch is the energy storage molecule in plants. (Hugh Spencer)

13.1 Classification

Carbohydrates are classified according to the size of the molecule. **Monosaccharides** are carbohydrates that cannot be broken into smaller units upon hydrolysis. **Disaccharides** yield two monosaccharides upon hydrolysis, and **polysaccharides** yield three or more monosaccharides upon hydrolysis (and can contain as many as three thousand monosaccharide units).

Monosaccharides, also called simple sugars, can be further classified by the number of carbons in the molecule.

3 carbons—triose	5 carbons—pentose
4 carbons—tetrose	6 carbons—hexose, etc.

Monosaccharides may also be classified by the carbonyl functional group found in the molecule.

Aldose—aldehyde functional group
Ketose—ketone functional group

For example, ribose is an aldopentose (a five-carbon sugar molecule containing an aldehyde group) and fructose is a ketohexose (a six-carbon sugar molecule containing a ketone group).

Monosaccharides

13.2 Hexoses

Simple monosaccharides are white crystalline solids that are highly soluble in water as a result of their polar hydroxyl groups. (As you may suspect, therefore, they are not soluble in nonpolar solvents.) Most monosaccharides have a sweet taste, the reason for which is not totally understood. The most abundant monosaccharides are the hexoses. The hexoses that play important roles in human metabolism are glucose, fructose, and galactose (Figure 13.3).

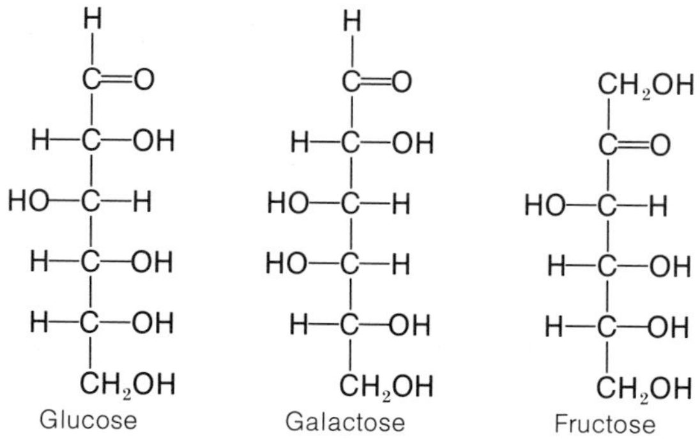

Figure 13.3 Hexoses that play an important role in human metabolism.

13.3 Glucose (Blood Sugar, Grape Sugar, Dextrose)

Glucose is the most abundant of the hexoses. It is an aldose that is found in the juices of fruits (especially grape juice), in the saps of plants, and in the blood and tissues of animals. Glucose is about as sweet as table sugar. It is the immediate energy reservoir for energy-requiring cellular reactions such as tissue repair and synthesis, muscle contraction, and nerve transmission. The average adult has five to six grams of glucose in his blood (about one teaspoon). This much glucose will supply the energy needs of your body for only about 15 minutes, so you must continuously replace the glucose in your blood from compounds stored in the liver. The level of glucose in the blood of a normal adult is fairly constant, rising after each meal and falling after periods of fasting.

Glucose is a component of many polysaccharides, and can be produced by the hydrolysis of these polysaccharides. (It is produced commercially by the hydrolysis of cornstarch.) Since glucose is found in most living cells, its chemistry is central to the carbohydrate chemistry of the body.

13.4 Structure of Glucose

The structure of glucose can be depicted in a straight chain form (as in Figure 13.3).

This open chain structure, however, does not help explain many of the properties of glucose, which actually is found in three forms in water solution. These three forms exist in equilibrium, and are readily converted one into another. The straight chain form of glucose makes up less than 0.5% of these molecules. The two other forms are ring compounds that result from the formation of a structural arrangement called an internal hemiacetal. A hemiacetal will form between an aldehyde and an alcohol group.

$$CH_3-\overset{\overset{\displaystyle O}{\|}}{C}-H + HOC_2H_5 \xrightarrow{HCl} CH_3-\overset{\overset{\displaystyle OH}{|}}{\underset{\underset{\displaystyle OC_2H_5}{|}}{C}}-H$$

| Aldehyde | Alcohol | A hemiacetal |

In glucose, the hemiacetal will form between the aldehyde group on carbon number 1, and the alcohol group on carbon number 5.

Ketones will react with alcohols to form a compound much like the hemiacetal, called a hemiketal.

$$CH_3-\overset{\overset{\displaystyle O}{\|}}{C}-CH_3 + HOC_2H_5 \xrightarrow{HCl} CH_3-\overset{\overset{\displaystyle OH}{|}}{\underset{\underset{\displaystyle OC_2H_5}{|}}{C}}-CH_3$$

| Ketone | Alcohol | A hemiketal |

The ring structure of glucose can be given a different representation by means of Haworth projections. These projections orient the hexagonal ring formed by the carbons of the glucose molecule perpendicular to the plane of the paper. The thickened side of the ring is the one closest to you, and the groups attached to the carbon atoms are then shown either above or below the ring.

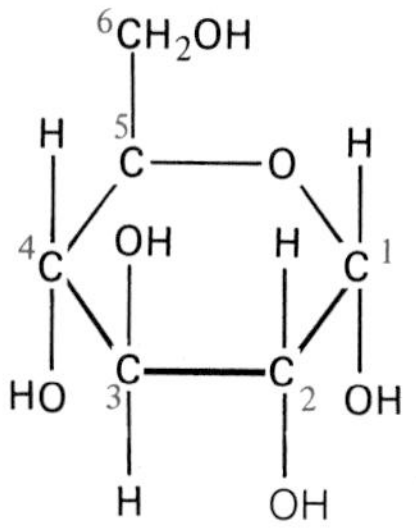

Haworth projection 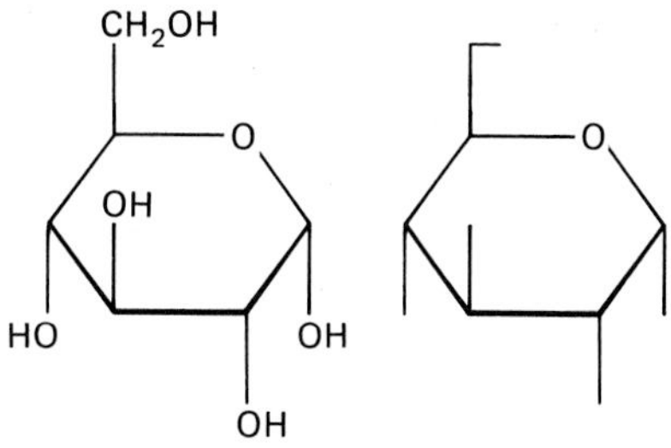   Shorthand forms

The two ring forms of glucose depend on the placement of the hydrogen and hydroxyl groups on carbon number 1. If the hydroxyl group is below the plane of the ring, this is the alpha (α) form; if it appears above the ring it is the beta (β) form (Figure 13.4).

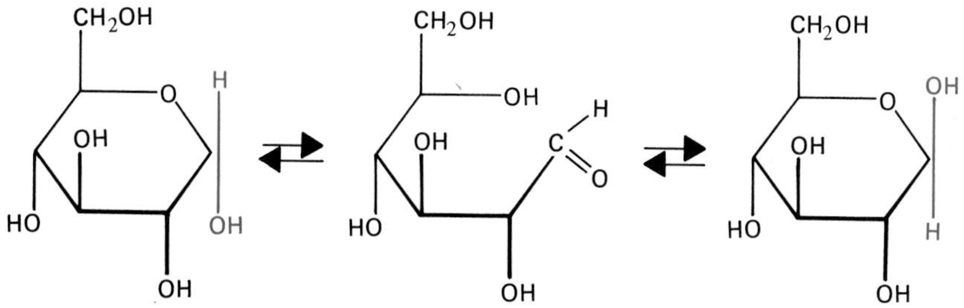

α–glucose   open chain β–glucose
(36%)   (0.02%)   (64%)

Figure 13.4 Three forms of glucose will exist in equilibrium in water solution: the alpha ring, the open chain, and the beta ring.

You may wonder what difference the position of that one hydroxyl group on the ring could possibly make. As we study the metabolism of living organisms, we will see that such small differences can determine whether a cell will be able to utilize a molecule or not. In this case, the difference between starch (a digestible glucose polymer) and cellulose (an indigestible glucose polymer) is in the position of the hydroxyl group on carbon number 1 of the glucose molecule.

13.5 Optical Isomerism*

In our study of biochemistry we will find that the position of other functional groups on the carbon atom is often critical. A carbon atom can have four

* This section on optical isomerism is optional, and may be skipped without loss of continuity.

groups attached to it, each directed toward one of the corners of a tetrahedron. If these four groups are all different, the carbon atom will possess a property called **asymmetry.** An asymmetric carbon atom, as with other asymmetric objects, has a mirror image that is not superimposable on itself. That is, if you take a three-dimensional model of an asymmetric carbon atom and a model of its mirror image, there is no way that you can superimpose the two models so that all of the groups match up (Figure 13.5).

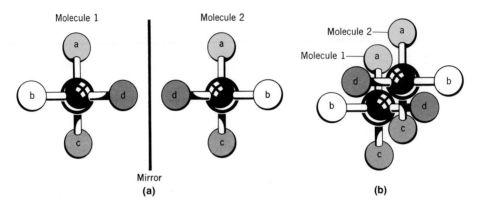

Figure 13.5 Molecules 1 and 2 are optical isomers. a) Molecule 1 contains an asymmetric carbon atom, and molecule 2 is the mirror image of molecule 1. b) The two molecules cannot be superimposed.

Consider a common example: your hand is asymmetric. Your right and left hands are not superimposable; there is no way that you can put a right-hand glove on your left hand (except by turning the glove inside out) (Figure 13.6). What difference will an asymmetric carbon atom make in a molecule?

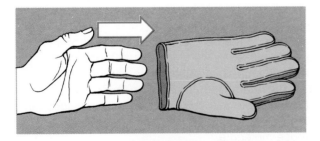

Figure 13.6 Your hand is asymmetric; there is no way that you can put a right-hand glove on your left hand (without turning it inside out).

If molecules are mirror images, they will have identical physical properties (such as boiling point, density, and vapor pressure) except one: the way in which they interact with polarized light, light in which the waves vibrate in only one plane. When placed in solutions, substances that interact with polarized light (by rotating the direction of this plane in either a clockwise or counterclockwise direction) are called optically active. **Optical isomers** are optically active compounds having the same molecular formula, but having the ability to rotate the plane of polarized light in opposite directions.

The system of classification of optical isomers was developed by Emil Fischer, and is based on the three-carbon compound glyceraldehyde.

D-Glyceraldehyde   L-Glyceraldehyde

The D-family of carbohydrates is represented as having the hydroxyl group located to the right of the asymmetric carbon farthest from the carbonyl group, and the L-family has the hydroxyl group to the left of the carbon. Most naturally occurring carbohydrates belong to the D-family (Figure 13.7). Of the 16 possible optical isomers in the aldohexoses (eight in the D-family and eight in the L-family), the three most abundant are D-glucose, D-mannose, and D-galactose.

Again, the small difference in arrangement of atoms between the D and L isomers may seem trivial to you, but to your body and its cells it is critical. Cells can recognize this difference, and often can use only one of the isomers. For example, yeast can ferment D-glucose to produce alcohol, but not L-glucose. As we will see in Chapter 15, our cells can use only L-amino acids to build proteins. The reason for this is that the enzymes that catalyze reactions in cells are asymmetrical compounds themselves, just as your shoes are asymmetrical. To catalyze a reaction, the enzyme must fit the reactant—just as your right shoe can fit only your right foot.

13.6 Fructose (Levulose, Fruit Sugar)

Fructose is a ketohexose found in many fruit juices and in honey. It is the sweetest sugar known, much sweeter than table sugar. Fructose is a component of the disaccharide sucrose (or cane sugar), and is produced

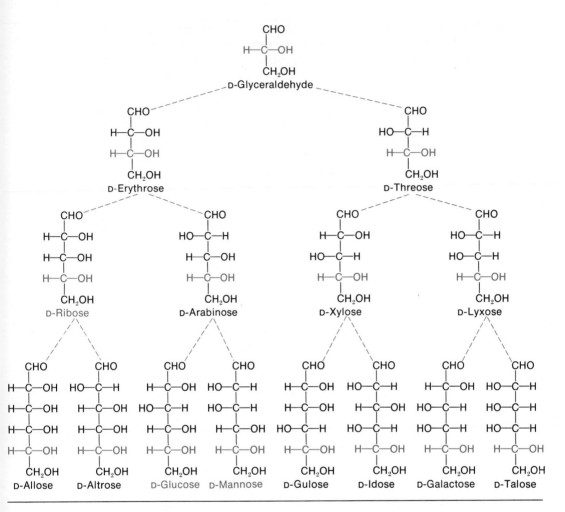

Figure 13.7 The D-family of aldoses.

by hydrolysis of the polysaccharide inulin. Fructose molecules, like glucose molecules, exist in two ring forms.

α-Fructose

β-Fructose

13.7 Galactose

Galactose is not found in nature as a free monosaccharide, but can be formed upon hydrolysis of larger carbohydrates. It is a component of lactose, the sugar found in milk, and as such is a component of our dietary

α-Galactose   β-Galactose

intake from birth. It is also found in glycolipids, fatlike substances that are components of the brain and nervous system. Agar-agar, a polysaccharide extracted from certain types of seaweed, is a polymer of galactose, and cannot be digested by man. It is used as a thickener in sauces and ice creams, and in nutrient broths used in microbiology.

13.8 Sugar Acids

Sugar acids are derivatives of carbohydrates in which the aldehyde group (or a hydroxyl group of a monosaccharide) has been oxidized to the carboxyl group.

Glucose   Gluconic acid   Glucuronic acid

Gluconic acid is one of the intermediates in the breakdown of glucose in some organisms. Glucuronic acid plays a special role in the body. By combining with unwanted toxic materials that have been accidentally swallowed, absorbed as medicine, or produced in normal metabolism, glucuronic acid makes these chemicals less toxic and more water soluble and, thus, easier for the body to excrete.

13.9 Pentoses

Some important five-carbon sugars are arabinose, which is formed by hydrolysis of gum arabic and the gum of a cherry tree, and xylose, which is a component of wood, straw, corncobs, and bran. Pentoses that play a role in human metabolism are ribose and deoxyribose, which are components of nucleic acids (the subject of Chapter 17). Both the alpha and beta forms of ribose exist in solution, but only the beta form is found in nucleic acids and in other metabolically active compounds.

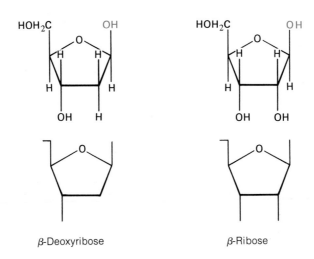

β-Deoxyribose   β-Ribose

13.10 Tests for Carbohydrate Identification

Molisch Test
This test is a general test for the presence of a carbohydrate. An alcoholic solution of α-naphthol is mixed with the unknown solution in a test tube. The tube is held at an angle, and cold concentrated sulfuric acid is poured into the tube. If a red-violet ring forms at the juncture of the two solutions, then a carbohydrate is present.

Seliwanoff's Test
This test detects the presence of a polyhydric aldehyde or ketone. The unknown solution is mixed with hot hydrochloric acid and resorcinol. Ketoses will produce a brignt red color, while aldoses will produce a pink color in the same period of time. This test is frequently used to detect fructose, but sucrose will also give a bright red color since it is hydrolyzed during the test.

$$
\begin{array}{c}
\text{CHO} \\
| \\
\text{H}\!-\!\text{C}\!-\!\text{OH} \\
| \\
\text{HO}\!-\!\text{C}\!-\!\text{H} \\
| \\
\text{H}\!-\!\text{C}\!-\!\text{OH} \\
| \\
\text{H}\!-\!\text{C}\!-\!\text{OH} \\
| \\
\text{CH}_2\text{OH} \\
\text{Glucose}
\end{array}
\qquad\qquad
\begin{array}{c}
\text{CH}_2\text{OH} \\
| \\
\text{C}\!=\!\text{O} \\
| \\
\text{HO}\!-\!\text{C}\!-\!\text{H} \\
| \\
\text{H}\!-\!\text{C}\!-\!\text{OH} \\
| \\
\text{CH}_2\text{OH} \\
\text{Xylulose}
\end{array}
$$

Figure 13.8 The urine of patients suffering from pentosuria will give a false positive Benedict's test for glucose. The urine contains the pentose sugar xylulose rather than glucose.

Test for Reducing Sugars

Carbohydrates that contain a free, or potentially free, aldehyde or ketone group will reduce alkaline solutions of mild oxidizing agents such as Cu^{2+} or Ag^+ **Benedict's test** is a widely used test for the detection of reducing sugars. The reagent contains an alkaline solution of copper sulfate, $CuSO_4$ This blue reagent solution is mixed with the unknown solution and is heated. If a reducing sugar is present, the Cu^{2+} ions will be reduced to Cu^+ ions, and a brick-red precipitate of copper oxide, Cu_2O, will form. The general reaction can be written

$$
\underset{\substack{(CuSO_4)\\ \text{Blue}}}{Cu^{2+}} + \underset{\substack{\text{(Free aldehyde or}\\ \text{ketone)}}}{\text{Reducing sugar}} \xrightarrow[\text{Base}]{\text{Heat}} \underset{\text{Brick-red}}{Cu_2O} + \text{Oxidized sugar}
$$

The amount of precipitate formed is proportional to the amount of reducing sugar present. All monosaccharides will give a positive Benedict's test. Clinitest tablets, which are widely used to test urine for sugar, are based on the same principle as Benedict's test. In this case a green color indicates very little sugar in the urine, while a brick-red color indicates more than 2 grams of reducing sugar per 100 ml of urine.

Note that a false positive Benedict's test for glucose, and an erroneous diagnosis of diabetes, can result in patients with a rare disease called pentosuria. These patients have the pentose sugar xylulose in their urine, but suffer no ill effects from it. The Tollens' pentose test can be used to determine if the sugar found in the urine is glucose or xylulose (Figure 13.8)

Disaccharides

13.11 Maltose (Malt Sugar)

Maltose is a disaccharide consisting of two glucose units. It is produced by the incomplete hydrolysis of starch, glycogen, or dextrins. Maltose produced from grains germinated under controlled conditions is called malt, and is used in the manufacture of beer.

Disaccharides are formed by a dehydration reaction between two monosaccharides. The reaction involves the formation of an acetal from a hemiacetal and an alcohol.

Hemiacetal   Alcohol   Acetal

In this reaction, one monosaccharide unit acts as the hemiacetal and the other as the alcohol. The bond that is formed is called an **acetal linkage** and is more stable than the hemiacetal. This acetal linkage will not react with bases, and requires acids or specific enzymes to cleave the bond.

The acetal linkage in maltose occurs between carbon 1 of a glucose molecule in the alpha form and carbon 4 on the other glucose; this bond is called an α(1-4) linkage.

Since the aldehyde group of the second glucose molecule is not involved in the acetal linkage, maltose can exist in either an alpha or beta form, and is a reducing sugar.

13.12 Lactose (Milk Sugar)

Lactose is found in the milk of mammals; it is synthesized by the mammary glands from glucose in the blood. Four to five percent of a cow's milk is lactose, while human milk contains 6 to 8%. Lactose is a colorless powder that is nearly tasteless. It can, therefore, be used in large amounts in special high calorie diets.

Lactose is formed by a dehydration reaction between glucose and galactose.

β-Galactose α-Glucose α-Lactose

As is the case with maltose, lactose has a potentially free aldehyde group in the second glucose unit, and is a reducing sugar.

13.13 Sucrose (Table Sugar, Cane Sugar, and Beet Sugar)

Sucrose is found in the juices of fruits and vegetables, and in honey. It is produced on a commercial scale from sugar cane or sugar beets, and is the sugar that we use in cooking. We each consume an average of 100 pounds of sucrose a year.

Sucrose is composed of one unit of glucose and one unit of fructose. The linkage occurs between the aldehyde group of glucose and the ketone group of fructose.

α-Glucose β-Fructose Sucrose

Since both the aldehyde group of glucose and the ketone group of fructose are involved in the linkage, sucrose does not have a potentially free aldehyde or ketone group, and is not a reducing sugar. Sucrose can be hydrolyzed by acids or enzymes found in the intestines and in yeast. The hydrolysis of sucrose produces a mixture of fructose and glucose called invert sugar.

Polysaccharides

13.14 Starch

Polysaccharides are polymers containing three or more monosaccharide units. They are used both as storage forms for energy and as structural components of the organism. Starch is the storage form of glucose used by plants. It is found in granules in the leaves, roots, and seeds of plants. These granules are insoluble in water, so their coating must be ruptured for the starch to mix with water. Heat will rupture the granules, producing a colloidal suspension whose thickness increases with heating. For this reason, cornstarch is widely used as a thickening agent in cooking.

Natural starches are a mixture of two types of polysaccharides: amylose and amylopectin. Amylose is a linear polysaccharide (molecular weight of about 50,000) whose glucose units are connected by α(1-4) linkages (Figure 13.9). Amylopectin (molecular weight of about 300,000) is a highly

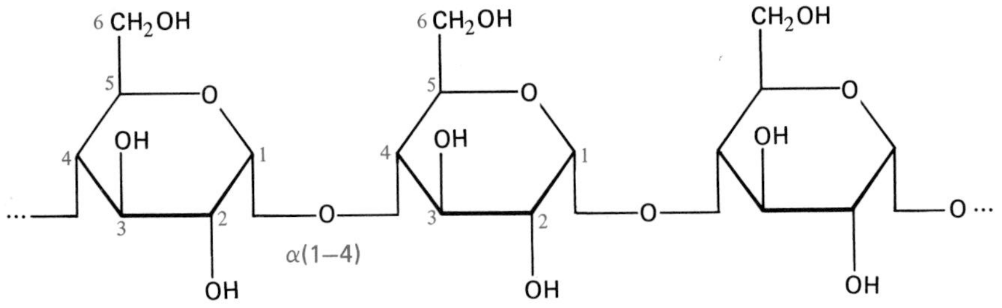

Figure 13.9 The structure of amylose. An α(1-4) linkage occurs between each of the glucose units.

branched glucose polymer. The nonbranching portion of the molecule consists of glucose units connected by α(1-4) linkages. The branching occurs every 20 to 24 glucose units, and is a result of α(1-6) linkages between the glucose units (Figure 13.10).

13.15 Dextrins

Dextrins are polysaccharides formed through the partial hydrolysis of starch by acids, enzymes, or dry heat. The golden color of bread crust results from the formation of dextrins. Dextrins get sticky when wet and are, therefore, used as adhesives on stamps and envelopes.

13.16 Glycogen

Glycogen is a heavily branched molecule that is the storage form of glucose in animals (Figure 13.11). It accounts for about 5% of the weight of the liver, and 0.5% of the weight of the muscle in the body. There is enough glucose stored in the form of glycogen in a well-nourished body to supply it with energy for about 18 hours.

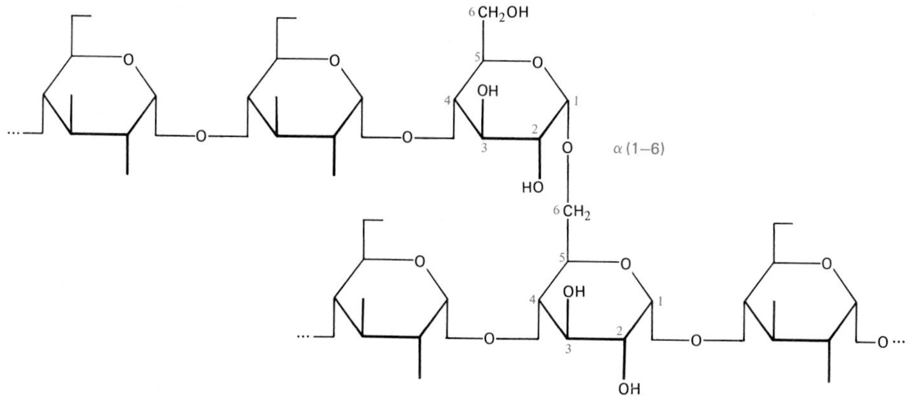

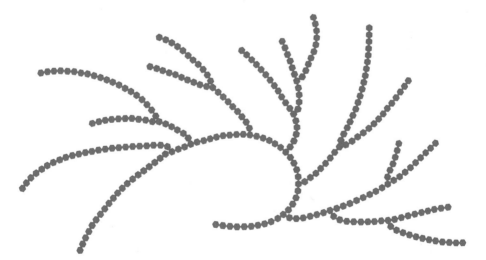

Figure 13.10 The structure of amylopectin and glycogen. The glucose units are connected by α(1-4) linkages and by α(1-6) linkages.

13.17 Cellulose

Cellulose is a glucose polymer (molecular weight from 150,000 to 1,000,000) produced by plants. It is the main structural component of plants, whose cells release this compound to form the exterior cell wall. Molecules of cellulose are insoluble in water because of their size and structure. The

Figure 13.11 The highly branched nature of the amylopectin and glycogen molecules is a result of the α(1-6) linkages that occur about every 20 glucose units. This diagram pictures a larger portion of the molecule than that shown in Figure 13.10. Each ● represents a glucose molecule.

strength and rigidity that cellulose gives to plants is the result of hydrogen bonding between the cellulose molecules (Figure 13.12).

The glucose units in cellulose are held together by an acetal linkage between carbon number 1 in the beta position on the first glucose and carbon number 4 on the second glucose.

Cellulose

This is a β(1-4) linkage, and human bodies do not possess the enzymes to cleave this bond. Therefore, any cellulose we eat passes through the digestive tract essentially undigested, constituting the roughage we need for proper elimination. Some microorganisms can digest cellulose; grass-feeding animals such as cows have extra stomachs to hold the grass for extended periods while these microorganisms break down the cellulose into glucose.

Over 50% of the total organic matter in the living world is cellulose. For example, wood is about 50% cellulose, and cotton is almost pure cellulose. When treated with a wide variety of chemicals, cellulose forms many useful products: guncotton, an explosive; celluloid; cellulose acetate, used in plastics, food wrapping films, and fingernail polish; rayon; methyl cellulose, used in fabric sizing, pastes, and cosmetics; and ethyl cellulose, used in plastic coatings and films.

13.18 Dextran

Dextran (molecular weight over 1,000,000) is a glucose polymer produced by bacteria. Partially hydrolyzed dextrans (molecular weight about 70,000) are used as blood plasma substitutes in the treatment of shock from low blood plasma volume. The dextrans so introduced are gradually eliminated through the urine.

13.19 Iodine Test

The iodine test is used to detect small traces of starch in solution. Starch will give an intense blue-black color when mixed with the iodine test reagent, a solution of potassium iodide containing iodine. This test can be used to monitor the hydrolysis of starch—the color will slowly change as the starch is progressively broken down to shorter carbon chain products (Table 13.1).

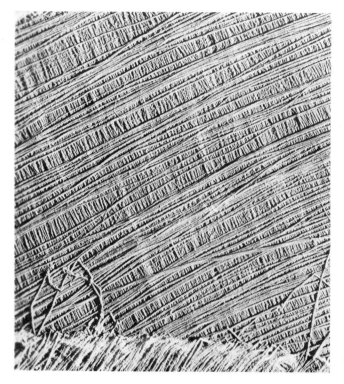

Figure 13.12 This electron micrograph of the cell wall of an alga shows the parallel arrangement of the cellulose fibers making up the wall. (Courtesy R. D. Preston)

Table 13.1   Colors Formed in the Iodine Test

$$Starch \xrightarrow{\text{Hydrolysis}} Dextrins \xrightarrow{\text{Hydrolysis}} Maltose$$

$$\textit{Blue-black} \longrightarrow \textit{Reddish} \longrightarrow \textit{Colorless}$$

13.20 Photosynthesis

Where does the energy stored in these polysaccharides originate? Ultimately, from the sun. Nuclear reactions occurring on the sun produce energy that radiates out into space. This radiant energy is trapped by plants growing on the earth, and is used to produce basic carbohydrates and certain amino acids. **Photosynthesis,** the process by which plants capture and use this energy, is quite complex and not totally understood.

The reactions of photosynthesis require the presence of light and molecules of chlorophyll. The overall equation, summarizing the many reactions of this process, shows the eventual formation of glucose from carbon dioxide and water:

$$6CO_2 + 6H_2O \xrightarrow[\text{Chlorophyll}]{\text{Light}} C_6H_{12}O_6 + 6O_2$$

The series of reactions producing these results can be divided into two categories: the light reactions and the dark reactions. The light reactions require the presence of chlorophyll to absorb radiant light energy and to use it in the production of oxygen and energy-rich molecules. The dark

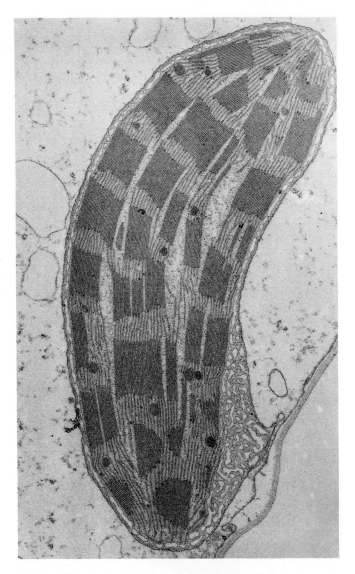

Figure 13.13 An electron micrograph of a chloroplast, the site of the light reactions of photosynthesis. The dense stacks in the chloroplasts are the grana, the main location of the chlorophyll molecules. (Courtesy T. Elliott Weier)

reactions then use these energy-rich molecules to reduce carbon dioxide to glucose and other organic products.

The reactions of photosynthesis take place in regions in the plant cells called chloroplasts. These chloroplasts contain chlorophyll and all the other enzymes necessary for the light reactions (Figure 13.13).

Metabolism of Carbohydrates

13.21 What Is Metabolism?

How many times have you seen this happen on television: a superbly trained athlete waits at the starting blocks for the gun, sprints 400 meters at top speed, breaks the tape in record time, and then collapses on the track infield (Figure 13.14). His legs feel like rubber, he painfully gasps for breath, and he feels nauseated. The athlete will require a rest of $1\frac{1}{2}$ hours

Figure 13.14 The tremendous exertion required to beat a world record has taken its toll. (Ken Regan, Camera 5)

to completely recover from the exhaustion of this extreme exertion. We might ask, where does the energy necessary for muscle contraction come from? Exactly how is it produced in the muscle cell? What causes the resulting muscle fatigue? And why does it take so long for the cell to recover after strenuous exercise? These are some of the questions we will answer in this section on carbohydrate metabolism.

Metabolism can be defined as all of the enzyme-catalyzed reactions in the body (Table 13.2). These reactions are of two types: **catabolic reactions,** in which molecules are broken down to yield smaller products and cellular energy, and **anabolic reactions,** in which the cell uses energy to produce molecules it needs for growth and repair. (These are also called biosynthetic reactions.)

Anabolic and catabolic reactions occur continuously in the cell, and each involves complicated series of carefully controlled enzyme-catalyzed reactions.

Table 13.2   Metabolism

METABOLISM	
Catabolic Reactions	Anabolic Reactions
Reactions that break down larger molecules.	Reactions in which larger molecules are produced from smaller components — biosynthetic reactions.
Reactions that produce the energy needed by the cell.	Reactions that require energy to occur.

13.22 Digestion and Absorption of Carbohydrates

Carbohydrates produced by plants from the products of photosynthesis supply the nutritional needs of animals. Plants are the major source of carbohydrate in our diet. But exactly how does the starch in, say, a potato chip get to your muscle cells? **Digestion,** the process by which complex foods are broken down into simple molecules, must take place before this food can be absorbed and used by the body.

The digestion of carbohydrates begins in the mouth, where teeth break large pieces of food into smaller ones. This food is then mixed with saliva, which contains an enzyme that begins the breakdown of starch to maltose. When the food is swallowed it passes into the stomach, where these salivary enzymes are inactivated by the low pH. The stomach further reduces the size of the food particles. As the food then passes into the small intestine, the pancreas and intestinal wall secrete juices containing enzymes that completely hydrolyze the polysaccharides and disaccharides contained in the food. Galactose, glucose, and fructose—the monosaccharides formed by this process—are taken up directly through the intestinal wall into the bloodstream and are transported to the liver. The galactose is converted to glucose by liver enzymes, while the fructose may be converted to glucose or may enter other metabolic reactions. Glucose may be stored in the liver as glycogen. The liver normally contains about 100 grams of glycogen (about $\frac{1}{2}$ cup), but it may store as much as 400 grams (Table 13.3).

Table 13.3   Digestion of Carbohydrates

Site of Digestion	Food Digested	End Products	Enzyme Source
Mouth	Starch	Dextrins, maltose	Saliva
Small intestines	Starch	Maltose	Pancreatic juice
	Lactose	Glucose and galactose	Pancreatic juice and intestinal secretions
	Maltose	Glucose	Pancreatic juice and intestinal secretions
	Sucrose	Glucose and fructose	Pancreatic juice and intestinal secretions

13.23 Cellular Energetics

We must now briefly interrupt this journey of the potato chip to your muscle cells for a brief discussion of cellular energetics and the compounds that assist in the anabolic and catabolic reactions in the cell.

The body has many energy needs, among which is the need for heat to maintain body temperature. This heat is produced by the controlled combustion of compounds such as glucose in the cell. The cell, however, cannot operate like a steam engine — it needs forms of energy other than heat in order to do work such as muscle contraction, transport of nerve impulses, and biosynthesis. For these tasks the cell requires "high-energy" or "energy-rich" compounds such as adenosine triphosphate, ATP (Figure 13.15). A molecule of ATP (and other molecules similar to ATP) contains two oxygen-to-phosphate bonds that are called "high-energy" phosphate bonds, and are often represented by a wavy line.

The reason that such phosphate bonds are called "high-energy" bonds is that hydrolysis of compounds such as ATP is a highly exothermic reaction, yielding about twice as much energy as hydrolysis of compounds containing low-energy phosphate bonds. In energy-releasing reactions in the cell, ATP is hydrolyzed to form ADP and an inorganic phosphate (denoted P_i).

$$ATP + H_2O \longrightarrow ADP + P_i + 7.3 \text{ kcal}$$

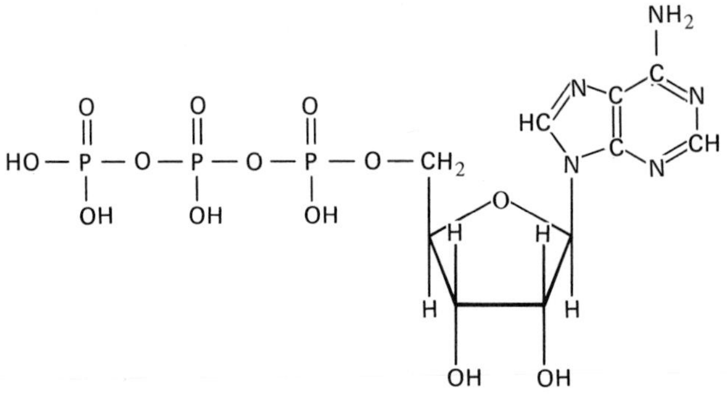

Figure 13.15 The structure of adenosine triphosphate, ATP.

If the hydrolysis of ATP is coupled with an endothermic reaction in the cell, such as the contraction of a muscle fiber or the synthesis of a large molecule, the hydrolysis of ATP will supply the energy necessary for the other reaction to occur.

$$\text{Relaxed muscle} + \text{ATP} + H_2O \longrightarrow \text{Contracted muscle} + \text{ADP} + P_i$$

Molecules other than ATP are also required to assist in the breakdown and synthesis processes of the cell. These molecules are members of a class of compounds called coenzymes, which will be discussed in detail in Chapter 16. There are three coenzymes, however, that we will mention here. Two are derivatives of the vitamin niacin, or nicotinic acid, and are called NAD (nicotinamide adenine dinucleotide) and NADP (nicotinamide adenine dinucleotide phosphate) (Figure 13.16). The other coenzyme, a derivative of riboflavin, or vitamin B_2, is FAD (flavin adenine dinucleotide). These three molecules serve as important hydrogen carriers in metabolic reactions. We can indicate how these coenzymes serve as hydrogen carriers by writing the following equation.

$$\text{NAD}^+ + \text{XH}_2 \xrightarrow{\text{Enzyme}_1} \text{NAD-H} + \text{H}^+ + \text{X}$$

(A molecule of NAD is positively charged, as shown in Figure 13.16; the reactant molecule is represented by XH_2.) The NADH produced by this reaction can then serve as a hydrogen donor in a reaction requiring hydrogen. In this way, one molecule of NAD^+ can be used over and over again in the cell.

$$\text{NADH} + \text{H}^+ + \text{Y} \xrightarrow{\text{Enzyme}_2} \text{NAD}^+ + \text{YH}_2$$

13.24 Glycogenesis and Glycogenolysis

Glycogenesis is the name given to the conversion of glucose to glycogen. This process occurs in the liver when excess glucose is present in the blood. It involves the complex series of reactions shown in Figure 13.17,

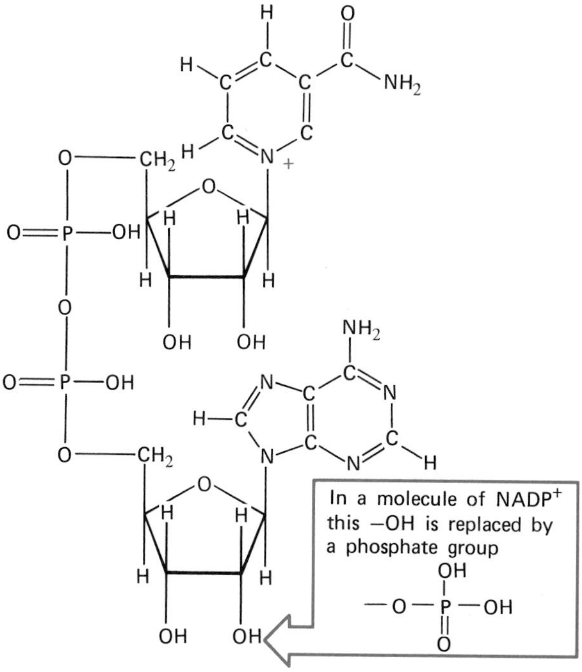

Figure 13.16 The structure of the two coenzymes NAD^+ and $NADP^+$. The difference between the structures of the two molecules is that a hydroxyl group on NAD^+ is replaced by a phosphate group on $NADP^+$. Note that both molecules carry a positive charge.

In a molecule of $NADP^+$ this —OH is replaced by a phosphate group

requiring several enzymes and two high-energy molecules, ATP and UTP (uridine triphosphate).

As the cells in the body use up the glucose in the blood, a process called **glycogenolysis** occurs in the liver. This consists of a series of reactions by which glycogen is broken down so that glucose can be released into the bloodstream to maintain the blood sugar level. Note that glycogenolysis is not the exact reverse of glycogenesis—they have only one step in common (Figure 13.17).

13.25 Oxidation of Carbohydrates

We can now return to the journey of our potato chip which, you will remember, we left in the form of glucose in the bloodstream. Once the glucose enters a cell of the body it can be used in biosynthetic reactions, or it can be broken down to yield useful cellular energy by a process called **cellular respiration.** In cellular respiration, glucose is oxidized to form carbon dioxide and water, and ATP is formed. This process is highly

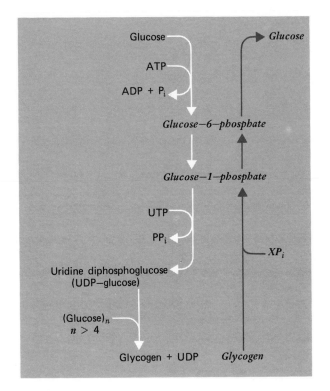

Figure 13.17 The reactions of glycogenesis (the synthesis of glycogen from glucose) and glycogenolysis (the breakdown of glycogen to glucose). Note that these processes have only one reaction in common. (PP_i = pyrophosphate)

complex, involving many steps and requiring many enzymes and coenzymes. A specific description of each step must be left to other biochemistry courses, but a general overview of this process is possible here.

The complete oxidation of glucose requires oxygen, and if carried out in the laboratory will yield 686 kilocalories per mole of glucose.

$$C_6H_{12}O_6 + 6O_2 \longrightarrow 6CO_2 + 6H_2O + 686 \text{ kcal}$$

If carried out in a living cell under enzyme control, the reaction will yield 36 ATP's per mole of glucose.

$$C_6H_{12}O_6 + 6O_2 \xrightarrow{\text{Enzymes}} 6CO_2 + 6H_2O + 36ATP$$

Each ATP can be hydrolyzed to yield 7.3 kcal of energy, so 36 ATP's represents the trapping of about 263 kcal of useful energy from the oxidation of 1 mole of glucose. These figures mean that the cellular processes of respiration are only about 38% efficient. What happens to the rest of the energy? It is given off as heat to maintain body temperature. The reactions of cellular respiration are self-regulating, and will occur only when the cellular supply of ATP is low.

13.26 Glycolysis

The cellular oxidation of glucose can be divided into two stages. The first, an anaerobic stage, requires no oxygen and occurs in the cytoplasm of the cell. This stage is called **glycolysis,** and involves the breakdown of glucose to form two molecules of lactic acid. This process yields two ATP's, and can be summarized by the following overall reaction.

$$\text{Glucose} + 2\text{ADP} + 2\text{P}_i \longrightarrow 2 \text{ Lactic acid} + 2\text{ATP} + 2\text{H}_2\text{O}$$

(The specific steps of this process are outlined in Figure 13.18.) The second phase of glucose oxidation, the aerobic stage, requires oxygen; it is the

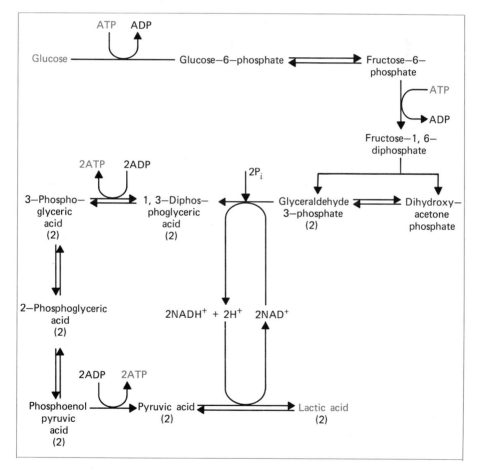

Figure 13.18   The steps of glycolysis, the anaerobic reactions in the oxidation of glucose. Shown here are the steps in the conversion of glucose to lactic acid. Note that enzymes are required for each of these reactions, but for simplicity we have not listed them here.

stage where the majority of energy is released in the breakdown of glucose. This second phase also serves as the final stage of oxidation for other compounds used by the cell to supply energy. It involves two series of reactions. The first series is called the **citric acid cycle** [or Krebs cycle, or tricarboxylic (TCA) cycle], and involves the final breakdown of the fuel molecule to carbon dioxide. This breakdown consists of oxidation steps that yield hydrogen atoms. These hydrogen atoms are then used in the second series of reactions, called the **electron transport chain.** This sequence of reactions requires oxygen, and yields ATP and water (Figure 13.19).

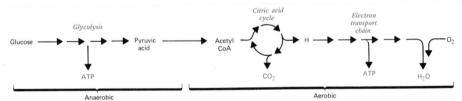

Figure 13.19 The cellular oxidation of glucose consists of an anaerobic stage (glycolysis) and an aerobic stage, linking the reactions of the citric acid cycle and the electron transport chain. It is in the aerobic stage that most of the energy from the oxidation of glucose is released.

The citric acid cycle and the electron transport chain occur in organelles in the cell called the mitochondria. Often also called the "power plants" of the cell, the mitochondria are located near the structures that require energy, such as the contractile filaments of muscle cells (Figure 13.20).

13.27 Citric Acid Cycle

When sufficient oxygen is present in the cell, the last step of glycolysis, the conversion of pyruvic acid to lactic acid, does not take place (Figure 13.18). Instead, pyruvic acid is oxidized to acetic acid in the form of acetyl coenzyme A—often referred to as acetyl CoA, acetic acid bound to the carrier molecule coenzyme A. This reaction requires the hydrogen carrier molecule NAD^+ in addition to coenzyme A, and yields $NADH + H^+$ along with one molecule of CO_2.

$$\text{Pyruvic acid} + NAD^+ + \text{Coenzyme A} \xrightarrow[\text{and Coenzymes}]{\text{Other Enzymes}}$$

$$\text{Acetyl CoA} + NADH + H^+ + CO_2$$

The acetyl CoA formed in this reaction can enter the citric acid cycle. This series of reactions, shown in Figure 13.21, produces two molecules of

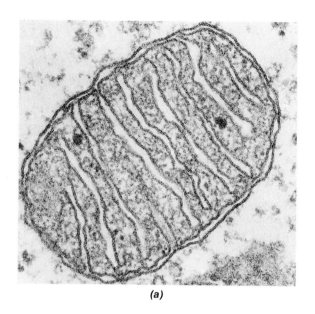

(a)

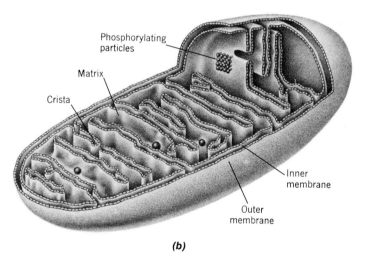

(b)

Figure 13.20 A mitochondrion. (a) An electron micrograph of a mitochondrion. (b) A model of a mitochondrion. The inner folds of the mitochondrion, called the cristae, contain the enzymes of the citric acid cycle and the electron transport chain.

carbon dioxide and four pairs of hydrogen atoms. Three pairs of these hydrogen atoms will be carried by NAD^+, and one by FAD. It is important that you note the cyclic nature of the reactions occurring in the citric acid

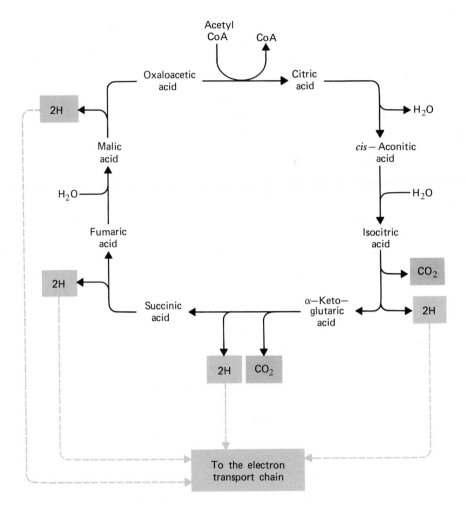

Figure 13.21 The citric acid cycle. Two carbon atoms enter the cycle in the form of acetyl CoA, and are joined to oxaloacetic acid to form citric acid. During the series of reactions that then occur, two molecules of carbon dioxide are formed, and four pairs of hydrogens are produced and carried to the electron transport chain. At the end of the cycle, a molecule of oxaloacetic acid is generated, ready for another turn of the cycle.

cycle. The first reaction involves the coupling of acetyl CoA with oxaloacetic acid to form citric acid, and the last reaction regenerates oxaloacetic acid that can then combine with another acetyl CoA. This cycle, however, cannot occur independently of the electron transport chain.

13.28 The Electron Transport Chain and ATP Formation

The electron transport chain is a series of reactions that produces most of the ATP molecules formed in the oxidation of glucose by the cell. In this series of reactions the hydrogen atoms released in the oxidation of glucose, and later just their electrons, are passed between a series of compounds until they are combined with oxygen to form water. Each transfer is part of an oxidation-reduction reaction in which some energy is released. As the electrons flow through this series of reactions they can do work, just as the electrons flowing through a copper wire can do work. The kind of work done by the electrons in the cell, however, is the production of ATP from ADP and inorganic phosphate. The exact mechanisms for this process are not well understood, but it is known at what points in the transport chain the ATP is produced (Figure 13.22).

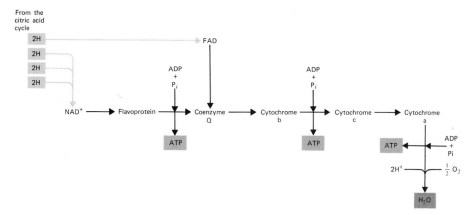

Figure 13.22 The electron transport chain. The hydrogens and their electrons produced in the citric acid cycle are passed between a series of compounds until they are combined with the oxygen we breathe to form water. It is this series of transfers that produces most of the ATP molecules formed in the oxidation of glucose.

13.29 Lactic Acid Cycle

When undergoing only moderate exercise, muscle cells have enough oxygen to carry out the respiration process aerobically. But during strenuous exercise the blood cannot supply oxygen to the muscles fast enough, and the muscle cells must rely upon a backup system—the production of energy by glycolysis. The lactic acid produced by glycolysis builds up in the muscle cells to the point where it hampers muscle performance, causing muscle fatigue and exhaustion. The lactic acid build-up produces

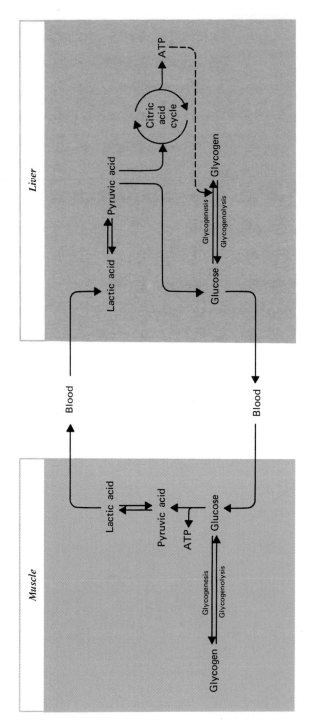

Figure 13.23 The lactic acid cycle. Lactic acid formed in strenuous exercise must be removed either by conversion to pyruvic acid in the muscle cells or by migration from these cells to the liver where it is converted to glycogen. Some of the lactic acid (about 25%) must be oxidized via the citric acid cycle to provide the energy necessary for this conversion.

mild acidosis, which causes nausea, headache, lack of appetite, and impairment of oxygen transport—resulting in the difficult, painful gulping of air experienced by athletes after maximum physical efforts. Muscle cells are slow to recover from this condition. The lactic acid must be removed either by conversion to pyruvic acid in the muscle cells, or by migration from these cells to the liver. There, about 25% of it is oxidized via the citric acid cycle and the remainder is converted to glycogen. The heavy breathing that occurs after strenuous exercise helps supply the oxygen necessary to oxidize the lactic acid (Figure 13.23). A major difference between a trained athlete and a nonathlete is that the trained athlete is able to supply his muscle cells with enough oxygen to maintain aerobic respiration for a longer period than can the nonathlete. This means that his muscle cells will require glycolysis for their supply of ATP only at much higher levels of exertion. Carbohydrate metabolism is summarized in Figure 13.24.

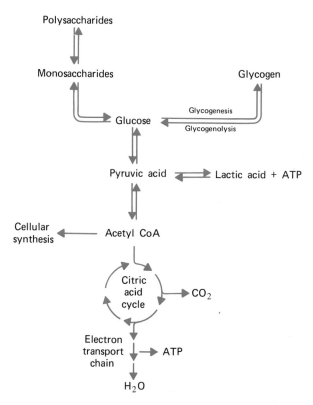

Figure 13.24 A summary of carbohydrate metabolism.

13.30 Fermentation

Microorganisms such as yeast are able to live quite well without oxygen. Glycolysis occurs in yeast cells in much the same way as it does in animal cells. The process differs in yeast cells only in the last step, where pyruvic acid is converted to ethanol and carbon dioxide rather than to lactic acid. The name given to this form of glycolysis in yeast cells is **fermentation,** and the overall reaction is as follows.

$$\underset{\text{Glucose}}{C_6H_{12}O_6} \xrightarrow[\text{Enzymes}]{\text{Yeast}} \underset{\text{Ethanol}}{2CH_3CH_2OH} + 2CO_2 + \text{Energy}$$

Control of Carbohydrate Metabolism

13.31 Cellular Control

Each cell maintains a sophisticated system for regulating carbohydrate metabolism. Each stage of carbohydrate metabolism has regulatory enzymes—enzymes that are turned on or off by fluctuations in the concentrations of specific substances in the cell. The overall reactions occur at a high rate only when the cell is low in ATP.

13.32 Insulin

Hormones are the body's second level of control of carbohydrate metabolism. Insulin is a hormone that acts on cell membranes to facilitate the passage of glucose from the blood into cells of muscle and fatty tissue. It also increases the production of glycogen and inhibits the synthesis of glucose in the liver. Insulin is synthesized by specialized cells in the pancreas called beta cells, which are sensitive to the level of glucose in the blood. These cells secrete insulin when the glucose level is high. This acts to increase the absorption of glucose by the cells, and to increase the conversion of glucose to glycogen in the liver, thereby lowering the blood sugar level.

13.33 Blood Sugar Level

In a normal adult, the blood sugar level remains fairly constant. Eight to twelve hours after a meal the level is 60 to 100 mg glucose/100 ml blood. This is called the normal fasting level. The brain, which maintains no storehouse of glucose, is totally dependent upon the glucose in the blood for its energy requirements. Hence the glucose level is critical for normal brain function (Figure 13.25).

Hypoglycemia is a condition resulting from blood sugar levels that are below normal. Under such conditions the brain becomes starved for glucose. Mild hypoglycemia produces irritability, dizziness, lethargy, grogginess, and fainting. Severe hypoglycemia can produce convulsions,

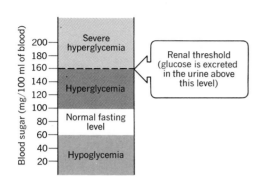

Figure 13.25 Blood sugar levels, and the conditions of hypoglycemia and hyperglycemia.

shock, and coma. Such severe hypoglycemia can be produced by overproduction or overinjection of insulin, a condition called hyperinsulinism, in which the blood sugar level drops precipitously and the person goes into convulsions and coma called "insulin shock." Mild hypoglycemia can be caused by too much carbohydrate in the diet. A meal, especially breakfast, which is rich in carbohydrates can overstimulate the pancreas to produce insulin, causing the blood sugar level to fall below normal. The diet can be an important factor in controlling mild hypoglycemia. Breakfast should be rich in proteins with some fats and carbohydrates. The proteins and fats reduce the speed with which the glucose enters the bloodstream and will, therefore, prevent the overstimulation of the pancreas.

Hyperglycemia is a condition resulting from blood sugar levels that are too high. Mild hyperglycemia occurs after meals; the body counteracts this higher level of glucose by storing it in the form of glycogen or fat, and by oxidizing it. In severe hyperglycemia, the blood sugar level can rise so high that the glucose level exceeds the amount tolerated by the kidneys (the renal threshold, 160 mg/100 ml blood), and glucose is excreted in the urine. Glycosuria, sugar in the urine, can be detected by using the Benedict's test. Glycosuria can result from conditions such as diabetes mellitus, emotional stress, kidney failure, or the administration of certain drugs.

13.34 Diabetes Mellitus

Diabetes mellitus is a complicated disease whose causes and metabolic pathways are not totally understood. It is a disorder that causes the blood sugar level to rise above normal. In diabetes, the beta cells in the pancreas respond slowly or not at all to the level of glucose in the blood. As we saw at the beginning of this chapter, there are two types of diabetes. Severe juvenile diabetes, whose onset is rapid and whose symptoms are readily apparent, will occur in individuals who lack the ability to produce insulin. When untreated, the patient will excrete large amounts of glucose in the urine. To supply the body with energy, the untreated diabetic will metabolize large amounts of fats, resulting in the production of high amounts of compounds called ketone bodies (Table 13.4).

Table 13.4   Ketone Bodies

Acetoacetic acid	$$CH_3\overset{\displaystyle O}{\overset{\|}{C}}CH_2\overset{\displaystyle O}{\overset{\|}{C}}OH$$
β-Hydroxybutyric acid	$$CH_3\overset{\displaystyle OH}{\overset{\|}{C}H}CH_2\overset{\displaystyle O}{\overset{\|}{C}}OH$$
Acetone	$$CH_3\overset{\displaystyle O}{\overset{\|}{C}}CH_3$$

Acetone can be detected in the breath, and acetoacetic acid and β-hydroxybutyric acid will be found in high concentrations in the blood and urine. These conditions can result in severe acidosis, coma, and death. Juvenile diabetes is treated by diet and daily injections of insulin.

Adult diabetes, the second form of this disorder, strikes many more people than the juvenile form (Figure 13.26). It is slow in developing, and

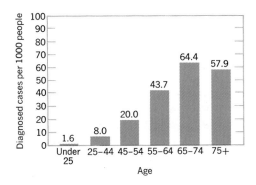

Figure 13.26 The incidence of diabetes increases with age.

will occur in individuals after the age of 30 to 40. One out of every five people is genetically disposed to diabetes. The beta cells of an adult diabetic will respond normally until put under an unusual or prolonged stress, such as an infection of the pancreas or obesity (Figure 13.27). Often this type of diabetes goes undetected. When detected, it can be treated by diet alone, diet and oral drugs that stimulate the pancreas to produce insulin, or diet and insulin injections.

The high concentrations of glucose in the blood of a diabetic result in serious disruption of the normal chemical balance in cells. Tissues in which insulin is not required for the uptake of glucose will contain very high concentrations of glucose, approaching that of the blood. Such excess glucose is converted to sorbitol and fructose, compounds that are retained

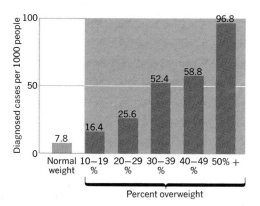

Figure 13.27 The incidence of diabetes also increases with certain stresses such as obesity.

in the cell and that cause electrolyte imbalance, water retention, and other abnormal conditions. Accumulation of sorbitol in the lens of the eye can cause cataracts. In the blood vessels, this chemical imbalance can cause atherosclerosis; in the eye it can cause retinopathy, a form of blindness. Diabetes is the second leading cause of blindness in the United States.

13.35 Glucose Tolerance Test

The glucose tolerance test is a useful tool in the diagnosis of diabetes. A patient who has fasted for 8 hours is given 50 to 100 grams of glucose by mouth. His blood sugar level is measured initially, and then for the next

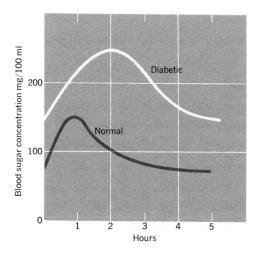

Figure 13.28 Glucose tolerance test.

several hours. A normal person will reach a maximum blood sugar level of about 160 mg/100 ml during the first hour, and this level will fall to normal by the end of the second hour. A diabetic will begin the test with a blood sugar level higher than normal; the sugar level will peak at the end of the second hour, and not return to the original level for several hours thereafter (Figure 13.28).

13.36 Other Hormones

There are other hormones that can affect the blood sugar level. Epinephrine will stimulate the release of glucose from the liver, and will increase the blood sugar level to a point that often exceeds the renal threshold. For this reason, glucose will often be found in the urine of individuals under stress. Glucagon is another hormone produced by the pancreas. It has the opposite effect of insulin, increasing the blood sugar level by increasing the rate of glycogenolysis in the liver. There are also several hormones produced by the pituitary gland with effects opposite to those of insulin. Just these few examples make it obvious that the control of carbohydrate metabolism by the body is an extremely complex process, and can be affected by many different factors in one's life.

Additional Reading

Books
1. J. Baker and G. Allen, *Matter, Energy and Life*, Addison-Wesley, Reading, Massachusetts, 1970.

2. A. L. Lehninger, *Bioenergetics*, W. A. Benjamin, 1965.

3. A. L. Lehninger, *Short Course in Biochemistry*, Worth Publishers, 1973. This text is an excellent reference text for Chapters 12 to 16.

4. R. H. Williams, *Disorders in Carbohydrate and Lipid Metabolism*, Saunders, Philadelphia, 1962.

Articles
1. W. Guild, "Theory of Sweet Taste," *Journal of Chemical Education*, March 1972, page 171.

2. A. Kappas and A. P. Alvares, "How the Liver Metabolizes Foreign Substances," *Scientific American*, June 1975, page 22.

3. H. Leese, "Ins and Outs of the Small Intestine," *New Scientist*, November 22, 1973, page 562.

4. T. H. Maugh II, "Diabetes: Epidemiology Suggests Viral Connection," *Science,* April 25, 1975, page 347.

5. R. Margaria, "The Sources of Muscular Energy," *Scientific American,* March 1972, page 84.

Questions and Problems

1. Define the following terms:

    (a) Carbohydrate
    (b) Monosaccharide
    (c) Disaccharide
    (d) Polysaccharide
    (e) Aldose
    (f) Ketose
    (g) Benedict's test
    (h) Reducing sugar
    (i) Sugar acid
    (j) Acetal linkage
    (k) Starch
    (l) Glycogen
    (m) Cellulose
    (n) Iodine test

    (o) Metabolism
    (p) Catabolic reaction
    (q) Anabolic reaction
    (r) Photosynthesis
    (s) ATP
    (t) Glycogenesis
    (u) Glycogenolysis
    (v) Glycolysis
    (w) Citric acid cycle
    (x) Electron transport chain
    (y) Lactic acid cycle
    (z) Fermentation
    (aa) Hyperglycemia
    (bb) Hypoglycemia

2. Describe four functions of carbohydrates in living organisms.

3. Identify each of the following as (a) an aldose or ketose, and (b) a triose, tetrose, pentose, hexose, or heptose.

(a)
$$
\begin{array}{c}
H \\
| \\
C=O \\
| \\
H-C-OH \\
| \\
H-C-OH \\
| \\
H-C-OH \\
| \\
CH_2OH
\end{array}
$$

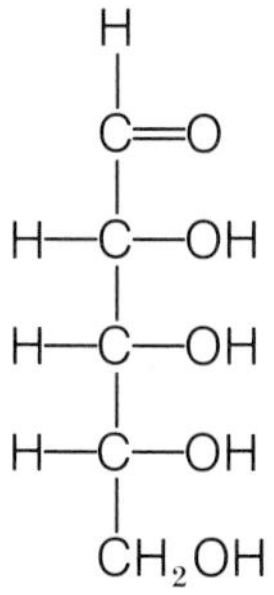

(b)
$$
\begin{array}{c}
CH_2OH \\
| \\
C=O \\
| \\
HO-C-H \\
| \\
H-C-OH \\
| \\
H-C-OH \\
| \\
CH_2OH
\end{array}
$$

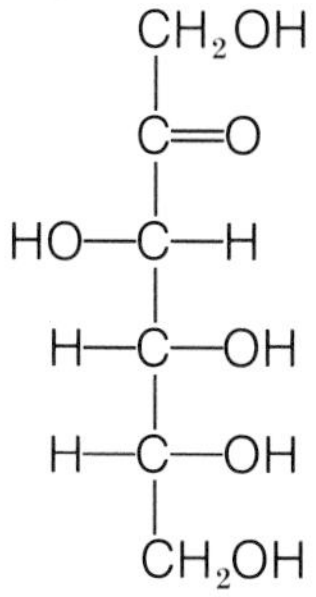

(c)
$$CH_2OH$$
$$|$$
$$C=O$$
$$|$$
$$HO-C-H$$
$$|$$
$$H-C-OH$$
$$|$$
$$H-C-OH$$
$$|$$
$$H-C-OH$$
$$|$$
$$CH_2OH$$

(d)
$$H$$
$$|$$
$$C=O$$
$$|$$
$$H-C-OH$$
$$|$$
$$CH_2OH$$

4. Draw the linear structure of glucose.

5. Given three unknown solutions labeled A, B, and C, propose a method for determining which solution contains starch, which contains glucose, and which contains fructose.

6. Which of the following sugars will give a positive Benedict's test? Give the reason for your answer.

(a) Fructose   (d) Maltose
(b) Ribose   (e) Sucrose
(c) Lactose

7. Suppose a sample of urine from an infant gives a positive test with a Clinitest tablet. Is it correct for the analyst to report glucose in the urine? Why or why not?

8. The liver protects the body from drugs and foreign materials (such as insecticides) by inactivating them and joining them with a natural substance such as glucuronic acid to facilitate excretion. For example, the liver metabolizes salicylic acid by forming an ester bond with glucuronic acid. Write a possible equation for this reaction.

9. Starch is composed of two polysaccharides.

(a) What are they?
(b) Describe the difference in their structures.
(c) Design an experiment for the hydrolysis of starch. How would you monitor the progress of the hydrolysis?

10. What is the storage form of glucose in animals? How does its structure compare with starch?

11. Celery and potato chips are both composed of molecules that are polymers of glucose. Explain why celery makes a good snack for individuals on a diet, while potato chips do not.

12. Write a general reaction for each of the following processes.

   (a) Photosynthesis
   (b) Glycogenesis
   (c) Glycogenolysis
   (d) Glycolysis
   (e) Fermentation
   (f) Citric acid cycle

13. In general terms describe the possible metabolic reactions that the starch in a piece of toast might undergo from the time you take a bite until the ATP formed from the toast is used in muscle contraction.

14. (a) Draw the structure of AMP, ADP, and ATP.
    (b) Describe what is meant by a "high-energy" phosphate bond.
    (c) Why are molecules such as ATP so critical to the survival of a cell?

15. The oxidation of glucose by a cell can be divided into two separate stages, an anaerobic and an aerobic stage.

    (a) What is the name given to the anaerobic stage?
    (b) Write the equation for the overall reaction that occurs in the anaerobic stage.
    (c) Where in the cell does each stage occur?
    (d) Which stage yields more energy?
    (e) Name the two series of reactions that occur in the aerobic stage, and describe in general the reactions that occur in each.
    (f) Write the equation for the overall reaction that occurs in the aerobic stage.

16. Your alarm doesn't go off and you wake up late for a final exam. You jump out of bed, dress quickly, run across campus, and race up three flights of stairs to your classroom. You collapse in your seat gasping for breath with your legs feeling like rubber. Describe the events that have occurred in your muscle cells during this experience. Why do you gasp for breath, and why are your muscles so weak?

17. Your brain requires a constant supply of glucose. Explain how your body maintains a fairly constant blood sugar level even though you eat foods that supply glucose only several times a day.

18. What is "insulin shock"? How does it occur? Explain why severe hyperinsulinism is more dangerous to a patient than is the lack of insulin over a short period.

19. Propose a diet for a patient suffering from hypoglycemia.

20. Does glycosuria result only during diabetes mellitus? How would you confirm diabetes mellitus in a patient who is suffering from glycosuria?

chapter 14

Lipids

Learning Objectives

By the time you have finished this chapter, you should be able to:

1. Describe the difference between
   (a) A simple and compound lipid.
   (b) A simple and mixed fat.
   (c) A saturated and unsaturated fatty acid.
   (d) A saponifiable and nonsaponifiable lipid.

2. Explain the meaning of "essential fatty acid."

3. Describe the process by which butter becomes rancid.

4. Given the iodine number of a lipid, describe the physical properties and probable source of the lipid.

5. Draw the general structure of a triglyceride, and write the equations for its hydrolysis and saponification.

6. Explain how soap is able to remove grease from your hands.

7. Give the general structure for the following compound lipids:
   (a) Phosphatides   (d) Sphingomyelins
   (b) Lecithins   (e) Glycolipids
   (c) Cephalins

8. Describe the function of each of the compound lipids listed immediately above.

9. Give three examples of nonsaponifiable lipids.

10. Describe the steps of lipid digestion and absorption.

11. Define "β-oxidation" and "ketosis."

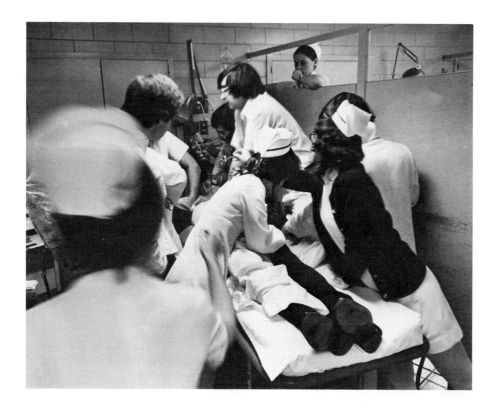

Dan was a hard-driving man of 34 who, through a great deal of work and self-sacrifice, had recently been promoted to branch manager of his bank. What little spare time he took off from his job was spent playing tennis and hiking with his family. During the last year Dan had been bothered occasionally by an intense chest pain that would develop during particularly strenuous games of tennis. But this pain was always brief and went away quickly, so Dan never paid much attention to it. One afternoon while Dan was working on plans for a Board of Trustees meeting, the pain began again. This time, however, it didn't go away, but rather gained in intensity and extended to his left arm. Dan found himself sweating profusely, and became extremely short of breath. His alert secretary recognized the symptoms and called an ambulance, and Dan was rushed to the hospital where he was treated for a heart attack. Luckily for Dan, he reached the hospital in time. He remained in guarded but stable condition for several days in the intensive care unit, and then began a slow recovery.

Dan kept asking himself, "Why me? Heart attacks happen only to older men, or to people in poor physical shape." Upon posing this question to his doctor, Dan was surprised to learn that coronary disease is responsible for over half of all the deaths reported for people between the ages of 35 and 64. In this particular case, the doctors had determined that Dan was suffering from atherosclerosis, the most common form of arteriosclerosis — disease of the arteries. Atherosclerosis is a slow, progressive disease that may begin in childhood, showing no symptoms for 20 to 50 years. It affects primarily the larger arteries, especially the aorta and the arteries that feed the heart, brain, and kidneys.

Atherosclerosis begins as a yellowish fatty streak that is present in the aortas of most children by the age of three. Such streaks are actually an accumulation of cholesterol and various fats, and they appear in most of the other major arteries by the age of 35. The streaks never cause symptoms themselves, but may in certain individuals develop into plaques, the major cause of concern and study in atherosclerosis. A plaque contains a core of fats (mainly cholesterol) covered by scar tissue (Figure 14.1). Older plaques, which may contain large deposits of calcium, can break open, releasing their contents and then remaining as open sores. Plaques may cause serious damage in several ways: they may grow in size and restrict the blood flowing to the tissues through the artery so severely that the tissue is damaged or destroyed. Or a blood clot may form on the rough surface of the plaque and then break off and block essential arteries. Or, as we just mentioned, the plaque itself may open up and release its contents, which can themselves block blood flow to critical areas of the body. As the plaque develops, the artery may become "hardened," or lose flexibility, which can cause a ballooning of the artery, called an aneurysm. Such an aneurysm may then burst, causing hemorrhaging, severe tissue damage, or death.

If atherosclerosis affects the arteries of the heart, as was the case with Dan, a person may suffer the first symptoms of angina pectoris, a short-term, intense pain in the chest. In such cases the plaques in the arteries leading to the heart have significantly reduced the blood flow, although there is sufficient flow to maintain normal activities. However, any activity requiring increased blood flow will result in an undernourished heart, and an intense pain. This condition can be treated with rest and nitroglycerin (which relaxes the muscles in the coronary arteries), and can occur for extended periods of time without harming the heart if the atherosclerosis doesn't progress. However, atherosclerosis can cause heart attacks by completely blocking arteries supplying the heart muscle. Recovery from such heart attacks depends on the amount and the location of damage to the heart muscle.

Atherosclerosis of the arteries of the brain can cause sudden attacks of dizziness, loss of balance, double vision, and weakness in the arms and legs. Blockage of these arteries results in a stroke, with recovery again depending upon the extent of the damage and the area of the brain that is affected. Atherosclerosis may affect the aorta that supplies the kidneys, resulting in reduced kidney function and high blood pressure; or it may affect the arteries supplying the legs, resulting in cramps, pain and, in extreme cases, gangrene.

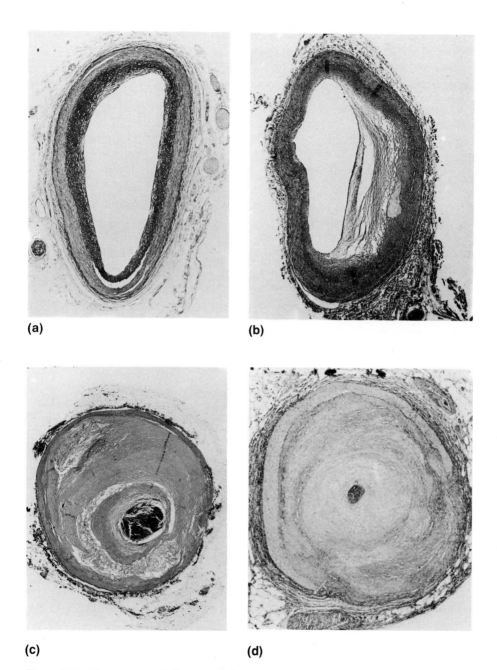

(a)

(b)

(c)

(d)

Figure 14.1 The progress of atherosclerosis. (*a*) A near-normal
artery. (*b*) Plaque forms on the inner lining of the artery. (*c*) The
narrowed channel within the artery is blocked by a blood clot.
(*d*) Atherosclerosis has progressed to the point that this artery is
completely blocked. (*a* and *d*, Courtesy National Institutes of
Health, National Heart & Lung Institute; *b* and *c*, Courtesy
American Heart Association)

Lipids

14.1 What Are Lipids?

Cholesterol, the material playing such an important role in atherosclerosis, belongs to a class of compounds called lipids. **Lipids** are oily or waxy substances that are insoluble in water and may be extracted from tissues using nonpolar solvents. Their major functions in living cells are to form structural components of membranes and to store energy for the cell.

Lipids are a diverse group of compounds that can be categorized in several ways. We can divide the lipids into two major classes: those that can be saponified (hydrolyzed by a base) and those that are nonsaponifiable. The saponifiable lipids can be further subdivided into **simple lipids,** those that yield fatty acids and an alcohol upon hydrolysis, and **compound lipids,** which yield fatty acids, alcohol, and some other compounds upon hydrolysis (Table 14.1).

Table 14.1   The Lipids

	LIPIDS		
Saponifiable		Nonsaponifiable	
Simple Lipids	Compound Lipids	Steroids	Terpenes
Neutral fats	Phospholipids	Cholesterol	Carotene
Waxes	Glycolipids	Cortisone	Vitamin A

Saponifiable Lipids: Simple Lipids

14.2   Fats and Oils

The simplest and most abundant of the lipid compounds are the neutral lipids, which are also called fats, triglycerides, or triacylglycerols. These compounds are esters of glycerol and three fatty acids, and are the major component of fat storage in plants and of the adipose cells (or fat cells) of vertebrates.

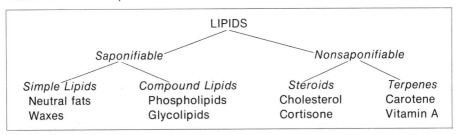

Fat or Triglyceride

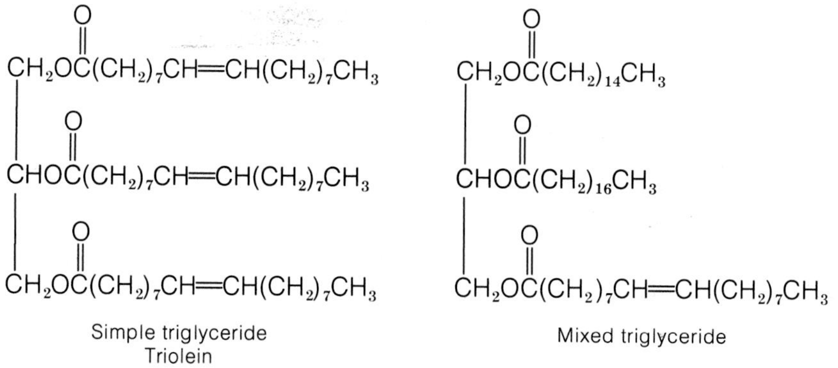

$$CH_2OC(CH_2)_7CH=CH(CH_2)_7CH_3$$

$$CHOC(CH_2)_7CH=CH(CH_2)_7CH_3$$

$$CH_2OC(CH_2)_7CH=CH(CH_2)_7CH_3$$

Simple triglyceride
Triolein

$$CH_2OC(CH_2)_{14}CH_3$$

$$CHOC(CH_2)_{16}CH_3$$

$$CH_2OC(CH_2)_7CH=CH(CH_2)_7CH_3$$

Mixed triglyceride

Figure 14.2 Simple triglycerides, such as triolein, contain only one type of fatty acid. Mixed triglycerides contain two or more different fatty acids.

Simple triglycerides contain the same fatty acid in all three positions on the glycerol molecule, while **mixed triglycerides** contain two or more different fatty acids. Natural fats are a mixture of simple and mixed triglycerides (Figure 14.2). Triglycerides have extensive commercial use in soaps, paints, varnishes, linoleum, printing inks, ointments, and creams.

14.3 Fatty Acids

Fatty acids are long-chain organic acids that normally are not found in the free state in cells, but are formed upon the hydrolysis of lipids. The naturally occurring fatty acids have an even number of carbon atoms and are generally nonbranching. The most abundant fatty acids have 16 or 18 carbon atoms in the chain. The most common fatty acids are palmitic, stearic, and oleic acids, with oleic acid making up more than half the total fatty acid content of many fats. **Saturated** fatty acids have only carbon-to-carbon single bonds, are unreactive, and are waxy solids at room temperature. **Unsaturated** fatty acids have one or more carbon-to-carbon double bonds, and are liquids at room temperature (Figure 14.3 and Table 14.2).

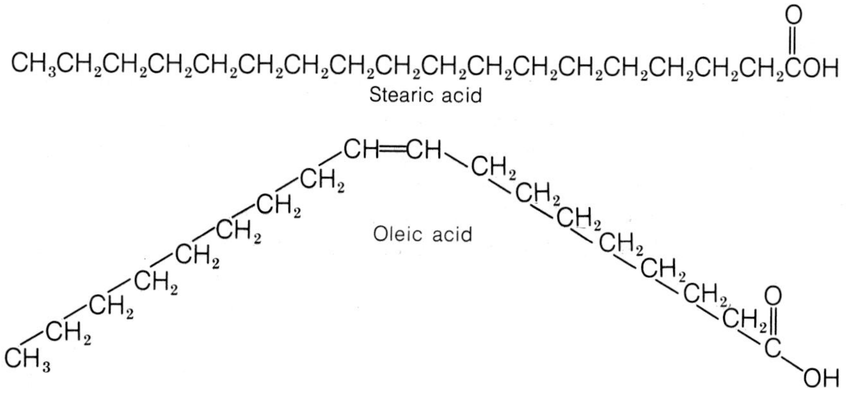

$$CH_3CH_2CH_2CH_2CH_2CH_2CH_2CH_2CH_2CH_2CH_2CH_2CH_2CH_2CH_2CH_2CH_2COH$$
Stearic acid

Oleic acid

Figure 14.3 Stearic and oleic acids are both fatty acids with 18 carbon atoms. Oleic acid is unsaturated; it contains one double bond in the *cis* configuration which gives the molecule a rigid bend.

Table 14.2 Some Common Fatty Acids

Name (Carbon Atoms)	Formula	Melting Point (°C)
SATURATED		
Butyric (4)	$CH_3(CH_2)_2\overset{\displaystyle O}{\overset{\|}{C}}OH$	− 4.2
Lauric (12)	$CH_3(CH_2)_{10}\overset{\displaystyle O}{\overset{\|}{C}}OH$	44.2
Myristic (14)	$CH_3(CH_2)_{12}\overset{\displaystyle O}{\overset{\|}{C}}OH$	53.9
Palmitic (16)	$CH_3(CH_2)_{14}\overset{\displaystyle O}{\overset{\|}{C}}OH$	63.1
Stearic (18)	$CH_3(CH_2)_{16}\overset{\displaystyle O}{\overset{\|}{C}}OH$	69.6
Arachidic (20)	$CH_3(CH_2)_{18}\overset{\displaystyle O}{\overset{\|}{C}}OH$	76.5
UNSATURATED		
Oleic (18)	$CH_3(CH_2)_7CH{=}CH(CH_2)_7\overset{\displaystyle O}{\overset{\|}{C}}OH$	13.4
Linoleic (18)	$CH_3(CH_2)_4CH{=}CHCH_2CH{=}CH(CH_2)_7\overset{\displaystyle O}{\overset{\|}{C}}OH$	− 5
Linolenic (18)	$CH_3CH_2CH{=}CHCH_2CH{=}CHCH_2CH{=}CH(CH_2)_7\overset{\displaystyle O}{\overset{\|}{C}}OH$	−11
Arachidonic (20)	$CH_3(CH_2)_4CH{=}CHCH_2CH{=}CHCH_2CH{=}CHCH_2CH{=}CH(CH_2)_3\overset{\displaystyle O}{\overset{\|}{C}}OH$	−49.5

The difference between fats and oils is the number of unsaturated fatty acids that are present. Animal fats, lard, tallow, and butter are mixed fats in which the number of saturated fatty acids is greater than the number of unsaturated fatty acids; they are waxy, white solids at room temperature. Vegetable oils, olive oil, corn oil, and cottonseed oil contain a higher concentration of unsaturated fatty acids, and are liquids at room temperature (Table 14.3).

Table 14.3   Some Common Fats and Oils and Their Fatty Acid Composition.*

ANIMAL FATS	Melting Point °C	Percent Composition of the Most Abundant Fatty Acids								Iodine Number
		Saturated				Unsaturated				
		Myris-tic	Pal-mitic	Stearic	Ara-chidic	Palmit-oleic	Oleic	Lino-leic	Lino-lenic	
Butter	32	11	29	9	2	5	27	4	–	36
Lard	30	1	28	12	–	3	48	6	–	59
Tallow	N/A	6	27	14	–	–	50	3	–	50
Human fat	15	3	24	8	–	5	47	10	–	68
PLANT OILS										
Corn	−20	1	10	3	–	2	50	34	–	123
Cottonseed	− 1	1	23	1	1	2	23	48	–	106
Linseed	−24	–	6	2	1	–	19	24	47	179
Olive	− 6	–	7	2	–	–	84	5	–	81
Peanut	3	–	8	3	2	–	56	26	–	93
Safflower	N/A	←————— 7 —————→				–	19	70	3	145
Soybean	− 16	–	10	2	–	–	29	51	6	130

* Values in this table are averages. Extreme variation may occur in the values depending upon the source, treatment, and age of the fat or oil.

14.4 Essential Fatty Acids

Our bodies can synthesize saturated fatty acids and unsaturated fatty acids containing one double bond. However, we can't form linoleic, linolenic, or arachidonic acids, known as the **essential fatty acids.** Infants lacking these fatty acids in their diets will lose weight and develop eczema. The essential fatty acids are used by the body in the formation of prostaglandins, compounds that are found in most mammalian tissues and that have a wide range of physiological effects (see Section 16.14). They are involved in the body's defenses against many sorts of change; in particular, they are powerful inducers of fever and inflammation. Aspirin seems to work as an antipyretic by regulating the synthesis of prostaglandins by the temperature-regulating tissues of the brain.

Reactions of Fats

14.5 Acrolein Test

The acrolein test will detect the presence of glycerol. When heated to high temperatures, fats will hydrolyze; the glycerol so produced will react to form acrolein. Acrolein vapors are irritating and tear-producing, and cause the unpleasant odor you recognize when oil or fat is burned. Acrolein is irritating to the digestive tract, and may be responsible for the stomach upset caused by deep-fried foods.

$$\begin{array}{c} H \\ | \\ H-C-OH \\ | \\ H-C-OH + Heat \\ | \\ H-C-OH \\ | \\ H \end{array} \longrightarrow \begin{array}{c} H \\ | \\ C=O \\ | \\ H-C \\ \| \\ CH_2 \end{array} + 2H_2O$$

Glycerol   Acrolein

14.6 Iodine Number

The unsaturated bonds in a fatty acid will react to add iodine. The iodine number of a fat is the number of grams of iodine that will react with 100 grams of fat or oil, and is an indication of the degree of unsaturation. The higher the iodine number, the more unsaturated the fat. Fats generally have an iodine number below 70, and oils have an iodine number above 70 (Table 14.2).

14.7 Hydrogenation

Oils can be converted to solid fats by hydrogenation, the addition of hydrogen to the double bonds of the molecule.

$$CH_3(CH_2)_4CH{=}CHCH_2CH{=}CH(CH_2)_7COOH + 2H_2 \longrightarrow CH_3(CH_2)_{16}COOH$$

Linoleic acid   Stearic acid

Vegetable shortenings such as Crisco®, Spry®, and Fluffo® are commercially produced by the partial hydrogenation of soybean, corn, or cottonseed oil. (The complete hydrogenation of these oils would yield a hard, brittle product.) These shortenings are usually a mixture of unsaturated oils and hydrogenated oils. Margarine is a mixture of unsaturated oils, hydrogenated oils, flavorings, coloring agents, and vitamins A and D.

14.8 Rancidity

Fats and oils often develop an objectionable odor and taste, and are then called rancid. There are two causes of rancidity: hydrolysis and oxidation. For example, when left at room temperature too long butter will become rancid. This occurs because some of the fat in the butter will have undergone hydrolysis, accelerated by enzymes produced by microorganisms in the air. This hydrolysis produces the fatty acid butyric acid, causing the odor of rancid butter. Oxygen in the air can oxidize unsaturated fats or oils to produce short-chain acids or aldehydes having disagreeable odors and tastes. Such oxidation is retarded in manufactured products such as crackers, potato chips, and pastries by the addition of chemicals referred to as antioxidants.

14.9 Hydrolysis

The hydrolysis of fats can occur in the presence of superheated steam, hot mineral acids, or specific enzymes. Hydrolysis under these conditions yields glycerol and three fatty acids. In general,

$$
\begin{array}{c}
\underset{\substack{\text{O} \\ \|}}{} \\
\text{CH}_2\text{OCR} \\
\underset{\substack{\text{O} \\ \|}}{} \\
\text{CHOCR}' \quad + 3\text{H}_2\text{O} \longrightarrow \\
\underset{\substack{\text{O} \\ \|}}{} \\
\text{CH}_2\text{OCR}'' \\
\text{Fat}
\end{array}
\qquad
\begin{array}{c}
\text{CH}_2\text{OH} \\
| \\
\text{CHOH} \quad + \\
| \\
\text{CH}_2\text{OH} \\
\text{Glycerol}
\end{array}
\qquad
\begin{array}{c}
\underset{\substack{\text{O} \\ \|}}{} \\
\text{RCOH} \\
\underset{\substack{\text{O} \\ \|}}{} \\
\text{R}'\text{COH} \\
\underset{\substack{\text{O} \\ \|}}{} \\
\text{R}''\text{COH} \\
\text{Fatty acids}
\end{array}
$$

For example,

$$
\begin{array}{c}
\underset{\substack{\text{O} \\ \|}}{} \\
\text{CH}_2\text{OC(CH}_2)_{16}\text{CH}_3 \\
\underset{\substack{\text{O} \\ \|}}{} \\
\text{CHOC(CH}_2)_7\text{CH}=\text{CH(CH}_2)_7\text{CH}_3 + 3\text{H}_2\text{O} \longrightarrow \\
\underset{\substack{\text{O} \\ \|}}{} \\
\text{CH}_2\text{OC(CH}_2)_{14}\text{CH}_3
\end{array}
$$

$$
\begin{cases}
\begin{array}{c}
\text{CH}_2\text{OH} \\
| \\
\text{CHOH} \\
| \\
\text{CH}_2\text{OH} \\
\text{Glycerol}
\end{array} \\[2em]
+ \text{CH}_3(\text{CH}_2)_{16}\overset{\substack{\text{O} \\ \|}}{\text{C}}\text{OH} \\
\text{Stearic acid} \\[1em]
+ \text{CH}_3(\text{CH}_2)_7\text{CH}=\text{CH(CH}_2)_7\overset{\substack{\text{O} \\ \|}}{\text{C}}\text{OH} \\
\text{Oleic acid} \\[1em]
+ \text{CH}_3(\text{CH}_2)_{14}\overset{\substack{\text{O} \\ \|}}{\text{C}}\text{OH} \\
\text{Palmitic acid}
\end{cases}
$$

14.10 Saponification

When hydrolysis is carried out in the presence of a strong base such as sodium hydroxide, glycerol and the sodium salts of the fatty acids are produced. In general

$$
\begin{array}{l}
\text{CH}_2\text{OCR} \\
| \\
\text{CHOCR}' \quad + \quad 3\text{NaOH} \quad \longrightarrow \\
| \\
\text{CH}_2\text{OCR}'' \\
\text{Fat}
\end{array}
\qquad
\begin{array}{l}
\text{CH}_2\text{OH} \\
| \\
\text{CHOH} \quad + \\
| \\
\text{CH}_2\text{OH} \\
\text{Glycerol}
\end{array}
\qquad
\begin{array}{l}
\text{RCO}^-\text{Na}^+ \\
\\
\text{R}'\text{CO}^-\text{Na}^+ \\
\\
\text{R}''\text{CO}^-\text{Na}^+ \\
\text{Sodium salt} \\
\text{(soap)}
\end{array}
$$

For example,

$$
\begin{array}{l}
\text{CH}_2\text{OC(CH}_2)_{16}\text{CH}_3 \\
| \\
\text{CHOC(CH}_2)_{16}\text{CH}_3 \ + 3\text{NaOH} \ \longrightarrow \\
| \\
\text{CH}_2\text{OC(CH}_2)_{16}\text{CH}_3 \\
\text{Tristearin}
\end{array}
\qquad
\begin{array}{l}
\text{CH}_2\text{OH} \\
| \\
\text{CHOH} \ + 3\text{CH}_3(\text{CH}_2)_{16}\text{CO}^-\text{Na}^+ \\
| \\
\text{CH}_2\text{OH} \\
\text{Glycerol} \qquad \text{Sodium stearate}
\end{array}
$$

Such salts are called **soaps.** Sodium salts of fatty acids are used in bar soaps, and potassium salts are used in liquid soaps. A hard soap will contain a larger number of saturated fatty acids than a soft soap. Various additives are used to give commercial soaps their colors and odors; also, as you may suspect, floating soaps contain air bubbles, and scouring soaps contain abrasives.

14.11 Cleansing Action of Soap

Water is a poor cleansing agent for removing grease and oil since water molecules tend to stick together rather than penetrate the nonpolar grease. Soap greatly improves the cleansing power of water. A soap molecule has two portions: a nonpolar "tail" formed by the hydrocarbon chain, and a polar "head" formed by the carboxyl group.

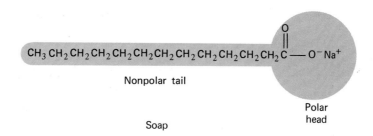

$$CH_3 CH_2 CH_2 CH_2 CH_2 CH_2 CH_2 CH_2 CH_2 CH_2 CH_2 CH_2 C - O^- Na^+$$

Nonpolar tail

Polar head

Soap

The nonpolar tail will readily dissolve in the nonpolar grease, while the polar head tends to remain dissolved in the water. In this way the soap breaks up the grease, forming small colloidal droplets — that is, emulsifying the grease — which can then be washed away (Figure 14.4).

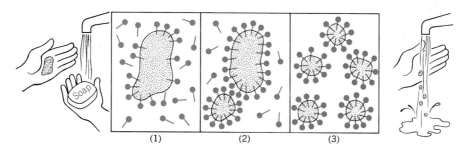

(1)   (2)   (3)

Figure 14.4 The cleansing action of soap. (1) The nonpolar tails of the soap molecules begin to dissolve in the nonpolar grease. (2) As the soap dissolves in the grease, small colloidal grease particles break off and are surrounded by the negatively-charged polar heads of the soap molecules. This keeps the particles in solution by preventing the particles from forming larger droplets. (3) In this manner the grease can be completely broken up and the colloidal droplets washed away with water.

The cleansing power of soap is affected by several factors. Hard water contains one or more of the metallic ions Ca^{2+}, Mg^{2+}, Fe^{2+}, or Fe^{3+}, ions that form insoluble salts with soap. These salts precipitate out to form soap scum and bathtub ring. This leaves less soap in the water to carry out the cleansing. Water softeners work by replacing these metallic ions with other ions, such as sodium, which do not interfere with the action of soap. Lowering the pH of the water will also decrease the cleansing action of soap by neutralizing the charge on the fatty acid ion.

Detergents, which are mixtures of sodium salts of sulfuric acid esters, have similar cleansing properties to soaps, but have significant advantages (Figure 14.5). Their calcium and magnesium salts are water soluble, and

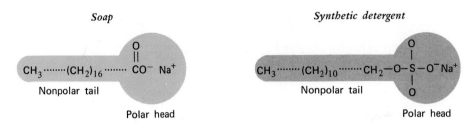

Soap   *Synthetic detergent*

Nonpolar tail   Polar head   Nonpolar tail   Polar head

Figure 14.5 Two organic salts that possess cleansing properties. Each has a nonpolar, hydrophobic (water-repelling) tail and a polar, hydrophilic (water-attracting) head.

they are not affected by pH. However, some early synthetic detergents containing branched chains in their hydrocarbon tails could not be naturally broken down by bacteria in sewage (that is, they were not biodegradable) in the same manner that soaps are, so they could not be removed from the water by sewage treatment plants. This resulted in rivers and streams being covered with foam and suds, creating a significant pollution problem. Newer detergents are partially or totally biodegradable, and the suds problem has been eliminated.

14.12 Waxes

Waxes are esters of long-chain fatty acids and long-chain alcohols. For example, beeswax is largely an ester of myricyl alcohol ($C_{30}H_{61}OH$) and palmitic acid. Waxes form protective coatings on skin, fur, feathers, leaves, and fruits (Table 14.4).

Table 14.4   Some Common Waxes

Name	Melting Point °C	Source	Uses
Beeswax	61–69	Honeycomb	Candles, polishes
Carnauba	83–86	Carnauba Palm	Floor waxes, polishes
Lanolin	36–43	Wool	Cosmetics, skin ointments
Spermaceti	42–50	Sperm whale	Cosmetics, candles

They have characteristics of water insolubility, flexibility, and nonreactivity that make them perfect protective coatings. Commercially produced waxes are used in cosmetics, furniture and car polishes, ointments, and creams.

Saponifiable Lipids: Compound Lipids

Phospholipids

The **phospholipids** are a class of waxy solids that form the structural components of cell membranes, and are important in the transport of lipids in the body. They can be divided into two general categories: glycerol-based phospholipids and sphingosine-based phospholipids.

14.13 Glycerol-Based Phospholipids: Phosphatides

The phosphatides are derivatives of phosphatidic acid. They contain glycerol, two fatty acids, phosphoric acid, and a nitrogen compound that can be choline, ethanolamine, serine, or inositol.

```
G ──Fatty acid (saturated)
l
y
c ──Fatty acid (unsaturated)
e
r
o ──Phosphoric acid─Nitrogen compound
l
```

Phosphatidic acid

Phosphatide

Cell Membranes. The phosphoric acid group forms a polar head on the phosphatide molecule, and the two fatty acids form two nonpolar tails.

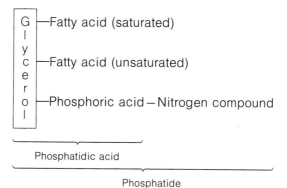

Nonpolar tails   Polar Head

This configuration gives the phosphatide good emulsifying properties and good membrane-forming properties. Cellular membranes are composed of proteins and phospholipids. The phospholipids form a double layer with their polar heads on the top and bottom, and their nonpolar tails in the middle. This double layer forms the framework for the membrane, and provides an anchorage for the protein (Figure 14.6).

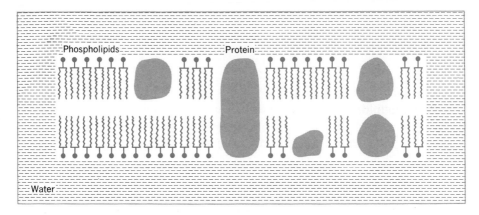

Figure 14.6 Cellular membranes are composed of phospholipids and proteins. The phospholipids are arranged in a bilayer
with their polar heads to the outside and their nonpolar tails
toward the middle. This bilayer serves as an anchor for the protein molecules.

Lecithins. Lecithins are phosphatides in which the nitrogen compound is
choline.

G —Fatty acid
l
y
c —Fatty acid
e
r
o —Phosphoric acid—Choline
l

$$CH_3-N-CH_2CH_2OH$$

with CH_3 and CH_3 on the nitrogen

Choline

Lecithin plays an important role in the metabolism of fats in the liver,
serves as a source of inorganic phosphate for tissue formation, and is an
excellent emulsifying agent. It is important in the transport of fats from one
part of the body to another, and is used commercially as an emulsifying
agent in such products as chocolate candies, margarine, and medicines.
Egg yolks contain a high amount of lecithin, and will emulsify salad oil
and vinegar in mayonnaise. Removal of one fatty acid from lecithin forms
lysolecithin, a compound that causes destruction of red blood cells and
spasmodic muscle contractions. The venom of poisonous snakes contains
enzymes to catalyze the formation of lysolecithin from lecithin.

Cephalins. When the nitrogen compound in a phosphatide is ethanolamine,
$H_2NCH_2CH_2OH$, the compound formed is called cephalin. Cephalins are
found in blood platelets and are important in the clotting of blood. They
also serve as a source of inorganic phosphate for the formation of new
tissue.

14.14 Sphingosine-Based Phospholipids: Sphingolipids

The alcohol of sphingolipids is not glycerol, but is sphingosine.

OH
|
CHCH=CH(CH$_2$)$_{12}$CH$_3$
|
CHNH$_2$
|
CH$_2$OH

Sphingosine

Sphingosine —Fatty acid

Sphingosine —Phosphoric acid – Choline

Sphingomyelin

The most common sphingolipid is sphingomyelin. Large amounts of sphingomyelins are found in brain and nervous tissue, and form part of the myelin sheath, the protective coating of nerves. The myelin sheath is very stable, due in part to interlocking of the long fatty acid chains of the sphingomyelins. Certain diseases such as Niemann-Pick disease and multiple sclerosis result in the production of defective myelin sheaths. Niemann-Pick disease is a congenital disease in which sphingomyelins build up in the brain, liver, and spleen, resulting in mental retardation and early death.

14.15 Glycolipids

The main difference between glycolipids and phospholipids is that the glycolipid contains a sugar group rather than a phosphate group. The sugar group is usually galactose, but may also be glucose. The alcohol is either glycerol or sphingosine.

Glycerol —Fatty acid
Glycerol —Fatty acid
Glycerol —Sugar

Sphingosine —Fatty acid
Sphingosine —Sugar

Glycolipids

Cerebrosides are glycolipids that contain the base sphingosine. They are found in high concentrations in the brain and nerve cells, especially in the myelin sheath. Several hereditary fat metabolism diseases have been linked to errors in the metabolism of the glycolipids. In Gaucher's disease, the glycolipids contain glucose rather than galactose, and they collect in the spleen and kidneys. In Tay-Sachs disease, the infant lacks an enzyme necessary to break down glycolipids, and they collect in the tissues of the brain and eyes, causing muscular weakness, mental retardation, seizures, blindness, and death by the age of three.

Nonsaponifiable Lipids: Terpenes and Steroids

Terpenes and steroids are high-molecular-weight lipids that are not cleaved by alkaline hydrolysis. The terpenes are polymers of isoprene, and were discussed in Chapter 9.

14.16 Steroids

Steroids are compounds whose structure is based on a complicated four-ring framework consisting of three cyclohexane rings and one cyclopentane ring.

Steroid nucleus

Steroids have a wide range of physiological functions, from the emulsification of fats to the determination of the sex characteristics of an organism. The structure and function of some familiar steroids are shown in Table 14.5.

Table 14.5   The Structure and Function of Some Steroids

Steroid	Structure	Function
Cortisone		One of many hormones produced in the adrenal glands. It is important in controlling carbohydrate metabolism and is used therapeutically to relieve symptoms of inflammation, especially in rheumatoid arthritis.

Table 14.5   The Structure and Function of Some Steroids (cont'd.)

Steroid	Structure	Function
Vitamin D$_2$		Irradiation of the steroid hormone ergosterol with ultraviolet light breaks open one of the rings in the steroid nucleus, producing vitamin D$_2$. This vitamin is essential to prevent rickets, a disease of calcium metabolism.
Digitoxigenin		Extracted from the digitalis plant, this steroid is used in small doses to regulate a diseased heart. In large doses it causes death.
Testosterone		This male sex hormone regulates the development of the male reproductive organs.
Progesterone		This is the female sex hormone that is produced in pregnancy and acts on the uterine lining, preparing it to receive the embryo.

14.17 Cholesterol

Sterols are steroid alcohols. The most abundant sterol is cholesterol.

Cholesterol

Cholesterol is found in most animal tissue, especially the brain, nervous tissue, cell membranes, and the fluids of the blood and bile. We each ingest some cholesterol daily from foods such as egg yolks and meat fats; the exact amount varies with our diet. In addition, many tissues, especially the liver, can synthesize cholesterol from acetyl CoA. Our bodies manufacture 3 to 5 grams of cholesterol a day, 80% of which is converted to cholic acid used in the formation of bile salts.

Cholesterol's functions in the body are not well understood. As we have seen, it may be a contributing factor in the development of atherosclerosis. It is known that cholesterol is an important precursor in the synthesis of steroids such as bile acids, sex hormones, and vitamin D.

Lipid Metabolism

14.18 Digestion and Absorption

About one-fifth of the solid matter in the food we eat is fat, although this figure will vary greatly with diet. Fats are not acted upon in the digestion process until they reach the small intestine. There they are mixed with bile salts, which emulsify the fats and allow enzymes in the pancreatic and intestinal juices to hydrolyze the fat into glycerol, fatty acids, and mono- and diglycerides.

Bile is a fluid that is continuously manufactured in the liver and stored in the gall bladder. Bile is composed of bile salts, bile pigments, and cholesterol (Figure 14.7). In some people, the mucous membranes of the gall bladder will absorb water, concentrating the bile. Under these conditions the cholesterol in bile, which is not very water-soluble, will crystallize out of solution along with bile salts and bile pigments, forming gall stones. These stones can cause infection and pain, and can obstruct the flow of bile, resulting in jaundice.

Figure 14.7 Bile contains the sodium salts of cholic acid and
other closely related compounds. These salts can act like soap
to emulsify fats and oils, preparing them for digestion.

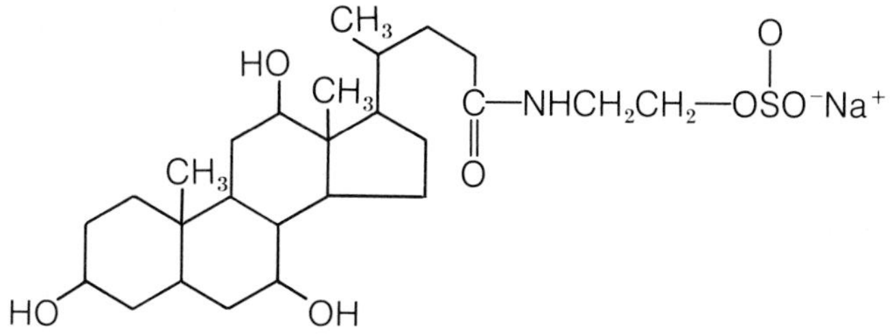

Taurocholic bile salt

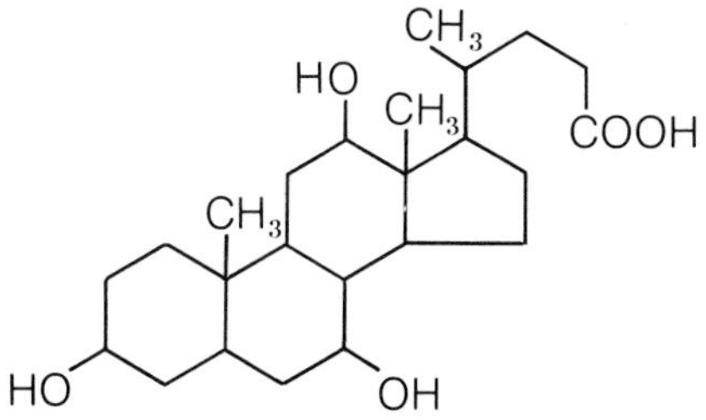

Cholic acid

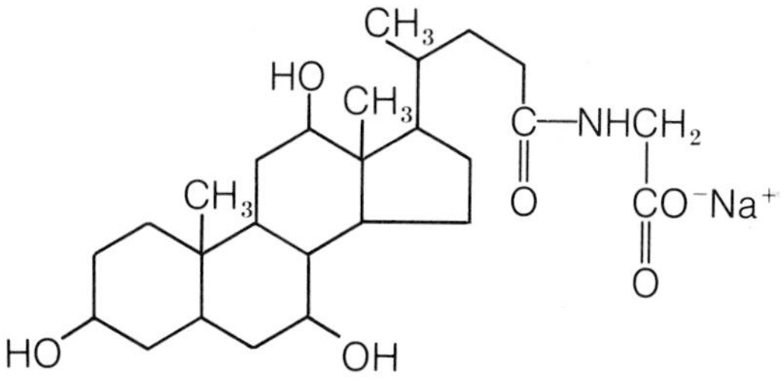

Glycocholic bile salt

Bile pigments are the waste products from the breakdown of red blood
cells, and are what give color to the feces. Bile salts are formed from bile
acids and cholic acid, and function to emulsify fat and aid in the transport
of the products of fat digestion across the intestinal barrier. The glycerol,
fatty acids, and mono- and diglycerides formed by the hydrolysis of
triglycerides cross the intestinal barrier, are reformed into glycerides, and
are transported to the blood via the body's lymph system. They enter the
blood as microdroplets and are joined with proteins for transport in the
blood. After a meal, the blood is rich in fat, and has a milky, opalescent
appearance.

14.19 Lipid Storage

Lipids and carbohydrates that are eaten in excess of energy requirements
are stored as triglycerides under the skin and around major organs in the
form of adipose tissue. Such storage fat has several functions: it is an
energy reserve, a support and shock absorber for inner organs, and heat
insulation for the body. While the glycogen reserves of our bodies are
sufficient to last only a few hours, the average adult male has enough
stored triglyceride to sustain him for 30 to 40 days if there is water
available.

The lipids in adipose tissue are in dynamic equilibrium with lipids in
the blood; stored fatty acids are constantly being exchanged with food

fatty acids. The particular composition of the storage fat differs for different organisms, but it can be altered by controlling the diet. For example, experiments are now underway to determine if cattle raised on special diets will develop higher concentrations of unsaturated fats in their meat.

14.20 Oxidation of Fatty Acids

Upon oxidation, fats yield much more energy than do carbohydrates; fats yield nine kilocalories per gram, while glycogen and starch yield four kcal per gram. In vertebrates, oxidation of fatty acids provides at least half the energy required by the liver, kidneys, heart, and resting skeletal muscles. In hibernating animals and migrating birds, fat is the sole energy source.

The process of fat oxidation starts when fats stored in the adipose tissues are hydrolyzed to fatty acids and glycerol. These compounds are then carried to energy-requiring tissues where they will be oxidized. The glycerol enters the glycolysis pathway (Chapter 13), and fatty acids are oxidized in the mitochondria by means of a sequence of reactions called beta oxidation.

β-Oxidation

In 1904, research by a chemist named Franz Knoop suggested that fatty acids were oxidized by the removal of successive two-carbon fragments. After many years of research, this observation has now been confirmed. Fatty acid oxidation starts with the oxidation of the beta carbon, and the successive removal of two-carbon fragments in the form of acetyl CoA (Figure 14.8).

The β-oxidation of the 16-carbon fatty acid palmitic acid will produce eight acetyl CoA's that can enter the citric acid cycle, and 14 pairs of hydrogen atoms that will enter the electron transport chain. Overall, one molecule of palmitic acid will yield 130 ATP's.

$$CH_3(CH_2)_{14}COOH + 23O_2 + 130ADP + 130P_i \longrightarrow$$

Palmitic acid

$$16CO_2 + 16H_2O + 130ATP + Energy$$

The heat of reaction for this reaction is 2338 kcal/mole. Since 130 ATP's will yield about 910 kcal of energy, the process in the cell is $\frac{910}{2338}$, or about 39%, efficient.

The acetyl CoA's produced in the oxidation of fatty acids can be put to many uses in the metabolism of the cell: they can enter the citric acid cycle; they can be used in biosynthesis of materials required by the cell, such as cholesterol, amino acids, or other fatty acids; or they can be used in the synthesis of ketone bodies (Figure 14.9).

14.21 Ketone Bodies

When acetyl CoA is produced in excess by the liver, it is converted into acetoacetate and two other compounds, all called **ketone bodies** (Table

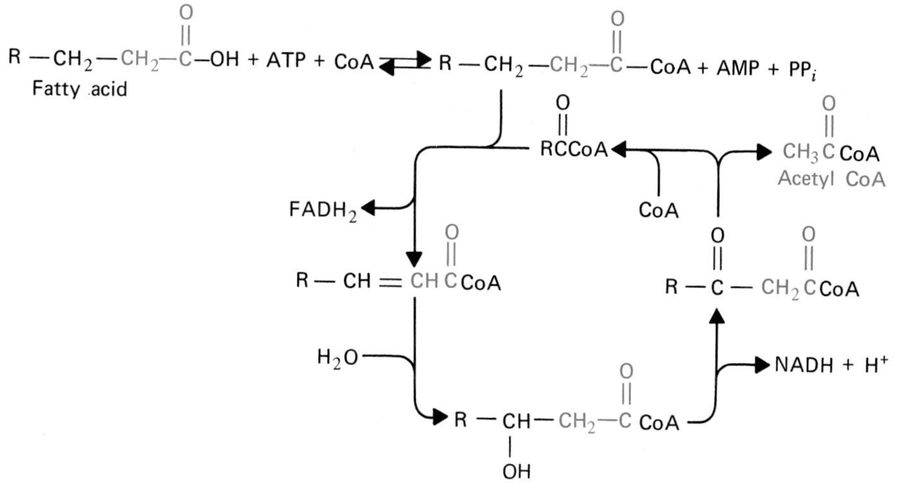

$$R-CH_2-CH_2-\overset{\overset{\displaystyle O}{\|}}{C}-OH + ATP + CoA \rightleftharpoons R-CH_2-CH_2-\overset{\overset{\displaystyle O}{\|}}{C}-CoA + AMP + PP_i$$

Fatty acid

RCCoA

$CH_3\overset{\overset{\displaystyle O}{\|}}{C}CoA$
Acetyl CoA

FADH$_2$

CoA

$$R-CH=CH\overset{\overset{\displaystyle O}{\|}}{C}CoA$$

$$R-\overset{\overset{\displaystyle O}{\|}}{C}-CH_2\overset{\overset{\displaystyle O}{\|}}{C}CoA$$

H$_2$O

NADH + H$^+$

$$R-\underset{\underset{\displaystyle OH}{|}}{CH}-CH_2-\overset{\overset{\displaystyle O}{\|}}{C}CoA$$

Figure 14.8 β-oxidation of fatty acids is a cyclic process in which one acetyl CoA is removed from the fatty acid molecule each time the molecule passes through the cycle.

13.4). These compounds, which cannot be oxidized by the liver, are transported to other tissues, where they are oxidized in the citric acid cycle or excreted by the kidneys. Their concentration in the blood is normally very low—about 1 mg/100 ml.

Any disruption of normal metabolism, such as liver damage, diabetes mellitus, starvation, or diets causing a restriction or decrease in glucose metabolism will increase the level of fat metabolism and, therefore, increase the production of ketone bodies. If the level of ketone bodies in the blood exceeds the amount that can be used by the tissues and excreted by the

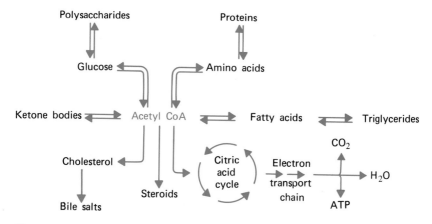

Figure 14.9 Acetyl CoA plays a central role in the anabolic and catabolic reactions of cellular metabolism.

kidneys, a condition known as **ketosis** occurs. Since two of the ketone bodies are acids, the pH of the blood will drop and acidosis will occur. As we have seen, acidosis can lead to nausea, depression of the central nervous system, dehydration and, in extreme cases, coma and death.

Additional Reading

Books
1. *Arteriosclerosis,* Volume I, National Institutes of Health, 1971.

2. B. F. Miller and L. Galton, *Freedom from Heart Attacks,* Simon and Schuster, New York, 1972.

Articles
1. R. O. Brady, "Hereditary Fat-Metabolism Diseases," *Scientific American,* August 1973, page 88.

2. R. A. Capaldi, "A Dynamic Model of Cell Membranes," *Scientific American,* March 1974, page 27.

3. C. F. Fox, "The Structure of Cell Membranes," *Scientific American,* February 1972, page 30.

Questions and Problems

1. Define the following terms:

    (a) Atherosclerosis
    (b) Simple lipid
    (c) Compound lipid
    (d) Triglyceride
    (e) Saponifiable and non-saponifiable lipid
    (f) Saturated and unsaturated fatty acid
    (g) Essential fatty acid
    (h) Acrolein test
    (i) Iodine number
    (j) Hydrogenation
    (k) Saponification
    (l) Soap
    (m) Wax
    (n) Phospholipid
    (o) Glycolipid
    (p) Steroid
    (q) Bile
    (r) β-Oxidation
    (s) Ketosis

2. Write the formula for the following triglycerides:

    (a) Tripalmitin
    (b) The mixed triglyceride containing glycerol, palmitic, oleic and linoleic acids.

3. Which triglyceride in question 2 would

   (a) Have the higher melting point?
   (b) Be more likely to be found in animal fat?
   (c) Have the higher iodine number?

4. Write the equation for the complete hydrogenation of the triglyceride in part (b) of question 2.

5. The following triglyceride is found in lard:

$$CH_2-O-\overset{\overset{\displaystyle O}{\|}}{C}(CH_2)_{14}CH_3$$
$$CH-O-\overset{\overset{\displaystyle O}{\|}}{C}(CH_2)_{16}CH_3$$
$$CH_2-O-\overset{\overset{\displaystyle O}{\|}}{C}(CH_2)_7CH=CH(CH_2)_7CH_3$$

   (a) Write the structure of all the products that would result from the digestion of this triglyceride.
   (b) How does the digestion and absorption of the fats in a strip of bacon differ from the digestion and absorption of the starch in a piece of toast?

6. Before commercially produced soap was readily available, housewives made soap from lard and lye that was extracted from wood ashes with a small amount of water. This lye solution contained basic substances such as KOH, Na_2CO_3, and K_2CO_3. Write the equation for the formation of soap from the triglyceride shown in question 5 and KOH.

7. In what ways are detergents superior to soaps?

8. Why does unrefrigerated butter turn rancid? What causes the odor associated with rancid butter?

9. Why would a diet lacking in linolenic acid be detrimental to a person's health?

10. Describe what is happening, on the molecular level, when you wash grease from your hands with soap.

11. Write the general formula for each of the following compounds, and describe the function of each in the human body.

   (a) Phosphatide
   (b) Lecithin
   (c) Cephalin

   (d) Sphingomyelin
   (e) Glycolipid

12. Advertising would lead us to believe that cholesterol is bad for us.

    (a) Give evidence to support this statement.
    (b) Then, explain why this statement is not entirely true.

13. Why is it more efficient for our bodies to store excess food calories in the form of body fat than as glycogen?

14. Describe the steps involved in the oxidation of fats by our bodies.

15. Why are diets that severely restrict the intake of carbohydrates potentially dangerous to a person's health?

chapter 15

Learning Objectives

By the time you have finished this chapter, you should be able to:

1. Write the general structure of an amino acid.

2. Describe the difference between

    (a) A simple and conjugated protein.
    (b) A globular and fibrous protein.

3. Explain how an amino acid can act as a buffer.

4. Define "zwitterion" and "isoelectric point."

5. Describe the types of bonding found in each of the following protein structures:

    (a) Primary   (c) Tertiary
    (b) Secondary   (d) Quaternary

6. State five methods for denaturing proteins.

7. Describe the steps of protein digestion.

8. Define "essential amino acid" and "adequate protein."

9. In general terms, give the reactants and products for each of the following reactions, and describe when they might occur during protein metabolism.

    (a) Transamination
    (b) Oxidative deamination
    (c) Urea cycle

10. Describe four metabolic uses of amino acids.

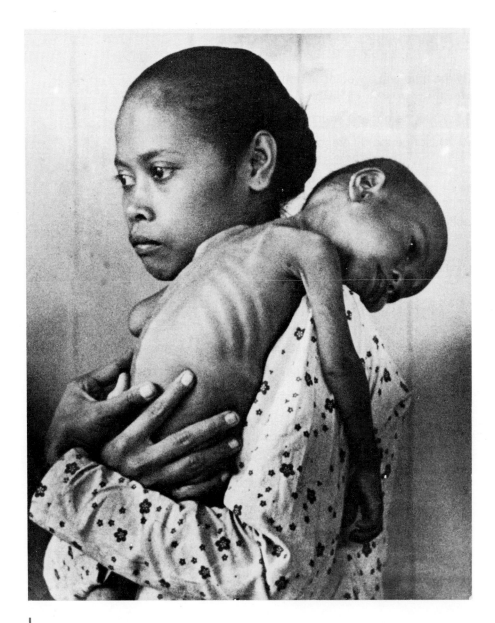

In a small village in western Africa, a young mother nurses her first child. He is a strong, healthy 18-month-old boy who will soon have to be weaned, for his mother is about to deliver her second baby. After the second child is born, the little boy is switched to a diet of mush made from corn. Slowly, day by day, he begins to lose energy. He becomes irritable and apathetic. His growth rate declines and he begins to lay miserably on the ground, crying and whimpering. After a year his appearance has changed drastically; his face and legs are swollen with fluid, his curly black hair has become soft and brown and is falling out, and his skin is flaky and covered with rashes. His mother has become greatly concerned, but has no way to obtain medical care. She watches helplessly as her son slowly sinks into a coma, and dies.

In a village to the south, another young boy has been more fortunate. His village has just been given powdered milk and a vegetable-protein mixture to supplement the normal diet of the inhabitants. Although this boy was suffering symptoms similar to the boy in the other village, he begins to recover. As he continues on his new diet, the swelling goes down, his rashes disappear, and his energy and appetite return.

Both of these children were suffering from Kwashiorkor, a disease that affects about 70% of the world's children under six years of age. Kwashiorkor is an African name meaning a disease that affects the first-born when he is displaced from the breast by the second child. The first child is then fed a diet of starchy foods that are inadequate sources of proteins essential for normal growth and development. The exact symptoms of Kwashiorkor will vary from country to country depending on the diet, but generally the children show appetite loss, growth retardation, edema (or swelling), disorders of the pigments of the skin and hair, and increased susceptibility to infection. More than 50% of the children afflicted with this disease will die. The disease can be cured by diets supplemented with milk or vegetable-protein mixtures high in the proteins missing from the child's regular diet. There is considerable debate as to whether protein deficiency in children causes mental retardation, but this seems possible since the brain and nervous system grow faster than any other tissues during the first four years of life. The widespread incidence of Kwashiorkor can be decreased only by supplementing the diets of a large portion of the world's population with adequate protein.

Proteins

15.1 Occurrence, Composition, and Function

Proteins are the most complex and varied class of macromolecules found in the cell. They occur in all living cells, and their biological significance can't be overemphasized. This fact was recognized by the German chemist G. T. Mulder in 1839, when he gave this class of compounds the name protein, which means "of prime importance."

All proteins are composed of the elements carbon, nitrogen, oxygen, and hydrogen. Most proteins also contain sulfur, and some have phosphorus and other elements, such as iron, zinc, or copper. Proteins are large polymers, and upon hydrolysis will yield monomer units called **amino acids.** Proteins can be divided into two major classes: **simple proteins,** which yield only amino acids upon hydrolysis, and **conjugated proteins,** which yield amino acids and other organic or inorganic components upon hydrolysis. These other components are called **prosthetic groups.**

The molecular weight of most proteins ranges from 12,000 to 1 million or more (Table 15.1).

Table 15.1 Proteins Have Very Large Molecular Weights Compared to Other Compounds

Compound	Molecular Weight
Inorganic Compounds	
Sodium chloride	58.5
Sulfuric acid	98
Organic Compounds	
Ethanol	48
Urea	60
Carbohydrates	
Glucose	180
Sucrose	342
Cellulose	about 500,000
Lipids	
Cholesterol	384
Triolein	836
Proteins	
Insulin	6,300
Ribonuclease	12,640
Hemoglobin	68,000
γ-Globulin	149,900
Fibrinogen	450,000
Glutanate dehydrogenase	1,000,000
Hemocyanin	9,000,000

This large size gives protein molecules colloidal properties. For example, they don't pass through differentially permeable membranes; the presence of proteins in urine, therefore, warns doctors of the possibility of damage to the membranes of the kidneys.

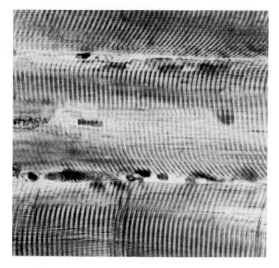

Figure 15.1 The structure of skeletal muscle. The muscle cells contain myofibrils which, when stained, exhibit cross banding or striations, hence the name striated muscle. The myofibrils consist of two types of contractile proteins, actin and myosin. (Courtesy J. Robert McClintic)

Because of their diverse natures, proteins can be classified in several ways. A second classification is based on the physical characteristics of the protein molecule. **Globular proteins** are soluble in water, are quite fragile, and have an active function, such as catalyzing reactions (in the case of enzymes), or transporting other substances (as, for example, hemoglobin). **Fibrous proteins** are insoluble in water, are physically tough, and have a structural or protective function (Figure 15.1). The keratin in hair and nails, and the collagen in tendons are examples of such fibrous proteins.

The biological importance of proteins results from their wide variety of functions (Table 15.2). Proteins are the body's main dietary source of nitrogen and sulfur. In addition to their catalytic and structural functions, they make up the contractile system of muscles. As antibodies they are the defense system of the body, and as hormones they regulate the body's glandular activity. In the blood they maintain fluid balance, are part of the clotting mechanism, and transport oxygen and lipids. They can act as poisons, such as venoms in animal bites and stings, or toxins, such as the bacterial toxin producing botulism in improperly produced foods. And some antibiotics that are secretions of bacteria and fungi are protein in nature.

Table 15.2 A Classification of Proteins Based on Their Functions
in Living Organisms

Class	Example
Enzymes	Pepsin Amylase
Structural proteins	Keratin Collagen
Storage proteins	Ferritin Casein
Transport proteins	Hemoglobin Myoglobin
Hormones	Insulin Parathormone
Contractile proteins	Actin Myosin
Protective proteins	Antibodies Fibrinogen
Toxins	Venoms Botulinus toxin

15.2 Amino Acids

The particular function of a given protein is determined by the sequence
of amino acids in the protein molecule. It is necessary, therefore, for any
study of proteins to include a thorough discussion of amino acids. Amino
acids are organic acids that have an amino group on the alpha carbon—
the carbon next to the carboxyl group. The general structure of an amino
acid is as follows:

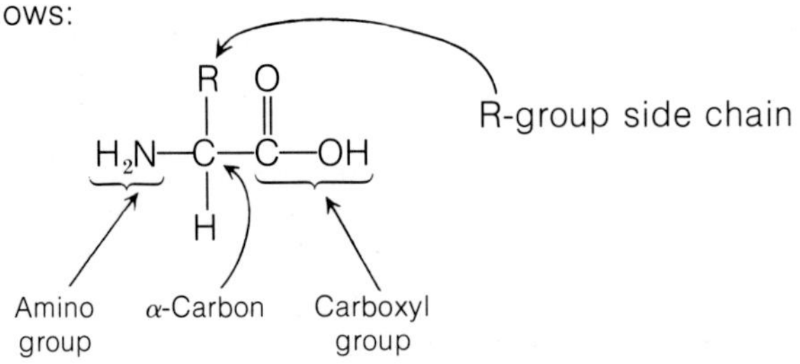

The different R-group side chains on the amino acids distinguish one
amino acid from another. Most naturally occurring proteins are composed
of the 20 amino acids shown in Figure 15.2, but there are a few specialized
types of proteins that contain other, more rare, amino acids. Still other
amino acids not found in proteins exist in a free or combined form; these
are, in general, derivatives of the 20 amino acids found in proteins (Fig. 15.3).

Figure 15.3 Citrulline and ornithine are non-protein amino acids that are involved in the reactions of the urea cycle.

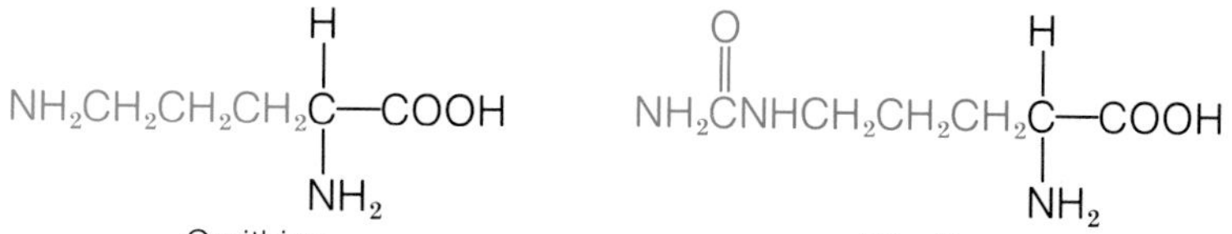

$$NH_2CH_2CH_2CH_2\overset{\overset{\displaystyle H}{|}}{\underset{\underset{\displaystyle NH_2}{|}}{C}}{-}COOH$$

Ornithine

$$NH_2\overset{\overset{\displaystyle O}{\|}}{C}NHCH_2CH_2CH_2\overset{\overset{\displaystyle H}{|}}{\underset{\underset{\displaystyle NH_2}{|}}{C}}{-}COOH$$

Citrulline

15.3 The L-Family*

All amino acids found in proteins, with the exception of glycine, are optically active and belong to the L-family. D-Family isomers of amino acids can be found in nature, but never occur in proteins.

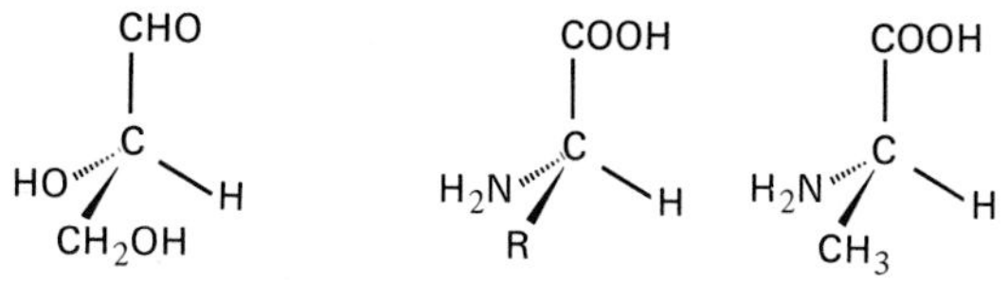

L-Glyceraldehyde   L-Amino acid   L-Alanine

15.4 Acid Base Properties

Since proteins can be viewed, in many cases, as very large amino acids, a knowledge of the acid-base properties of amino acids can be of great help in understanding some of the properties of proteins. Amino acids contain both an acid group (the carboxyl group) and a basic group (the amino group). In water, amino acids can act either as acids or bases; molecules having this property are called **amphoteric.**

Amino acids are soluble in water, and have very high melting points. This suggests that they do not exist as uncharged molecules, but rather are found in the form of the highly polar **zwitterion** or dipolar ion.

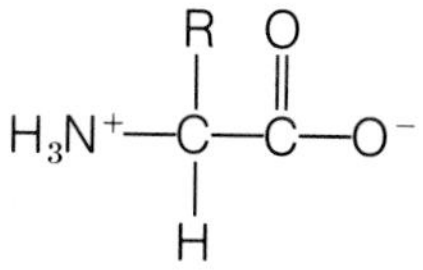

$$H_3N^+{-}\overset{\overset{\displaystyle R}{|}}{\underset{\underset{\displaystyle H}{|}}{C}}{-}\overset{\overset{\displaystyle O}{\|}}{C}{-}O^-$$

Zwitterion or dipolar ion

The zwitterion is formed when the acidic carboxyl group donates a hydrogen ion to the basic amine group. Although we may write the structure of the amino acid in the uncharged form, keep in mind that it will usually be found as a dipolar ion.

---

* Section 15.3, The L-Family, is optional and may be omitted without loss of continuity.

Figure 15.2 The 20 amino acids commonly found in proteins.

NONPOLAR R-GROUPS (HYDROPHOBIC)

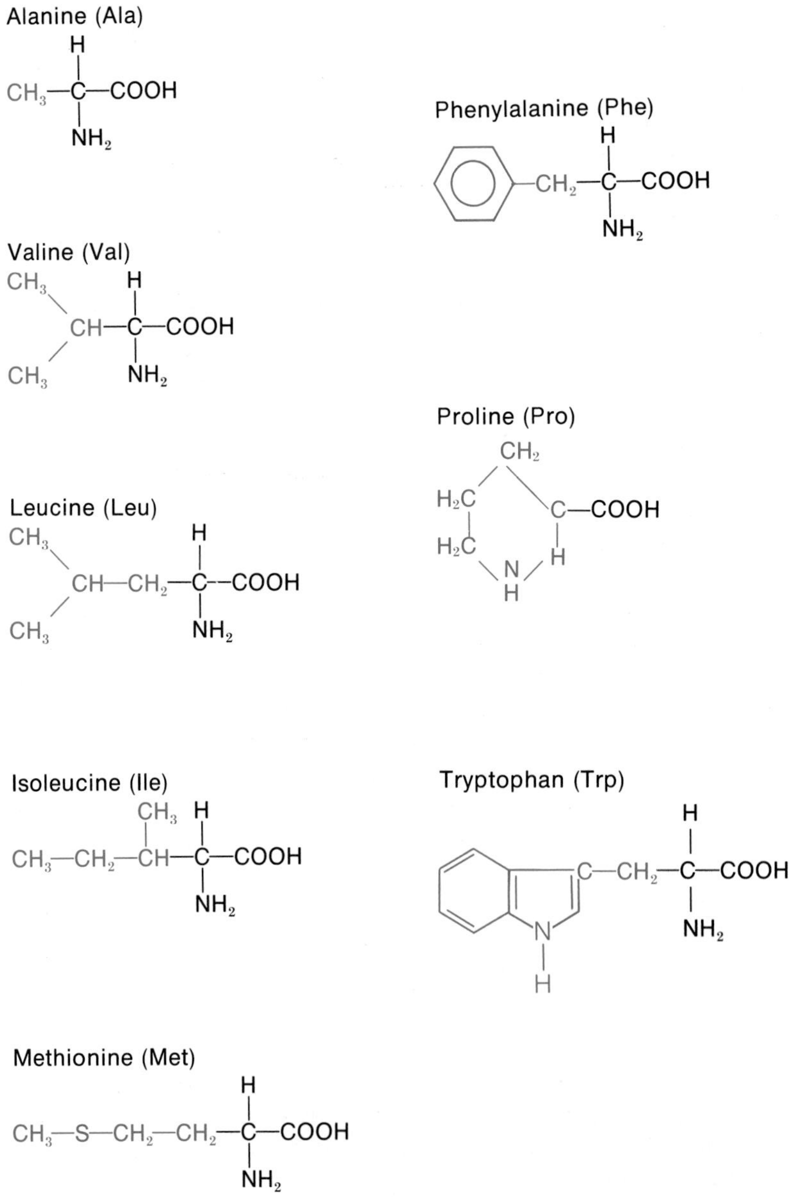

Alanine (Ala)

$$CH_3-\underset{\underset{NH_2}{|}}{\overset{\overset{H}{|}}{C}}-COOH$$

Phenylalanine (Phe)

Valine (Val)

Proline (Pro)

Leucine (Leu)

Isoleucine (Ile)

Tryptophan (Trp)

Methionine (Met)

$$CH_3-S-CH_2-CH_2-\underset{\underset{NH_2}{|}}{\overset{\overset{H}{|}}{C}}-COOH$$

Figure 15.2 The 20 amino acids commonly found in proteins. (cont).

POLAR R-GROUPS (HYDROPHILIC)

Glycine (Gly)

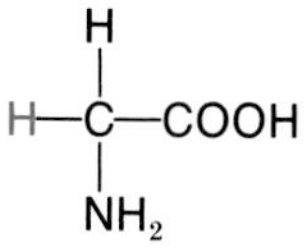

$$H-\underset{\underset{NH_2}{|}}{\overset{\overset{H}{|}}{C}}-COOH$$

Serine (Ser)

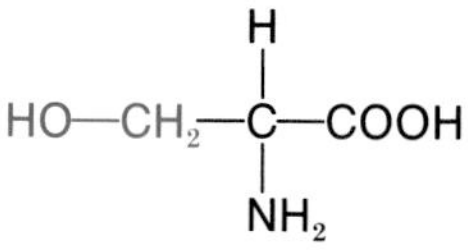

$$HO-CH_2-\underset{\underset{NH_2}{|}}{\overset{\overset{H}{|}}{C}}-COOH$$

Threonine (Thr)

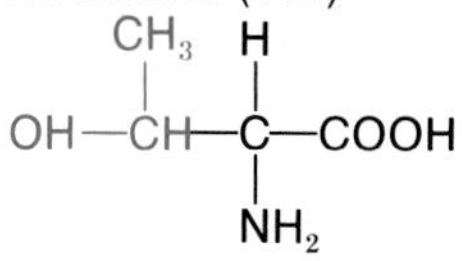

$$OH-\underset{}{\overset{CH_3}{\underset{}{CH}}}-\underset{\underset{NH_2}{|}}{\overset{\overset{H}{|}}{C}}-COOH$$

Tyrosine (Tyr)

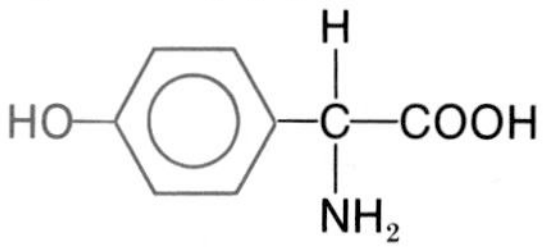

$$HO-\text{⟨benzene ring⟩}-\underset{\underset{NH_2}{|}}{\overset{\overset{H}{|}}{C}}-COOH$$

Cysteine (Cys)

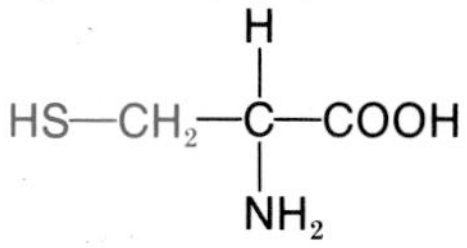

$$HS-CH_2-\underset{\underset{NH_2}{|}}{\overset{\overset{H}{|}}{C}}-COOH$$

Asparagine (Asn)

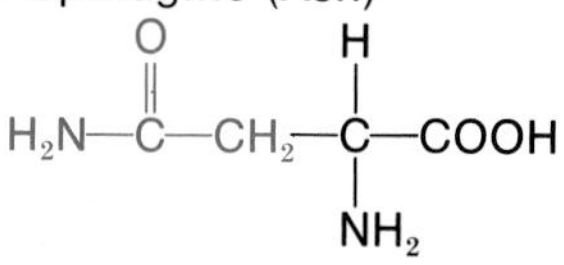

$$H_2N-\overset{\overset{O}{\|}}{C}-CH_2-\underset{\underset{NH_2}{|}}{\overset{\overset{H}{|}}{C}}-COOH$$

Glutamic acid (Glu)

$$HO-\overset{\overset{O}{\|}}{C}-CH_2-CH_2-\underset{\underset{NH_2}{|}}{\overset{\overset{H}{|}}{C}}-COOH$$

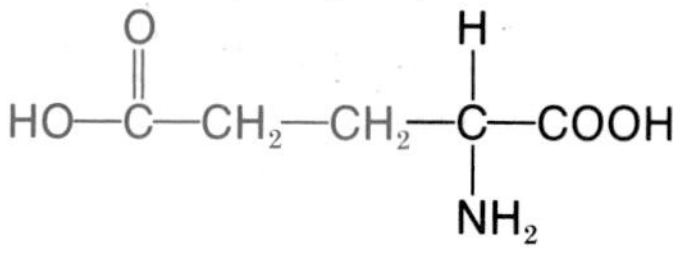

Basic R-Group

Lysine (Lys)

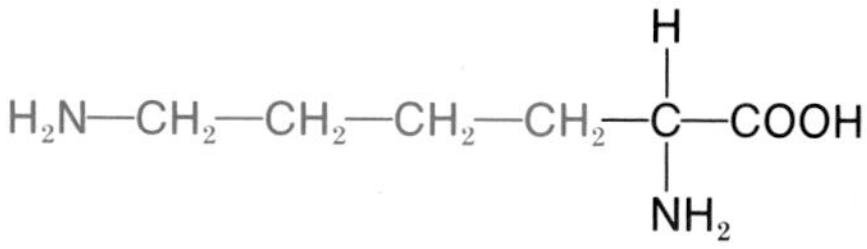

$$H_2N-CH_2-CH_2-CH_2-CH_2-\underset{\underset{NH_2}{|}}{\overset{\overset{H}{|}}{C}}-COOH$$

Arginine (Arg)

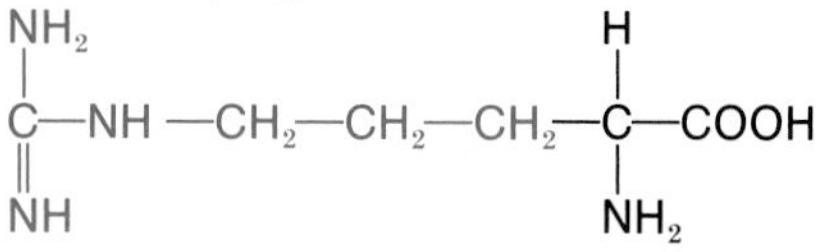

$$\underset{\underset{NH}{\|}}{\overset{\overset{NH_2}{|}}{C}}-NH-CH_2-CH_2-CH_2-\underset{\underset{NH_2}{|}}{\overset{\overset{H}{|}}{C}}-COOH$$

Histidine (His)

$$HC=\!\!=C-CH_2-\underset{\underset{NH_2}{|}}{\overset{\overset{H}{|}}{C}}-COOH$$

(imidazole ring: N, NH, C, H)

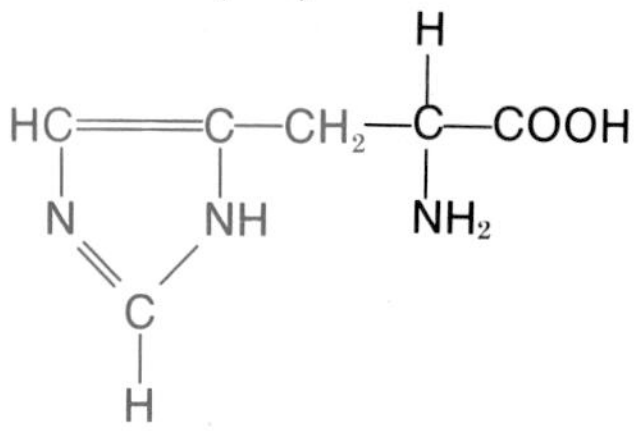

Acidic R-Group

Aspartic acid (Asp)

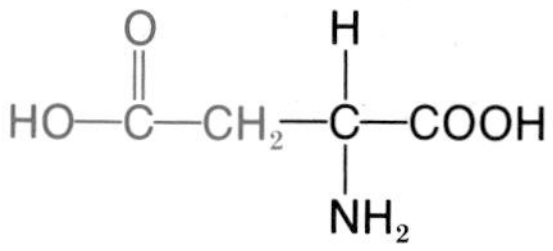

$$HO-\overset{\overset{O}{\|}}{C}-CH_2-\underset{\underset{NH_2}{|}}{\overset{\overset{H}{|}}{C}}-COOH$$

Glutamine (Gln)

$$H_2N-\overset{\overset{O}{\|}}{C}-CH_2-CH_2-\underset{\underset{NH_2}{|}}{\overset{\overset{H}{|}}{C}}-COOH$$

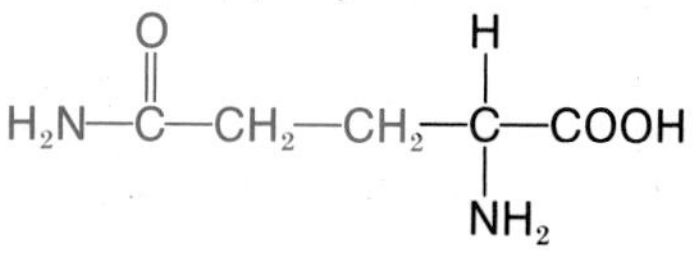

15.5 Isoelectric Point

There is a specific pH at which each amino acid and protein will be electrically neutral, and will not migrate in an electric field. This pH is called the **isoelectric point** for that molecule. Some amino acids have an ionizable R-group (Figure 15.2), and their isoelectric points will vary (Table 15.3).

Table 15.3 Isoelectric Points of Some Amino Acids and Proteins

Compound	Isoelectric Point
Amino acid	
Glutamic acid	4.0
Alanine	6.0
Lysine	10.5
Protein	
Egg albumin	4.6
Urease	5.0
Hemoglobin	6.8
Myoglobin	7.0
Chymotrypsin	9.5
Lysozyme	11.0

At a pH above the isoelectric point, the amino acid will have a net negative charge and will migrate toward the positive electrode. At a pH below the isoelectric point, the amino acid will carry a net positive charge and will migrate toward the negative electrode. Each protein has a characteristic isoelectric point. At this particular pH, the protein will exhibit minimum solubility; the protein molecules can cluster together and are most easily removed from solution. Casein, for example, is the protein found in cow's milk, and has an isoelectric point of pH 4.7. The normal pH of cow's milk is 6.3. However, in the production of cheese, bacteria produce lactic acid, which lowers the pH of the milk. This, then, lowers the solubility of the casein, causing the milk to curdle.

15.6 Buffering Properties

Since amino acids (and proteins) can act as either acids or bases, they are effective buffers in aqueous solution.

$$H_2N-\underset{\underset{H}{|}}{\overset{\overset{R}{|}}{C}}-\overset{\overset{O}{\|}}{C}-O^- \xleftarrow{+OH^-} H_2N-\underset{\underset{H}{|}}{\overset{\overset{R}{|}}{C}}-\overset{\overset{O}{\|}}{C}-OH \xrightarrow{+H^+} H_3N^+-\underset{\underset{H}{|}}{\overset{\overset{R}{|}}{C}}-\overset{\overset{O}{\|}}{C}-OH$$

One of the functions served by the proteins in the blood is to act as buffers, helping to keep the blood pH within its very narrow normal range (pH = 7.35 to 7.45).

Protein Structure

Proteins are complex molecules that can be classified according to specific primary, secondary, tertiary, and quaternary structural characteristics.

15.7 Primary Structure: The Amino Acid Sequence and the Peptide Bond

The **primary structure** of a protein is given by the sequence of amino acids in the protein molecule. These amino acids are coupled together by covalent bonds called **peptide bonds.** The peptide bond is an amide linkage, formed by joining the carboxyl group of one amino acid to the amino group of a second amino acid through elimination of water (a condensation reaction) (Figure 15.4).

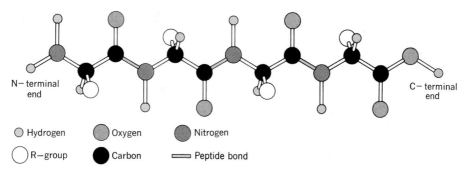

Amino acid₁   Amino acid₂   Peptide bond

N–terminal end   C–terminal end

○ Hydrogen   ◐ Oxygen   ● Nitrogen

○ R–group   ● Carbon   ▭ Peptide bond

Figure 15.4 Primary structure of proteins. This polypeptide chain contains four amino acids joined by peptide bonds. The end of the molecule having the free amino group is the N-terminal end, and the end of the molecule with the free carboxyl group is the C-terminal end.

The peptide bond is stable in the face of changes in pH, in solvents, or in salt concentrations; it can be broken only by acid or base hydrolysis, or by specific enzymes. Two amino acids held together by a peptide bond are called a **dipeptide;** three amino acids form a **tripeptide,** and more than three form a **polypeptide.** There is no precise division between polypeptides and proteins. For example, insulin, with 51 amino acids in its primary structure, is a very small protein, whereas glucagon, with 21 amino

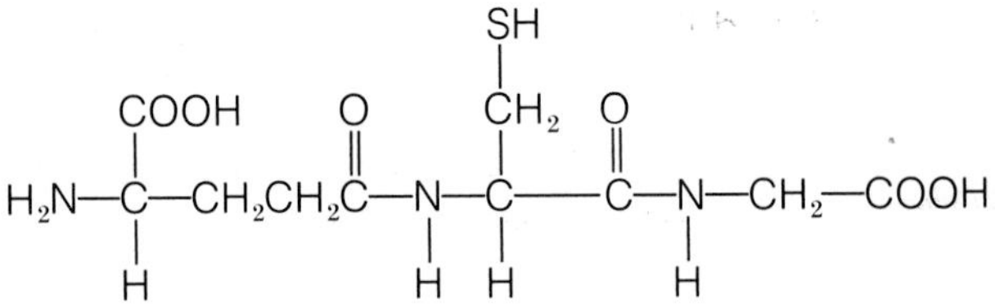

Glutathione

Figure 15.5 The structure of the tripeptide glutathione.

acids, is considered a large polypeptide. There are many small polypeptides that have important functions in biological systems. Glutathione, which plays a role in oxidation-reduction reactions, is a tripeptide; vasopressin, a hormone produced by the pituitary, is an octapeptide; and the antibiotic bacitracin is a polypeptide with 11 amino acids in its sequence (Figure 15.5).

Two different dipeptides can be formed between the amino acids glycine and alanine.

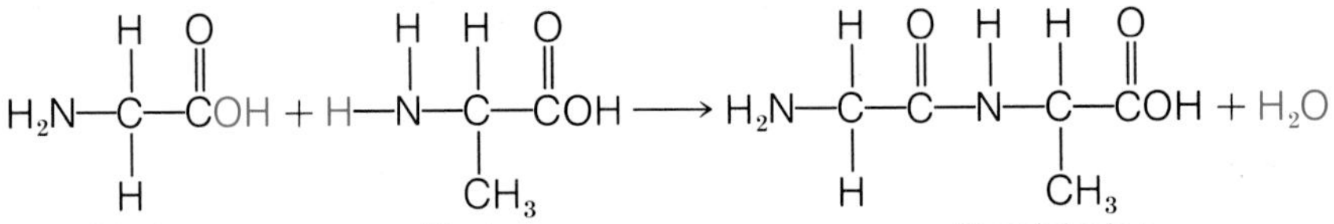

Glycine (Gly) — Alanine (Ala) — Glycylalanine (Gly-Ala)

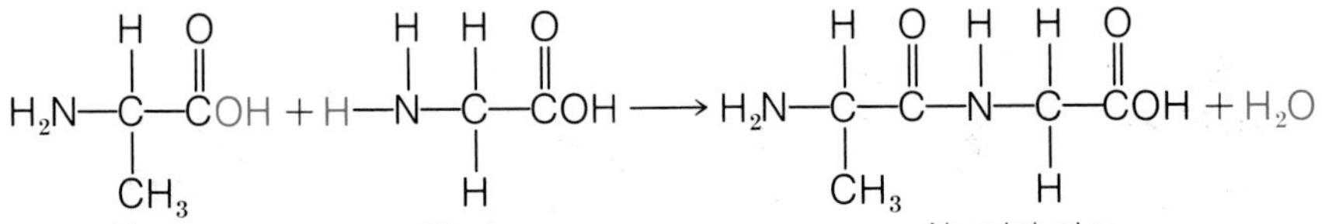

Alanine (Ala) — Glycine (Gly) — Alanylglycine (Ala-Gly)

There are some standard conventions for the naming of peptides and proteins. Since proteins contain very large numbers of amino acids, three-letter abbreviations of the amino acid names are used when writing the amino acid sequence. (See Figure 15.2 for the abbreviations.) The peptide bond is represented by a dash or dot between the amino acid names. One end of the protein will consist of an amino acid having a free amino group. This is called the N-terminal end of the protein, and the N-terminal amino acid is listed first in the sequence of amino acids. At the other end of the protein there will be an amino acid having a free carboxyl group; this is the

C-terminal amino acid, and it is the last amino acid listed in the sequence. The following are two examples of this naming convention:

Lysine–Aspartic acid–Serine–Asparagine–Glutamic acid   (I)
Lys–Asp–Ser–Asn–Glu

Valine · Phenylalanine · Alanine · Tryptophan · Leucine   (II)
Val · Phe · Ala · Trp · Leu

N-terminal end   C-terminal end

The order of amino acids in a protein will determine its function, and is critical to its biological activity. A change of just one amino acid in the sequence can disrupt the entire protein molecule. For example, hemoglobin, the molecule in the blood that carries oxygen, consists of four polypeptide chains having a total of 574 amino acid units. Changing just one specific amino acid in one of the chains results in the defective hemoglobin molecule found in patients having sickle cell anemia.

β-chain

Adult Hemoglobin   Val–His–Leu–Thr–Pro–Glu–Glu–Lys– . . .
(Hb-A)

Sickle Cell Hemoglobin   Val–His–Leu–Thr–Pro–Val–Glu–Lys– . . .
(Hb-S)

Determination of the amino acid sequence in a protein is a complex procedure that was first developed by F. Sanger in 1953. He determined the amino acid sequence of the protein insulin (Figure 15.6).

15.8 Secondary Structure: Noncovalent Bonding

The **secondary structure** of a protein is the specific geometric arrangement of the amino acids. These configurations, resulting from hydrogen bonding, were established by Linus Pauling using X-ray diffraction (Figure 15.7).

α-Helix

Pauling determined that the polypeptide chains in the protein keratin were curled in an arrangement called an α-helix. In this configuration, the amino acids form loops in which the hydrogen on the nitrogen atom in the peptide bond is hydrogen bonded to the oxygen attached to the carbon atom of a peptide bond farther down the chain (Figure 15.8). There are 3.6 amino acids in each turn of the α-helix, and the R-groups on these amino acids extend outward from the helix. Hair and wool are made up of several coils of keratin wound around each other and held together by disulfide bridges.

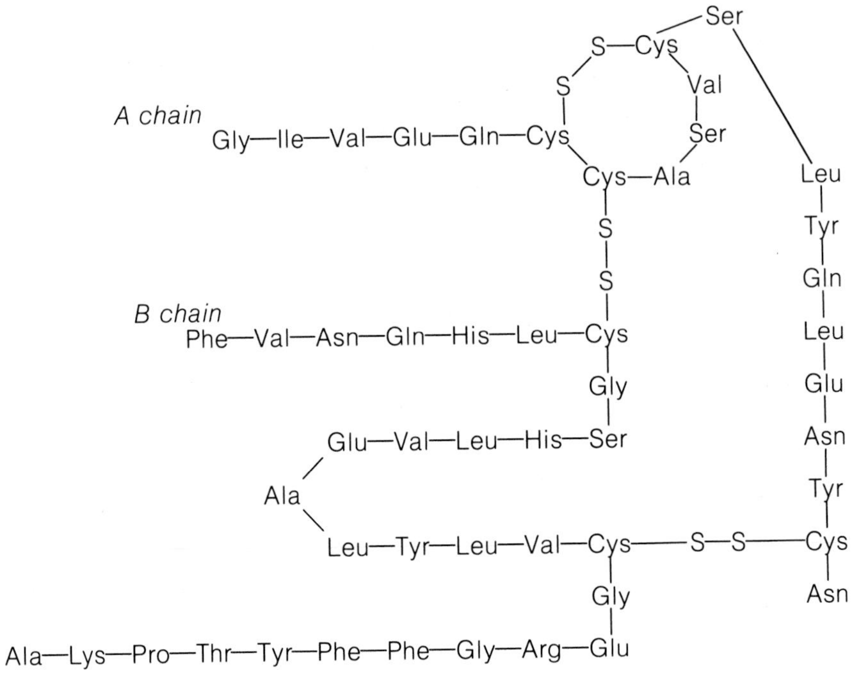

Figure 15.6 The complete amino acid sequence of bovine insulin was first determined by F. Sanger in 1953, for which he was awarded the Nobel prize. The insulin molecule contains 51 amino acids in two polypeptide chains.

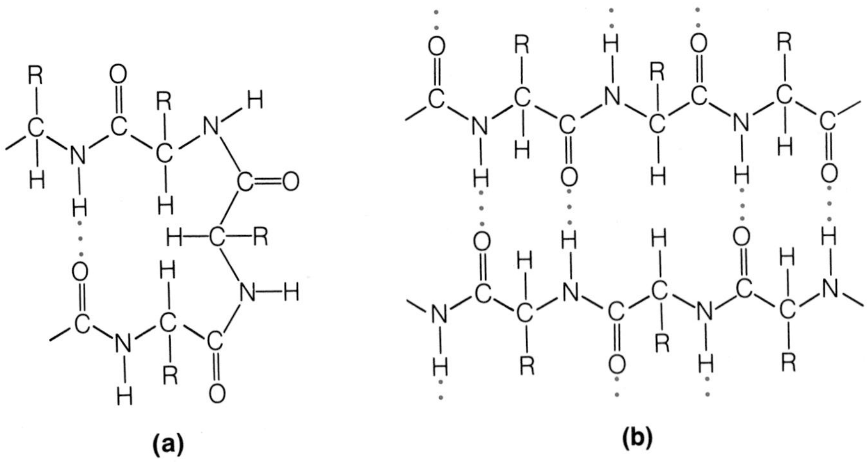

(a)　　　　　　　　　　(b)

Figure 15.7 Hydrogen bonds can form (a) between amino acids on the same polypeptide chain, forming a loop in the molecule, or (b) between amino acids on different polypeptide chains.

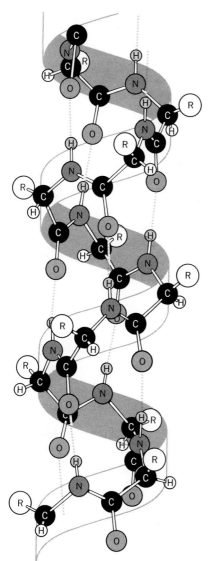

Figure 15.8 Model of a polypeptide chain in an alpha-helix configuration. The amino acids are coiled in a manner resembling a circular staircase, with the loops held together by hydrogen bonding (dotted lines). Note that the R-groups on the amino acids point away from the center of the helix. (Adapted from B. Low and E. T. Edsall, *Currents in Biochemical Research,* Wiley-Interscience, New York, 1956. Used by permission.)

β-Configuration, or Pleated Sheet

The fibrous protein of silk has a different secondary structure. In silk, several polypeptide chains going in different directions are located next to each other; this gives the protein a zigzag appearance, hence the name pleated sheet (Figure 15.9). The chains are held together by hydrogen bonds, and the R-groups extend above and below the sheet.

Each protein will have a specific secondary structure depending upon the amino acid sequence. For example, the α-helix is formed when the R-groups are small and uncharged. However, this configuration will be disrupted by the amino acid proline, which has no amide hydrogen and therefore can't form hydrogen bonds. Kinks or bends in the molecule are found at proline sites. It is possible in a large protein to have separate

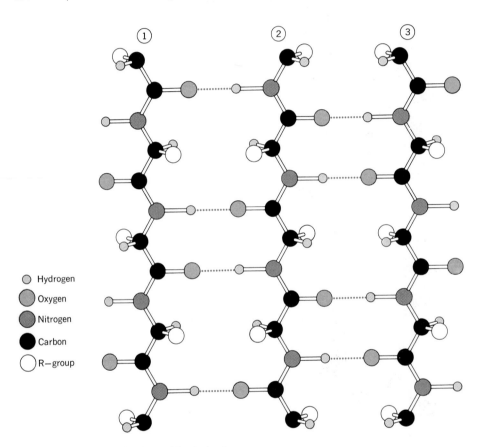

○ Hydrogen
◐ Oxygen
● Nitrogen
● Carbon
○ R—group

Figure 15.9 Model of polypeptide chains in a pleated sheet structure. In this configuration several polypeptide chains are held together by hydrogen bonds. The R-groups extend above and below the plane of the sheet.

areas of α-helix and pleated sheet configuration (Figure 15.10). Note that hydrogen bonding is weak noncovalent bonding, and is easily disrupted by changes in pH, temperature, solvents, or salt concentrations.

15.9 Tertiary Structure: Globular Proteins

Tertiary structure is the three-dimensional structure of globular proteins. In general, globular proteins are very tightly folded into a compact spherical form. This folding results from interactions between the R-group side chains of amino acids, and may involve hydrogen bonding and the following interactions.

Disulfide Bridges

Disulfide bridges are interactions that include disulfide linkages between molecules of the amino acid cysteine. Sulfhydryl groups (—S—H) are easily oxidized to form a disulfide (—S—S—). If this reaction occurs between two cysteines, the amino acid cystine is formed. (Unfortunately, cysteine and cystine have identical pronunciations, although many scientists go to great lengths to pronounce them differently.)

Cysteine   Cysteine   Cystine

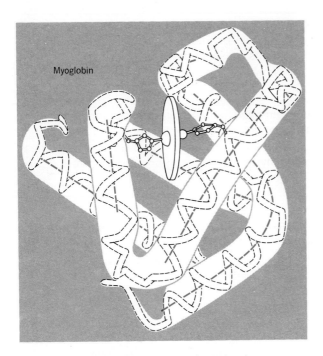

Figure 15.10 The tertiary structure of a molecule of myoglobin consists of eight sections of alpha-helix surrounding a heme group. The main polypeptide chain is shown here without the R-group side-chains. (From R. E. Dickerson and I. Geis, *The Structure and Action of Proteins*, W. A. Benjamin, Inc., Menlo Park. Copyright 1969 by Dickerson and Geis.)

This disulfide linkage can be formed between two cysteines on different amino acid chains, thereby linking the two chains together to form such tough, strong material as keratin, or can be found between cysteines on the same amino acid chain, creating a loop in the chain.

$$
\begin{array}{l}
\text{Gly—Arg—Pro—Cys—Asn} \\
\qquad\qquad\quad\; | \qquad\quad \backslash \\
\qquad\qquad\quad\; \text{S} \qquad\quad \text{Gln} \\
\text{Vasopressin} \qquad\; | \qquad\qquad | \\
\qquad\qquad\quad\; \text{S} \qquad\quad \text{Phe} \\
\qquad\qquad\quad\; | \qquad\quad / \\
\qquad\qquad \text{Cys—Tyr}
\end{array}
$$

These bridges are covalent linkages that can be ruptured by reduction, but that are stable in the face of changes in pH, solvents, or salt concentrations. The insulin molecule contains examples of both types of disulfide bridges (Figure 15.6).

The fact that cysteine reacts very easily to form cystine may play an important role in protecting the human body from the effects of free radicals. Free radicals can be produced by ionizing radiation or can be formed from compounds such as caffeine or sodium nitrite, a preservative used in certain meat products. Because of the abundant supply of sulfur-containing compounds such as cysteine, the free radicals are likely to be destroyed by reactions with these compounds rather than in interactions with biologically critical molecules. This theory is supported by the finding that mice injected with cysteine just before exposure to radiation are less likely to develop radiation sickness than mice receiving the same radiation dose without prior injection of cysteine.

Salt Bridges

Salt bridges result from ionic interactions between charged carboxyl or amino side chains found on amino acids such as aspartic acid, glutamic acid, lysine, and arginine. (See Figure 15.2.) These linkages are particularly vulnerable to changes in pH.

Hydrophobic Interactions

Hydrophobic interactions occur between the nonpolar side chains of the amino acids in the protein molecule. Since they are repulsed by the solvent water, these groups tend to be found on the inside of the molecule along with other hydrophobic groups.

At normal pH and temperature, each protein will assume a shape that is energetically the most stable given the specific sequence of amino acids and the various types of interactions we have mentioned. This shape is called the **native state** or native configuration of the protein.

15.10 Quaternary Structure: Two or More Polypeptide Chains

The **quaternary structure** of a protein is the manner in which separate polypeptide chains fit together in those proteins containing more than one chain. Hydrogen bonding, hydrophobic interactions, and salt bridges may be involved in holding the chains in position. Hemoglobin is just one example of a protein containing more than one polypeptide chain (Figure 15.11). It consists of four polypeptide chains—two alpha chains and two beta chains—arranged around an iron-containing heme group.

15.11 Denaturation

Various changes in the environment of a protein can disrupt the complex secondary, tertiary, or quaternary structure of the molecule. Disruption of the native state of the protein is called **denaturation,** and will cause the protein to lose its biological activity. Denaturation may or may not be permanent; in some cases the protein will return to its native state when the denaturing agent is removed. Denaturation may result in coagulation, with the protein being precipitated from solution. The process of denaturation

Figure 15.11 This model of hemoglobin shows the quaternary arrangement of the four polypeptide chains. The two alpha chains are represented by the light blocks, and the two beta chains by the dark blocks. The iron-containing heme groups that bind oxygen to the molecule are represented by the disks. (Courtesy M. F. Perutz, Medical Research Council, Laboratory of Molecular Biology.)

involves the uncoiling of the protein molecule into a random state (Figure 15.12). Denaturing agents come in many forms, a few of which are as follows:

pH

Changes in pH have their greatest disruptive effect on hydrogen bonding and salt bridges. The polypeptide polylysine is composed entirely of the amino acid lysine, which has an amino group on its side chain. In acidic pH the side chains will all be positively charged and will repel each other, causing the molecule to uncoil. In basic pH, however, the side chains will be neutral. Therefore, they do not repel, and the molecule will coil into an α-helix.

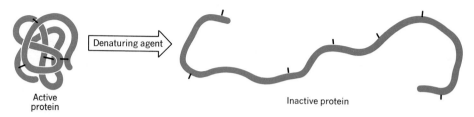

Figure 15.12 Denaturation of a protein disrupts the tertiary structure of the molecule, causing it to uncoil. This results in the loss of protein activity. Denaturation can be either temporary or permanent depending upon the denaturing agents used.

Heat

Heat causes an increase in thermal agitation of the molecule, disrupting hydrogen bonding and salt bridges. After gentle heating the protein can usually regain its native state, but violent heating will result in irreversible denaturation and coagulation of the protein (as, for example, in cooking an egg). Similarly, heat used to sterilize equipment coagulates the protein of microorganisms. Heat can also be used to detect the presence of protein in urine; heated urine that turns cloudy indicates the presence of protein.

Organic Solvents

Organic solvents such as alcohol or acetone are capable of forming hydrogen bonds themselves, which then compete with the hydrogen bonds naturally occurring in the protein. This causes denaturation and coagulation. A 70% alcohol solution is a good disinfectant since it will coagulate the protein in bacteria, thereby destroying these organisms.

Heavy Metal Ions

Heavy metal ions such as Pb^{2+}, Hg^{2+}, and Ag^+ may disrupt the natural salt bridges of the protein by forming salt bridges of their own with the protein molecule. This usually causes coagulation of the protein. The heavy metal ions also bind with sulfhydryl groups, disrupting the disulfide linkages in the protein molecule and denaturing the protein. Such heavy metals, therefore, are toxic to living organisms. As an antidote to heavy metal poisoning, substances high in protein, such as milk or egg whites, are given. These substances will bond with the metals while they are still in the stomach. The victim must then be made to vomit before the metals are again released by the process of digestion.

Alkaloidal Reagents

Reagents such as tannic or picric acid affect salt bridges and hydrogen bonding, causing proteins to precipitate. Tannic acid is used to precipitate proteins in animal hides; this is the process of tanning used in the manufacture of leather.

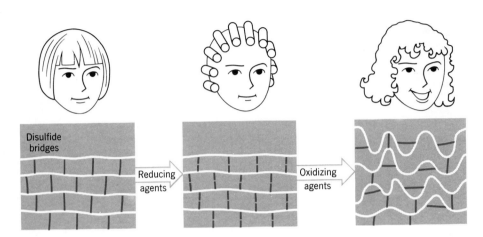

Figure 15.13 Permanent waves work by breaking the disulfide bridges between the cysteines in hair protein molecules, and then reforming the bridges in different locations.

Reducing Agents

Reducing agents disrupt disulfide bridges formed between cysteine molecules. For example, hair permanents work by reducing and disrupting the disulfide bridges in hair; when the hair is curled and oxidizing agents are applied, new disulfide bridges are formed (Figure 15.13).

Radiation

Radiation is similar to heat in its effect on proteins. For example, ultraviolet light will burn the skin, causing sunburn.

15.12 Identification of Proteins and Amino Acid Sequences

Many color tests and separation procedures are used in protein research. Two commonly used color tests are the biuret test and ninhydrin test.

Biuret Test

The biuret test detects the presence of two or more peptide linkages. The unknown solution is mixed with sodium hydroxide and a few drops of dilute copper sulfate. If a violet color appears, peptides larger than dipeptides are present. This color test is often used to follow the progress of protein hydrolysis.

Ninhydrin Test

The ninhydrin test is widely used to detect the presence of amino acids, peptides, and proteins. The sample solution is heated with excess ninhydrin, and a blue color indicates a positive test (except in the case of the amino acid proline, which gives a yellow color).

15.13 Separation of Proteins

Proteins and amino acids can be separated and identified using a variety of techniques. The process of fractional precipitation is based on the differences in protein solubility in different salt solutions at various pH levels. An ultracentrifuge can be used to separate proteins by their molecular weights. The technique of paper chromatography makes use of the differences in protein and amino acid solubility in various solvents.

Column chromatography, a technique widely used in biochemical research, can separate proteins or amino acids on the basis of electric charges or size of the molecule. In column chromatography, mixtures of amino acids or proteins are passed through a column of starch, cellulose powder, or ion exchange resins. The solutions that come off the column are collected in very small volumes, which are analyzed for their contents. If the conditions are varied, the amino acids or proteins can be made to come off the column one at a time (Figure 15.14).

The process of electrophoresis separates proteins and amino acids by means of their specific charge at the pH of the solution. In this technique, each protein will migrate toward the positive or negative electrode at a different rate depending upon the pH and voltage applied. Paper electrophoresis is a useful tool in analyzing the proteins in human blood serum (Figure 15.15).

Protein Metabolism

15.14 Digestion and Absorption

The digestion of proteins begins in the stomach. When food is swallowed, the stomach lining produces hydrochloric acid and inactive forms of protein-digesting enzymes called zymogens. The zymogens travel from the stomach lining to the stomach, where they are activated by HCl. This process prevents the enzyme from digesting the proteins of the stomach lining as it is traveling through the lining. Formation of peptic ulcers may be due to an increased production of hydrochloric acid or a decrease in the amount of protective mucous of the stomach lining. Protein digestion continues in the small intestines. The enzymes of the small intestines whose job it is to digest proteins are shown in Table 15.4 on page 453. At this time, the amino acids produced by digestion and some very small polypeptides are absorbed through the intestinal wall, and enter the bloodstream.

Amino acids have many uses in the body. They are used by cells to synthesize tissue protein for use in the formation of new cells or the repair of old cells. They may also be used by cells to synthesize enzymes, to form nonprotein nitrogen-containing compounds such as nucleic acids or heme groups, or to form new amino acids. The amino acids not used in such

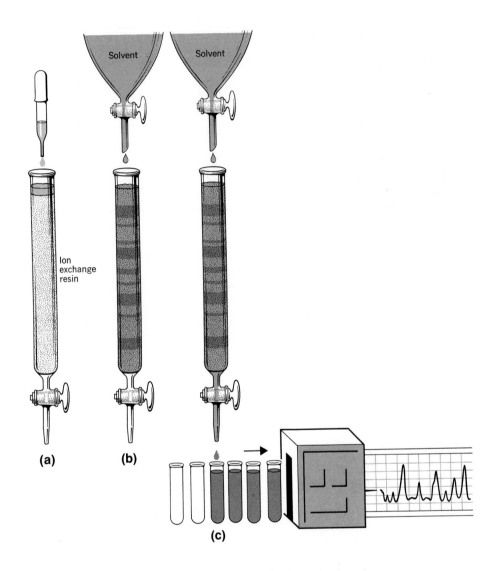

Figure 15.14 Separation of a mixture of amino acids or proteins by column chromatography. (a) The mixture to be separated is placed on the top of the column. (b) One or more solvents are passed through the column. (c) The effluent is collected in small volumes which are then analyzed for their contents.

synthesis can be catabolized, or broken down, for energy. The amine group of the amino acid will be removed in a process called oxidative deamination, and will enter the urea cycle (which we will discuss shortly). The remainder of the molecule can enter the citric acid cycle or be converted to glucose and stored as glycogen.

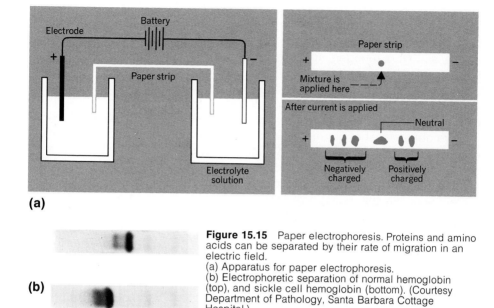

(a)

(b)

Figure 15.15 Paper electrophoresis. Proteins and amino acids can be separated by their rate of migration in an electric field.
(a) Apparatus for paper electrophoresis.
(b) Electrophoretic separation of normal hemoglobin (top), and sickle cell hemoglobin (bottom). (Courtesy Department of Pathology, Santa Barbara Cottage Hospital.)

Table 15.4 The Enzymes of Protein Digestion

Site	Enzyme	Action
Mouth	None	
Stomach	Rennin (infants)	Coagulates milk proteins
	Pepsin	Proteins ⟶ polypeptides
Small intestine		
Pancreatic juice	Trypsin, chymotrypsin, and carboxypeptidase	Polypeptides ⟶ dipeptides and amino acids
Intestinal juice	Erepsin	Dipeptides ⟶ amino acids

15.15 Sources of Nitrogen

We mentioned at the beginning of this chapter that proteins are our main dietary source of nitrogen, an element essential to life. The atmosphere is 80% nitrogen, but even though we inhale this molecule with every breath, we have no capacity to use elemental nitrogen. We are totally dependent upon microorganisms in the soil or in the roots of leguminous plants to fix nitrogen into forms that plants can use to form amino acids (Figure 15.16).

We obtain most of our nitrogen from the plant and animal protein that we eat. There is no storage form of amino acids in the body, as there is for carbohydrates (in the form of glycogen) and lipids (in the form of fat). Instead, there is a pool of amino acids that is constantly changing; tissue protein is continually being broken down and resynthesized (Figure 15.17). The turnover of amino acids in the liver and blood is relatively rapid—half will be replaced every six days. This turnover is much slower in muscles and supportive tissue—half of the amino acids are replaced every 180 days in muscles, and every 1000 days in the collagen of supportive tissue.

A healthy adult maintains a nitrogen balance within his body; he will excrete as much nitrogen as he takes in in his diet. Rapidly growing young children will be in a state of positive nitrogen balance. They will not excrete as much nitrogen as they take in, since they need the amino acids to synthesize new tissues. A negative nitrogen balance, in which more nitrogen is excreted than absorbed, occurs in starvation, malnutrition, wasting diseases, and fevers.

Figure 15.16 The nitrogen cycle.

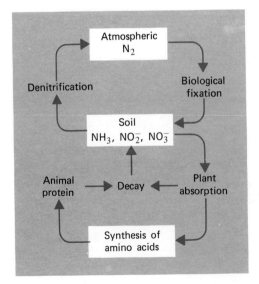

Synthesis of Proteins

In order to maintain a proper nitrogen balance, we need to eat an adequate supply of protein. Although the needs of different people vary considerably, the Recommended Daily Allowance (which exceeds the requirements of most individuals) is 0.8 grams of protein per kilogram of body weight. For the average adult this amounts to 55 to 70 grams of protein, or $\frac{1}{4}$ to $\frac{1}{3}$ pound of protein-rich food, each day. However, not only do we need to eat enough protein, but we must also be careful that it is the right type of protein.

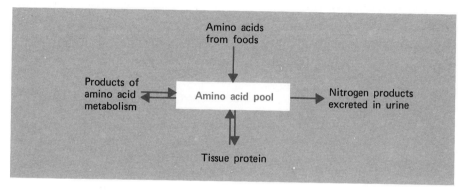

Figure 15.17 The amino acid pool in the human body is constantly changing.

15.16 Essential Amino Acids

There is strong evidence that man cannot synthesize eight amino acids. These amino acids are called the essential amino acids, and must be absorbed in the diet (Table 15.5).

Table 15.5   Essential Amino Acids*

Isoleucine	Phenylalanine
Leucine	Threonine
Lysine	Tryptophan
Methionine	Valine

* These amino acids cannot be synthesized by the human body or are not produced in sufficient quantities to meet the body's requirements for them; therefore, they must be supplied by foods in the diet.

Proteins that contain these amino acids are known as **adequate proteins.** Animal protein and milk are adequate proteins, but many vegetable proteins lack one or more of the essential amino acids and, therefore, are inadequate proteins. The protein in soybeans is adequate, but the protein in corn is too low in lysine and tryptophan to support growth in young children. Rice is low in lysine and threonine, and wheat is low in lysine. In parts of the world where plant protein is the major source of protein, deficiency diseases such as Kwashiorkor result. It is clear, then, that a major problem in our overpopulated world is to increase the supplies of adequate protein.

15.17 Transamination

Amino acids can be synthesized by the body in a process called transamination. Transamination involves the transfer of an amino group from one carbon to another. The donor of the amino group is an amino acid, and the receiving molecule is an α-keto acid.

| Alanine (Amino acid) | α-Ketoglutaric acid (α-Keto acid) | Pyruvic acid | Glutamic acid (New amino acid) |

15.18 Specific Syntheses

Amino acids play a role in the synthesis of many metabolic products. The following are just a few examples: Tyrosine is used to produce the hormones epinephrine, norepinephrine, and thyroxin, as well as the skin pigment melanin. Tryptophan is used in the synthesis of the nerve transmission chemical serotonin and the coenzymes NAD and NADP. Serine is converted to ethanolamine, which is found in lipid cephalins, and cysteine is used in the synthesis of bile salts.

Amino Acid Catabolism

Amino acids that are not used in synthesis are converted in the liver to ammonia, carbon dioxide, water, and energy.

15.19 Oxidative Deamination

Oxidative deamination is a process that occurs in the tissues, especially the liver and involves removing the amino group from an amino acid.

The α-keto acids that are formed in oxidative deamination can be used in several ways. They may be oxidized in the citric acid cycle to produce energy, or they may be converted to other amino acids through transamination. These α-keto acids may also be used in the synthesis of carbohydrates and fats.

15.20 Urea Cycle

The ammonia that is formed in oxidative deamination is toxic to cells; a concentration of 5 mg/100 ml of blood is toxic to man. The liver disposes of this ammonia by converting it to urea in a cyclic reaction called the **urea cycle,** or the Krebs ornithine cycle (Figure 15.18). The urea can then be

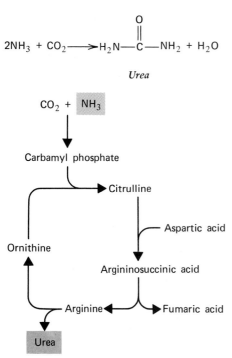

Figure 15.18 The urea cycle. Ammonia, produced in the breakdown of amino acids in the tissues, is converted to the less toxic urea by liver cells in this series of reactions. Ammonia enters the cycle as carbamyl phosphate, and is joined to the amino acid ornithine. In each turn of the cycle, one molecule of urea is produced and a molecule of ornithine is generated to begin the next turn.

safely transported through the body for elimination by the kidneys. Any condition that impairs the elimination of urea by the kidneys can lead to uremia, an accumulation of urea and other nitrogen wastes in the blood, which can be fatal. A person suffering from uremia will feel nauseated, irritable, and drowsy. His blood pressure will be elevated and he may be anemic. He may experience hallucinations, and in serious cases may lapse into convulsions and coma. To reverse this condition, either the cause of kidney failure must be removed or the patient must undergo hemodialysis to remove the nitrogen wastes from his blood.

Figure 15.19 summarizes the metabolic pathways of proteins, lipids, and carbohydrates.

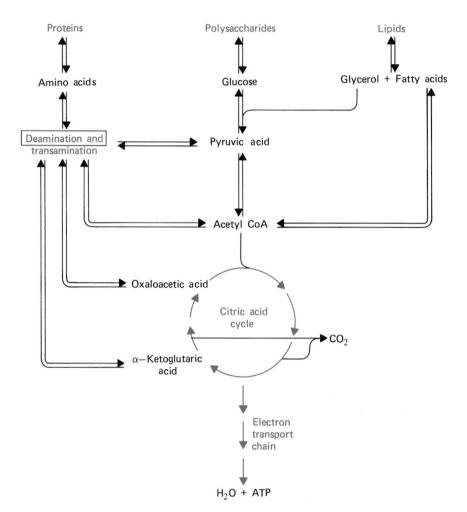

Figure 15.19 The metabolic pathways of proteins, lipids, and carbohydrates are intricately interrelated.

Additional Reading

Books
1. R. E. Dickerson and I. Geis, *The Structure and Action of Proteins,* Harper and Row, New York, 1969.

2. R. A. McCance and E. M. Widdowson, ed., *Calorie Deficiencies and Protein Deficiencies,* Little, Brown and Company, Boston, 1968.

3. D. S. McLaren, *Nutrition and its Disorders,* Churchill Livingstone, Edinburgh, 1972.

Articles
1. C. B. Anfinsen, "Principles that Govern the Folding of Protein Chains," *Science,* July 20, 1973, page 223.

2. R. E. Dickerson, "The Structure and History of an Ancient Protein: Cytochrome C," *Scientific American,* April 1972, page 58.

3. R. D. B. Fraser, "Keratins," *Scientific American,* August 1969, page 87.

4. R. Freedman, "What Makes Proteins Fold?" *New Scientist,* May 31, 1973, page 560.

5. F. M. Lappe, "Proteins from Plants," *Chemistry,* October 1973, page 11.

6. M. F. Perutz, "The Hemoglobin Molecule," *Scientific American,* November 1964, page 64.

7. J. C. Kendrew, "The Three-Dimensional Structure of a Protein Molecule: Myoglobin," *Scientific American,* December 1961, page 96.

8. D. C. Philips, "The Three-Dimensional Structure of an Enzyme Molecule," *Scientific American,* November 1966, page 78.

9. N. Sharon, "Glycoproteins," *Scientific American,* May 1974, page 78.

10. H. C. Trowell, "Kwashiorkor," *Scientific American,* December 1954, page 46.

Questions and Problems

1. Define the following terms:

   (a) Amino acid
   (b) Simple and conjugated protein
   (c) Globular and fibrous protein
   (d) Zwitterion
   (e) Isoelectric point
   (f) Peptide bond
   (g) Primary, secondary, tertiary, and quaternary structure
   (h) Polypeptide
   (i) α-Helix
   (j) β-Configuration
   (k) Disulfide bridge
   (l) Salt bridge
   (m) Hydrophobic interactions
   (n) Native state
   (o) Denaturation
   (p) Essential amino acid
   (q) Adequate protein
   (r) Transamination
   (s) Oxidative deamination
   (t) Urea cycle

2. Why is protein in the urine indicative of kidney disorders?

3. What accounts for the difference between the solubility of globular and fibrous proteins?

4. (a) Write the formula for leucine as it would be found in solution at a pH of 7.
   (b) Write the equation for the addition of acid to the solution in (a).
   (c) Write the equation for the addition of base to the solution in (a).

5. Describe the mechanism by which blood proteins help protect against pH changes.

6. If placed in an electric field, how would each of the following proteins migrate if the pH of the solution was 7?

   (a) Urease
   (b) Myoglobin
   (c) Chymotrypsin

7. What aspect of the structure of amino acids accounts for their high melting points?

8. Write the structure for the two polypeptides described on page 441.

   (a) Which polypeptide, I or II, would be more soluble in water? Explain why.
   (b) In which polypeptide, I or II, would hydrophobic interactions be more likely to occur? Explain why.
   (c) In which polypeptide, I or II, would salt bridges be more likely to occur? Explain why.

9. How do the processes of denaturation and digestion of a protein differ?

10. Why are eggs good antidotes for lead poisoning? Why is the administration of the egg antidote not sufficient to prevent poisoning?

11. Explain, on the molecular level, the process of hair straightening.

12. Describe two methods for separating the proteins in a blood sample.

13. Why can't insulin be administered to diabetics in the form of a pill?

14. (a) Is a diet consisting mainly of rice an adequate diet? Why or why not?
    (b) Suggest several ways in which a diet consisting mainly of corn can be supplemented to provide enough adequate protein.

15. Write the equation for the transamination reaction between valine and pyruvic acid.

16. Write the equation for the deamination of alanine.

17. Describe in general terms the metabolic fate of the ammonia produced in question 16.

18. Describe four possible metabolic uses of the amino acids contained in a hamburger.

chapter 16

Enzymes, Vitamins, and Hormones

Learning Objectives

By the time you have finished this chapter, you should be able to:

1. Identify the various components of an enzyme molecule.

2. Describe the method of enzyme action.

3. Indicate how enzymes differ from inorganic catalysts.

4. Explain the "lock-and-key" and "induced-fit" theories of enzyme activity.

5. Explain how changes in pH and temperature will affect enzyme activity.

6. Explain why vitamins are essential to normal cellular function.

7. Explain the mechanisms by which multienzyme systems are regulated.

8. State the difference between

    (a) Reversible and irreversible inhibition.
    (b) Competitive and noncompetitive inhibition.

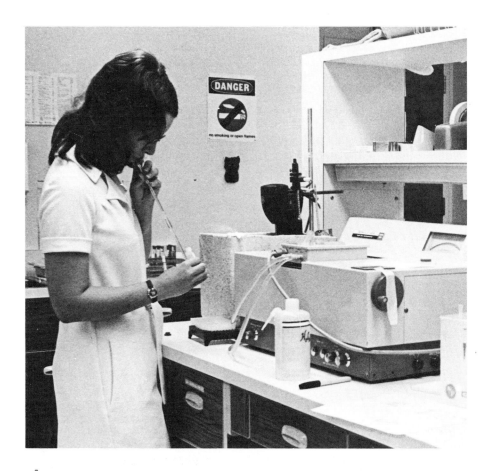

After evening rounds at the hospital, Dr. Feldman was discussing with two medical students some of the patients they had just seen. He was remarking on how far diagnostic techniques had progressed in the last 15 years as a result of increased knowledge of human biochemistry. In particular, he singled out two patients whom they had visited that evening as important illustrations of how enzymes have enabled doctors to design quick and precise diagnostic tests for certain diseases.

The patient in 204A, for example, had been admitted two days earlier suffering from a heart attack. Until recently, Dr. Feldman would have assessed the extent of damage to the patient's heart by the injection of radioactive dyes into the heart. However, it has now been found that certain enzymes are released into the blood by the damaged heart muscle, and that the concentration of these enzymes in the blood gives a direct indication of the extent of damage to the heart muscle. Therefore, on this patient the doctor was able to use a newly developed diagnostic technique requiring only an analysis of a sample of the patient's blood. In particular, one enzyme that he monitored at regular intervals was creatine phosphokinase (CPK), an enzyme found only in heart muscle cells, and not in the surrounding heart tissue. By evaluating the level of CPK in the patient's blood, the doctor had been able to prescribe immediate treatment and to quickly assess the chances for the patient's recovery.

A similar procedure had been used with the patient in 305C. By measuring the blood level of the enzyme glutamic pyruvic transaminase, the doctor had been able to quickly confirm the preliminary diagnosis of infectious hepatitis. Until this test had been developed, the only way to confirm such a diagnosis was by microscopic examination of a sample of the patient's liver tissue. Moreover, since the blood level of this enzyme rises even before any symptoms of the disease appear, Dr. Feldman had arranged to test other members of the patient's family to determine if anyone else had contracted the disease.

As one further example of how knowledge of specific enzyme functions and their levels in body fluids was bringing about new and important diagnostic tests, the doctor mentioned the enzyme acid phosphatase. This enzyme is an important indicator because its level in the blood increases in cases of cancer of the prostate and in certain types of bone diseases. The doctor concluded by speculating that the discovery of such enzyme "flags" was just beginning, and that doctors would immensely benefit from research results yet to come.

Enzymes

16.1 What Are Enzymes?

Enzymes are the largest and most highly specialized class of proteins. They function in the body as biological catalysts for the reactions involved in metabolism. We have previously described catalysts as substances that increase the rate of a chemical reaction by lowering the activation energy of the reaction. The reactions of metabolism would occur at extremely slow rates at normal body temperature and pH. Enzymes greatly increase this reaction rate, allowing cells to function under normal body conditions. Without the enzymes in our digestive tract, for example, it would take us about 50 years to digest a single meal! Since a catalyst is not consumed in the reaction, it can be used over and over again. Such compounds, therefore, need be present in only very small amounts.

Enzymes are water-soluble globular proteins that can vary from a molecular weight of 12,000 to over 1 million. The enzyme molecule can consist of a fairly simple single polypeptide chain, or it may be a more complex molecule composed of several polypeptide chains and other nonprotein subunits.

16.2 Some Important Definitions

Enzymes may be simple proteins composed entirely of amino acids, or may be conjugated proteins that require a nonprotein group for their biological activity. Before we continue with our discussion of enzymes, it will be helpful for us to define some special terms that are commonly used in the study of these complex molecules (Figure 16.1).

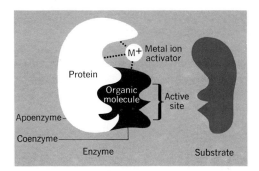

Figure 16.1   Some important terms used in the study of enzymes.

Apoenzyme.   Apoenzyme is the term used to describe the protein portion of the enzyme molecule.

Cofactor.   Cofactors are the additional chemical groups appearing in enzymes that are conjugated proteins. These cofactors are required for enzyme activity, and often take part in the reactions that occur. They may consist of metal ions or complex organic molecules; some enzymes require both types of cofactors.

Activator.   When the cofactor of an enzyme is a metal ion, such as the ions of magnesium, zinc, iron, or manganese, the cofactor is called an activator.

Coenzyme.   When the cofactor of an enzyme is a complex organic molecule other than a protein, the cofactor is called a coenzyme.

Prosthetic Group.   In the enzyme molecule the cofactor may be loosely associated with the apoenzyme, or may be tightly bound to the protein. If it is tightly bound, the cofactor is called a prosthetic group.

Zymogen.   Zymogen (or pre-enzyme) is the name applied to the inactive form of an enzyme. Enzymes, as we saw in the case of the digestive enzymes, are often secreted in inactive forms as zymogens. They are transported to the site where activity is desired, and are then converted to their active forms.

Substrate. The substrate is the chemical substance or substances upon which the enzyme acts.

Active Site. The active site is the specific area on the enzyme to which the substrate attaches during the reaction.

16.3 Enzyme Nomenclature and Classification

Since enzymes are the largest class of proteins, a great number of specific enzymes have been isolated and described. Initially, the only attempt at systematic nomenclature consisted of ending the enzyme name with the suffix -in to indicate a protein. Names such as trypsin, rennin, and pepsin are examples of such nomenclature, and are still used today. However, such names give no indication of the substrate involved or the type of reaction taking place. As the number of known enzymes increased, scientists began naming them by adding an -ase ending to the name of the substrate. For example, urease catalyzes the hydrolysis of urea; maltase catalyzes the hydrolysis of maltose; and phosphatases catalyze the hydrolysis of phosphoric acid esters.

In 1961 the Commission on Enzymes of the International Union of Biochemistry proposed a systematic classification of enzymes. They recommended that enzymes be named by the substrate or type of reaction involved, and that they be classified into six major divisions (Table 16.1). Many enzymes, however, are still commonly referred to by their original names, and we will use these more common names in this book.

16.4 Method of Enzyme Action

We have stated that enzymes catalyze reactions in cells by lowering the activation energy. They do this by forming a complex with the substrate (that is, by attaching to the substrate), thereby increasing the probability that the reaction will occur. We can roughly outline the steps of an enzyme-catalyzed reaction in the following manner (Figure 16.2).

1. The substrate (or substrates) becomes loosely bound to the enzyme surface, forming the enzyme-substrate complex.

$$E + S \rightleftharpoons ES$$

2. The substrate becomes activated; that is, the bonds in the substrate become polarized.

Indicates activated state

$$ES \rightleftharpoons ES^*$$

Table 16.1 Classes of Enzymes

Hydrolases: Enzymes that catalyze hydrolysis reactions.	
Example	Reaction Catalyzed
Carbohydrases:	Polysaccharides and disaccharides $\xrightarrow{+H_2O}$ Monosaccharides
Esterases:	Ester $\xrightarrow{+H_2O}$ Acid + alcohol
Proteases:	Protein $\xrightarrow{+H_2O}$ Peptides and amino acids
Nucleases:	Nucleic acids $\xrightarrow{+H_2O}$ Pyrimidines + purines + sugars + phosphoric acid

Oxido-Reductases: Enzymes that catalyze oxidation-reduction reactions.	
Example	Reaction Catalyzed
Oxidases:	Addition of oxygen to a substrate
Dehydrogenases:	Removal of hydrogen from a substrate

Transferases: Enzymes that catalyze reactions involved in the transfer of functional groups.	
Example	Reaction Catalyzed
Transaminases:	Transfer of $-NH_2$
Transmethylases:	Transfer of $-CH_3$
Transacylases:	Transfer of $-\overset{\overset{\textstyle O}{\|}}{C}-R$
Transphosphatases: (Kinases)	Transfer of $-O-\overset{\overset{\textstyle O}{\|}}{\underset{\underset{\textstyle OH}{\|}}{P}}-OH$

Lyases: Enzymes that catalyze the addition to double bonds.

Isomerases: Enzymes that catalyze the interconversion of isomers.

Ligases: Enzymes that, in conjunction with ATP, catalyze the formation of new bonds.

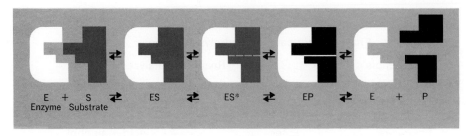

E + S ⇌ ES ⇌ ES* ⇌ EP ⇌ E + P
Enzyme Substrate

Figure 16.2 The steps of an enzyme-catalyzed reaction.

3. The products of the reaction form on the surface of the enzyme.

$$ES^* \rightleftharpoons EP$$

4. The products are released or "kicked off" the surface of the enzyme, making the enzyme available to catalyze further reactions.

$$EP \rightleftharpoons E + P$$

The rate at which an enzyme catalyzes a reaction will vary with different cellular conditions, and from enzyme to enzyme. Some enzymes, such as carbonic anhydrase, are extremely efficient in catalyzing a reaction. The efficiency of enzyme action is represented by the **turnover number,** which is the number of substrate molecules transformed per minute by one molecule of enzyme under optimum conditions of temperature and pH (Table 16.2).

Table 16.2   Turnover Numbers of Some Enzymes

Enzyme	Turnover Number (Molecules of Substrate/Minute)
Carbonic anhydrase	36,000,000
Sucrose invertase	1,000,000
Glutamic dehydrogenase	30,000
Phosphoglucomutase	1,240
Chymotrypsin	100
DNA polymerase	15

16.5 Specificity

Enzymes differ from inorganic catalysts in their specificity. For example, platinum will catalyze several different types of reactions, but a given enzyme will catalyze only one type of reaction and, in some cases, will limit its activity to only one particular type of molecule. Pancreatic lipase will hydrolyze the ester linkage between glycerol and fatty acids in lipids, but will have no effect on the hydrolysis of proteins or carbohydrates. Kidney phosphatase catalyzes the hydrolysis of esters of phosphoric acid, but at a different rate for each substrate. Urease is even more specialized; it will catalyze only the hydrolysis of urea. An extreme example of the specificity of enzyme action is given by the enzyme aspartase, which will catalyze only the following reversible reaction.

$$
\underset{\text{Fumaric acid}}{\overset{\text{H} \qquad \text{COOH}}{\underset{\text{HOOC} \qquad \text{H}}{\text{C}=\text{C}}}} \quad + NH_3 \underset{\text{Aspartase}}{\rightleftharpoons} \underset{\text{L-Aspartic acid}}{\overset{\text{COOH}}{\underset{\text{CH}_2\text{COOH}}{\text{H}_2\text{N}-\text{C}-\text{H}}}}
$$

This enzyme won't catalyze the addition of ammonia to any other unsaturated acid, not even maleic acid, which is the *cis*-isomer of fumaric acid.

$$
\underset{\text{Maleic acid}}{\overset{\text{HOOC} \qquad \text{COOH}}{\underset{\text{H} \qquad \text{H}}{\text{C}=\text{C}}}} \qquad \underset{\text{D-Aspartic acid}}{\overset{\text{COOH}}{\underset{\text{CH}_2\text{COOH}}{\text{H}-\text{C}-\text{NH}_2}}}
$$

Moreover, aspartase won't even catalyze the removal of ammonia from D-aspartic acid.

16.6 Lock-and-Key Theory

How can we explain the strict specificity of enzymes? There is a specific area on the surface of the enzyme molecule called the **active site,** to which the substrate attaches itself during a reaction. The configuration or geometry of this active site is specially designed for the specific substrate involved, and is determined by the amino acid sequence of the enzyme. The native configuration of the entire enzyme molecule must be intact for the active site to have the correct configuration. The substrate then fits into the active site of the protein in much the same way as a key fits into a lock. The configuration of the lock is specific for only one key; no other keys will turn the lock (Figure 16.3).

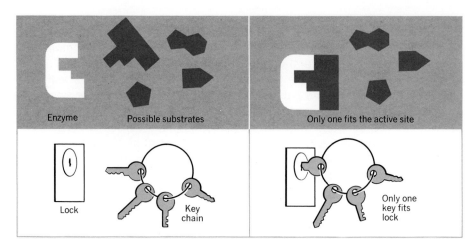

Figure 16.3 Many enzymes are very specific, often limiting their action to only one particular molecule (just as a lock is specific for only one key).

Not only is the geometry of the active site important, but so also is the arrangement of charged R-groups around the active site. There is an electrical attraction between the substrate and the enzyme, which draws them together. The products of the resulting reaction will have a different distribution of charges, and the repulsion between the product and the active site may cause the product to be "kicked off" the enzyme.

16.7 Induced-Fit Theory

The model of a lock-and-key fit, which is quite sufficient to explain the action of some enzymes, must be slightly modified for other enzymes. In these cases a more appropriate analogy would be a hand slipping into a glove, thereby inducing a fit. Enzyme molecules are flexible, and the active

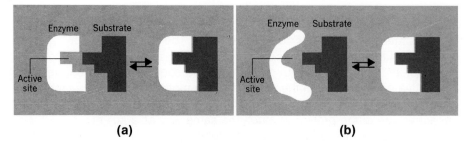

Figure 16.4 Theories of enzyme action. (a) In the lock and key theory, the active site conforms exactly to the substrate molecule. (b) In the induced fit theory, the substrate induces the active site to take on a shape complementary to the shape of the substrate molecule.

site of some enzymes may not initially conform to the substrate. However, the substrate itself, as it is drawn to the enzyme, may induce the enzyme to take on a shape complementary to the substrate. In such cases the active site is still quite specific to the substrate, just as a left-hand glove cannot be made to fit a right hand (Figure 16.4).

Enzymes that are secreted as zymogens have their active sites blocked. These sites must be unblocked by the hydrolysis of part of the molecule to activate the enzyme. Cofactors contribute to the activity of the enzyme either by providing the arrangement of molecules necessary for the active site, or by forming a bridge between the substrate and the enzyme (Figure 16.5).

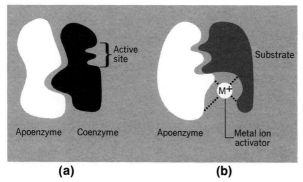

(a)   **(b)**

Figure 16.5 Cofactors may contribute to the activity of the enzyme by (a) providing the active site, or (b) forming a bridge between the enzyme and the substrate.

16.8 Factors Affecting Enzyme Activity

We have seen that factors such as pH, temperature, solvents, and salt concentrations can alter the structure of a protein. Such factors, therefore, will have an effect on the activity levels of enzymes.

pH
Changes in the pH of the surrounding medium can change the secondary or tertiary structure of an enzyme. This may alter the geometry of the active site or the surrounding charge distribution. Each enzyme has an optimum pH at which its activity is maximized. For example, pepsin has an optimum pH of 1.5, while trypsin has its maximum activity at pH 8. Increasing or decreasing the pH from these optimal levels will lower the activity of the enzyme (Figure 16.6).

Temperature
Temperature also affects the rate at which enzyme-catalyzed reactions occur. Most body enzymes exhibit maximum activity at temperatures

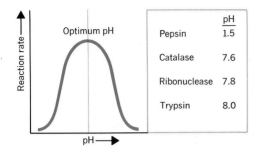

	pH
Pepsin	1.5
Catalase	7.6
Ribonuclease	7.8
Trypsin	8.0

Figure 16.6   Each enzyme has its optimum activity at a specific pH.

ranging from 35 to 45°C, above which the enzyme begins to denature and the reaction rate decreases. When you are running a high fever, you feel ill partially because your high body temperature inhibits enzyme activity. Above 80°C enzymes will become permanently denatured. Below 35°C the reaction rate slowly decreases until it essentially stops as the enzymes become inactive (Figure 16.7). Enzymes, however, are not denatured at low temperatures, and will resume activity if the temperature is again raised. It is this fact that allows researchers to preserve cell cultures, human tissues for transplants, and organisms such as bacteria for further studies. The preservation of food by refrigeration is possible because the enzyme-catalyzed reactions that cause spoilage are significantly slowed or stopped at low temperatures.

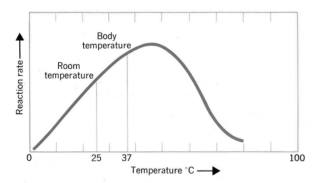

Figure 16.7   Temperature affects enzyme activity. Body enzymes will have a maximum activity at temperatures ranging from 35° to 45°C.

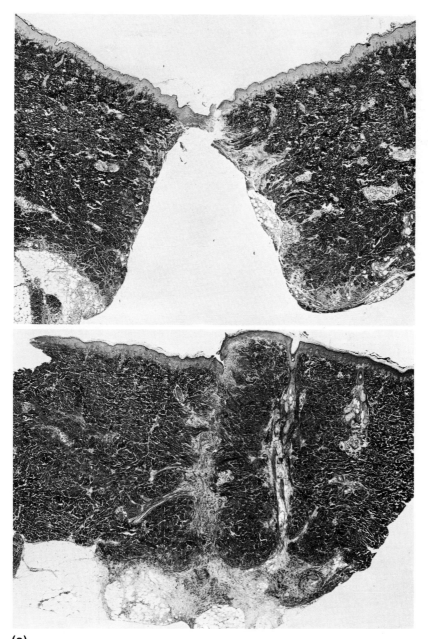

(a)

Figure 16.8
Illustrations of diseases caused by vitamin deficiencies. (a) These photographs show the first satisfactorily controlled experiment on human scurvy. The top photograph shows a biopsy 10 days after a wound was made in the midback region of a subject who had eaten an ascorbic acid-free diet for six months. The wound shows no healing except in the outer layer of skin cells. The lower photograph shows a biopsy taken after 10 days of treatment with ascorbic acid. The wound is healing, and shows abundant formation of connective tissue. (Courtesy J. H. Crandon and The Upjohn Company)

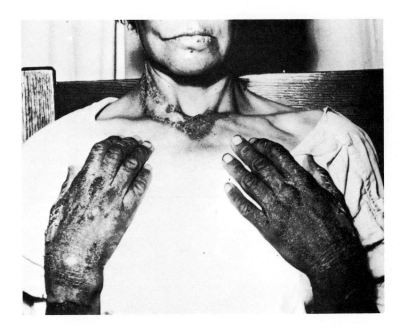

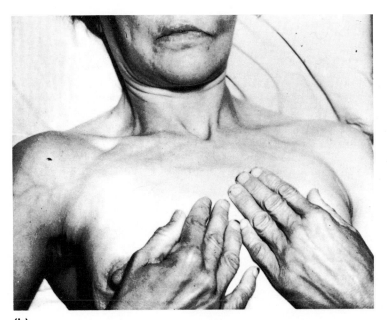

(b)

Figure 16.8
(b) The patient in the top photograph shows the skin lesions in advanced pellagra, the disease that results from a deficiency in the vitamin niacin. The lower photograph is of the same patient after niacin therapy. (Courtesy The Upjohn Company)

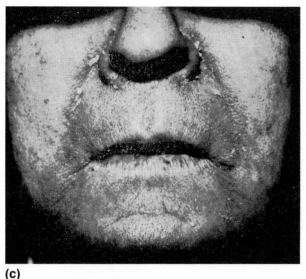

(c)

Figure 16.8
(c) The rough, red skin around the nose and chin,
and the greasy scaliness of the skin in this area
is caused by a deficiency of vitamin B_6. (Courtesy
R. W. Vilter, et al., *J. Lab. Med.* 42:355, 1953)

Vitamins

16.9 Vitamins as Coenzymes

We have stated that cofactors may be either metal ion activators or
nonprotein coenzymes. Many coenzymes are vitamins or derivatives of
vitamins. Exactly what are vitamins?

The need for vitamins has been recognized for over 200 years. British
sailors were given the slang name "limeys" from the lime juice they drank
to prevent scurvy. The substance in lime juice that prevents scurvy is
vitamin C, which was first isolated in 1930.

Vitamins are a class of organic nutrients that must be present in the
diet in trace amounts for proper cellular function, growth, and reproduction.
When vitamins are lacking in the diet, specific deficiency diseases result
(Figure 16.8). Most vitamins are present in all plant and animal cells. Some
organisms, however, are able to manufacture certain vitamins themselves,
and do not require these nutrients in their diet. Most vitamins function as
coenzymes in the cell, and can be classified as either water-soluble or
fat-soluble (Table 16.3).

Table 16.3   Vitamins

Name	Minimum Daily Requirements	Dietary Sources	Function	Symptoms of Deficiency
Water Soluble				
Vitamin B₁ (Thiamine)	1.5 mg	Whole grains; liver, brain, heart, kidney; legumes	Formation of enzymes involved in citric acid cycle	Beriberi, heart failure, mental disturbance
Vitamin B₂ (Riboflavin)	1 to 2 mg	Milk, eggs, liver, yeast, leafy vegetables	Coenzymes in electron transport chain	Fissures of the skin, visual disturbances
Vitamin B₆ (Pyridoxine)	1 to 2 mg	Whole grains, pork, glandular meats, legumes	Coenzyme for amino acid and fatty acid metabolism	Convulsions in infants; skin disorders in adults
Vitamin B₁₂ (Cyanocobalamin)	2 to 5 micrograms	Liver, kidney, brain. Synthesized by bacteria in gut	Synthesis of nucleoprotein	Pernicious anemia
Niacin (Nicotinic acid)	17 to 20 mg	Yeast, lean meat, liver, whole grains	NAD, NADP; coenzymes in hydrogen transport	Pellagra, skin lesions, diarrhea, dementia
Vitamin C (Ascorbic acid)	75 mg	Citrus fruits, green vegetables	Maintenance of normal connective tissue, carbohydrate metabolism	Scurvy, bleeding gums, loosened teeth, swollen joints
Folic acid	0.1 to 0.5 mg	Yeast, organ meats, wheat germ	Synthesis of nucleoprotein	Anemia, inhibition of cell division
Pantothenic acid	8 to 10 mg	Yeast, liver, kidneys, egg yolk	Forms part of Coenzyme A (CoA)	Neuromotor disorders, digestive disorders, cardiovascular disorders.

Table 16.3   Vitamins (Cont'd)

Name	Minimum Daily Requirements	Dietary Sources	Function	Symptoms of Deficiency
Biotin	0.15 to 0.3 mg	Liver, egg white, dried peas and lima beans, synthesized by bacteria in the gut	Protein synthesis, CO_2 fixation, transamination	Skin disorders
Inositol	Unknown	Found in many fruits and vegetables	Aids in fat metabolism	Fatty liver
Choline (essential factor; not strictly a vitamin)		Egg yolk, wheat germ	Aids in fat oxidation and transport	Inadequate fat absorption, fatty liver
Fat Soluble				
Vitamin A (A_1-retinol) (A_2-dehydroretinol)	5000 I.U. (1 I.U. = 0.3 micrograms of retinol)	Green and yellow vegetables and fruits; cod liver oil	Formation of visual pigments; maintenance of normal epithelial structure	Night blindness, skin lesions, eye disease (In excess—vitamin A toxicity, hyperirritability, skin lesions, bone decalcification, increased pressure on the brain)
Vitamin D (D_2-calciferol) (D_3-cholecalciferol)	400 I.U. (1 I.U. = 0.025 micrograms of cholecalciferol)	Fish oils, liver; provitamins in our skin activated by sunlight	Increase Ca^{2+} absorption from gut; important in bone and teeth formation	Rickets (defective bone formation) (In excess—growth retardation in infants above 2000 I.U. per day)
Vitamin E (Tocopherol)	10 to 40 mg depending upon intake of polyunsaturated fatty acids	Green, leafy vegetables	Maintain resistance of red cells to hemolysis	Increased fragility of red blood cells
Vitamin K (K_2-phylloquinone)	Unknown	Produced by bacteria in the intestines	Enables prothrombin synthesis in liver	Failure of coagulation

Regulation of Enzyme Activity

Enzymes give living systems the capacity to act. But if all enzymes were equally active in the cell all the time, the cell would probably short-circuit itself and die. Living systems, therefore, not only have the capacity to act, but also the capacity to control that action. For example, enzymes in the bloodstream catalyze the formation of blood clots when we are bleeding, but are not active and do not form clots under normal conditions. Enzymes catalyze the contraction of muscle fibers when we walk, but are inactive when we sit or rest. The control mechanisms of the living system involve control of enzyme concentrations and control of enzyme activity.

16.10 Regulatory Enzymes

We have seen that most biological chemical reactions occur in sequential chains that eventually accomplish a specific metabolic result. Each reaction in such a chain is catalyzed by a separate enzyme, and a series of reactions is therefore called a **multienzyme system.** For example, a multienzyme system is involved in the breakdown of glucose to lactic acid in muscle cells, and another multienzyme system is involved in the synthesis of specific amino acids.

In most multienzyme systems, the enzyme that catalyzes the first reaction of the series is the **regulatory** or **allosteric enzyme;** this enzyme controls the rate of the entire process. Regulatory enzymes are usually complex, high-molecular-weight molecules containing several polypeptide chains and cofactors. These enzymes usually have more than one site for attachment of molecules—one for the active site and one or more for regulatory molecules. A site for a regulatory molecule is called a **regulatory** or **allosteric** (meaning other space or location) **site.** The **regulatory molecule** can either inhibit or increase the activity of the enzyme. The allosteric enzyme is a flexible molecule, and the regulatory molecule causes a slight change in the shape of the enzyme; this changes the shape of the active site, making it either more or less receptive to the substrate.

Many regulatory enzymes are inhibited by the end product of the multienzyme system.

$$A \xrightarrow{E_1} B \xrightarrow{E_2} C \xrightarrow{E_3} D \xrightarrow{E_4} F \xrightarrow{E_5} G$$

In the above example, E_1 is the regulatory enzyme for the process, and it will be inhibited by high concentrations of product G (Figure 16.9).

All cells need energy in the form of ATP, and there are three multienzyme systems working to control the level of ATP in the cell (Figure 16.10). When ingested into the human body, glucose is either broken down in system 1 to form carbon dioxide, water, and ATP, or is converted to glycogen in system 2. System 3 breaks down glycogen to release glucose when demand for this sugar is great. The regulatory enzymes for these three systems are affected by cellular levels of ATP and AMP (see Chapter 13, Section 13.23).

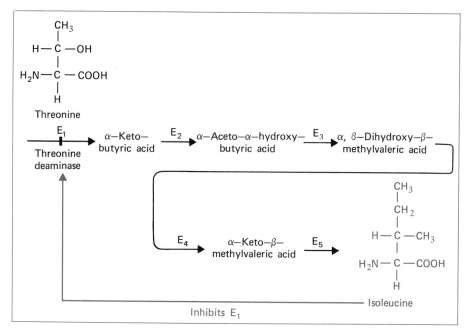

Figure 16.9 In the synthesis of the amino acid isoleucine from the amino acid threonine, the first enzyme, threonine deaminase, is the regulatory enzyme. It is inhibited by the end product of the multienzyme system — isoleucine.

High levels of AMP will stimulate systems 1 and 3, and inhibit system 2; similarly, high levels of ATP will inhibit systems 1 and 3, and stimulate system 2.

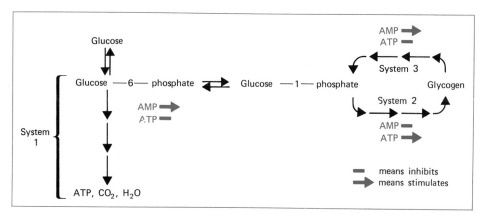

Figure 16.10 ATP and AMP have opposite regulatory effects on three multienzyme systems that control the level of ATP in the cell. System 1 involves the oxidation of glucose. System 2 involves glycogenesis, and system 3 glycogenolysis.

Table 16.4 The Principle Endocrine Glands and Tissues of Man, and Their Hormones*

Gland or Tissue	Hormone	Major Function of the Hormone
Thyroid	1. Thyroxin	1. Stimulates rate of oxidative metabolism and regulates general growth and development.
	2. Thyrocalcitonin	2. Lowers level of calcium in the blood.
Parathyroid	Parathormone	Regulates the levels of calcium and phosphorus in the blood.
Pancreas (Islets of Langerhans)	1. Insulin 2. Glucagon	1. Decreases blood glucose level. 2. Elevates blood glucose level.
Adrenal medulla	Epinephrine	Various "emergency" effects on blood, muscle, temperature.
Adrenal cortex	Cortisone and related hormones	Control carbohydrate, protein, mineral, salt, and water metabolism.
Anterior pituitary	1. Thyrotropic	1. Stimulates thyroid gland functions.
	2. Adenocorticotropic	2. Stimulates development and secretion of adrenal cortex.
	3. Growth hormone	3. Stimulates body weight and rate of growth of skeleton.
	4. Gonadotropic (two hormones)	4. Stimulate gonads.
	5. Prolactin	5. Stimulates lactation.

16.11 Regulatory Genes

As we will see in the next chapter, the function of a gene is to direct the synthesis of a protein molecule. Genes, therefore, also direct the synthesis of cellular enzymes. One way in which a cell can control its metabolic activities is to switch on and off the genes that direct the synthesis of cellular enzymes. By this means the cell can control the concentration of enzymes within itself.

Table 16.4 The Principle Endocrine Glands and Tissues of Man, and
Their Hormones* (Cont'd)

Gland or Tissue	Hormone	Major Function of the Hormone
Posterior pituitary	1. Oxytocin 2. Vasopressin	1. Causes contraction of some smooth muscles. 2. Inhibits excretion of water from the body by way of urine.
Ovary (follicle)	Estrogen	Influences development of sex organs and female characteristics.
Ovary (corpus luteum)	Progesterone	Influences menstrual cycle; prepares uterus for pregnancy; maintains pregnancy.
Uterus (placenta)	Estrogen and progesterone	Function in maintenance of pregnancy.
Testis	Androgens (testosterone)	Responsible for development and maintenance of sex organs and secondary male characteristics.
Digestive system	Several gastrointestinal	Integration of digestive processes.

*From G. E. Nelson, G. G. Robinson, and R. A. Boolootian, *Fundamental Concepts of Biology,* Second Edition, 1970. John Wiley and Sons, Inc., New York, page 114. Used by permission.

16.12 Regulatory Hormones

The body has control systems in addition to those within individual cells; these mechanisms coordinate the actions between cells in multicellular systems. This higher level of communication between cells, tissues, and organs involves the nervous system and the endocrine system. The endocrine system is a group of glands (Figure 16.11) that produce and secrete into the body fluids, particularly the blood, chemical messengers called **hormones** (Table 16.4).

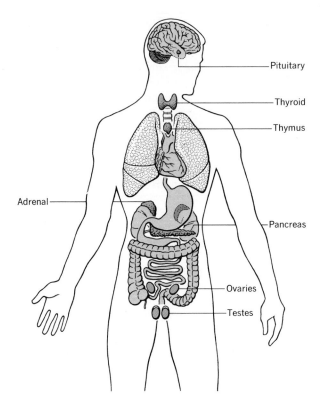

Figure 16.11 The endocrine glands secrete chemical messengers called hormones.

Each hormone has target organs or cells that are influenced by its presence. The secretion of a particular hormone by an endocrine gland may be triggered by the nervous system, as in the release of adrenalin triggered by hearing a nearby explosion, or by the concentration of a specific chemical compound, as in the release of insulin triggered by an increase in blood glucose levels.

Hormones can regulate cellular processes in one of several ways. Steroid hormones such as estrogen and testosterone are thought to activate genes. Adrenalin works by converting an enzyme from an inactive to an active form. Insulin, human growth hormone, and hormones that regulate gastric and kidney secretions are thought to work by causing changes in the cell membrane.

16.13 Cyclic AMP

Cyclic AMP is a molecule that acts within the cell as a secondary messenger, or intracellular hormone. Its formation is triggered by the attachment of a hormone molecule to a target cell. The system works as follows: a hormone, the primary messenger, is released into the blood-

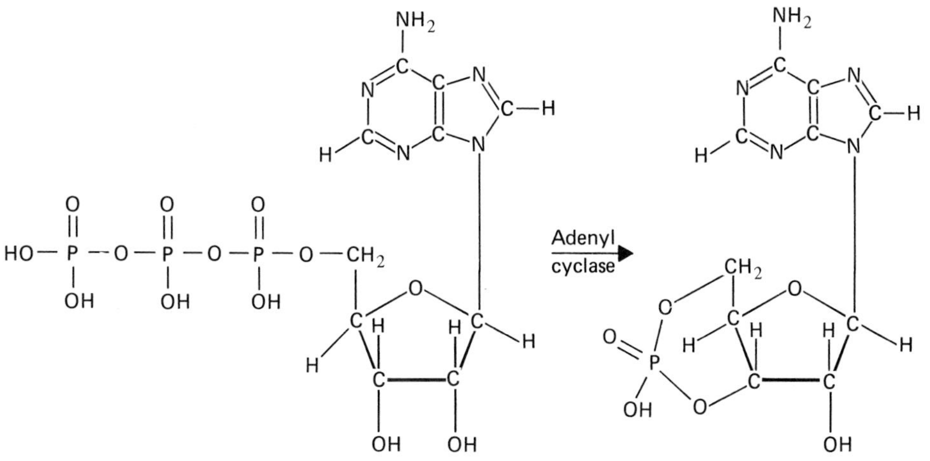

Figure 16.12 The formation of cyclic AMP from ATP.

stream. It travels to a target cell where it attaches to a specific receptor site on the outside of the cell membrane. This attachment causes a conversion of the enzyme adenyl cyclase from an inactive form to an active form on the inside of the cell membrane. This enzyme then catalyzes the formation of cyclic AMP from ATP (Figure 16.12). The cyclic AMP then diffuses through the cell as a secondary messenger, instructing the cell to respond to the hormone in the particular manner characteristic of that cell (Table 16.5 and Figure 16.13).

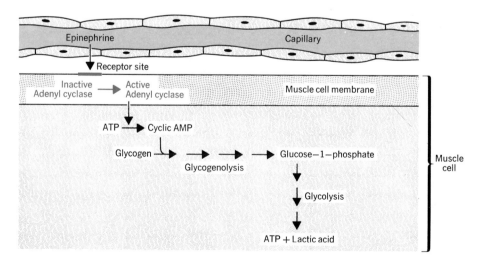

Figure 16.13 Epinephrine stimulates glycogen breakdown and glycolysis in the muscle cell by attaching to a receptor site on the muscle cell membrane. This triggers the activation of the enzyme adenyl cyclase, which catalyzes the formation of cyclic AMP from ATP. The cyclic AMP sets off a series of reactions that result in glycogenolysis and glycolysis, and the production of ATP for use in the contraction of the muscle cell.

Table 16.5 Some Hormones That Trigger Response in Target Tissues by Increasing the Level of Cyclic AMP in the Tissue

Thyrotropin	Thyroid	Thyroxin secretion
Luteinizing hormone	Ovary	Progesterone secretion
Epinephrine	Muscle	Glycogenolysis
	Fatty Tissue	Fat hydrolysis
	Liver	Glycogenolysis
Parathyroid hormone	Bone	Calcium resorption
	Kidneys	Phosphate excretion

For example, the hormone thyrotropin is secreted by the pituitary gland, and travels to target cells in the thyroid gland. There it triggers the formation of cyclic AMP, to which the thyroid cells respond by secreting more thyroxin.

16.14 Prostaglandins

Prostaglandins are a class of compounds that resemble hormones in their effects, but are chemically quite different. They are 20-carbon fatty acids that are synthesized in the cell membrane from unsaturated fatty acids such as arachidonic acid (Figure 16.14). Like cyclic AMP they are present in almost all cells and tissues, and function to carry out the messages that cells receive from hormones. They are among the most potent of all biological agents. They can raise or lower blood pressure, and can regulate gastric secretions. They cause uterine contractions, and so are used to induce labor; they open air passages to the lungs, and so may be used in the relief of asthma. Prostaglandins are now under careful study to determine their full range of physiological effects and possible therapeutic uses.

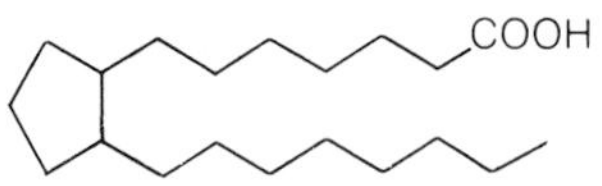

Prostanoic acid

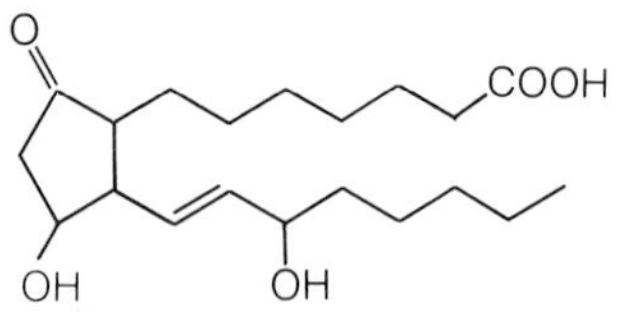

PGE₁

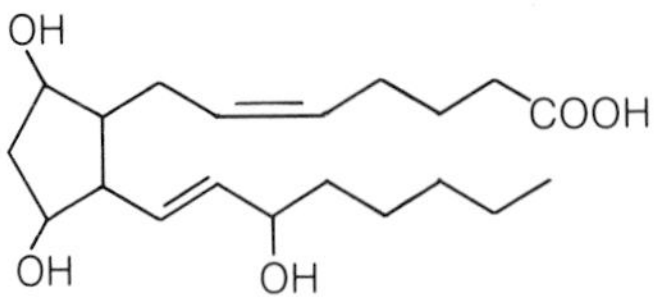

PGE₃

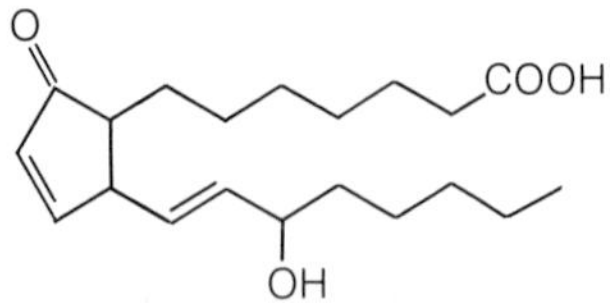

PGA₁

Figure 16.14 Some important prostaglandins. In general the prostaglandins are variants of the basic structure of prostanoic acid. Although there are only slight differences in structure, the prostaglandins have widely diverse biological effects.

Inhibition of Enzyme Activity

The activity of enzymes can be inhibited in several ways. Research into the mechanisms of enzyme inhibition has yielded a great deal of knowledge about the specificity of enzymes and the nature of their active sites. The action of many poisons and drugs is due to their ability to inhibit specific enzymes.

16.15 Irreversible Inhibition

Irreversible inhibition occurs when a functional group or cofactor required for the activity of the enzyme is destroyed or modified. For example, cholinesterase is an enzyme that catalyzes a reaction taking place between the juncture of nerve cells; the enzyme is necessary for normal transmission of nerve impulses. Components of nerve gases combine with the —OH group on a serine molecule that is vital to the active site of the cholinesterase enzyme. The enzyme then loses its ability to catalyze the reaction, so that animals poisoned by nerve gas are paralyzed (Figure 16.15). Cyanide poisons release the cyanide ion ($C\equiv N^-$), which binds very tightly to metal

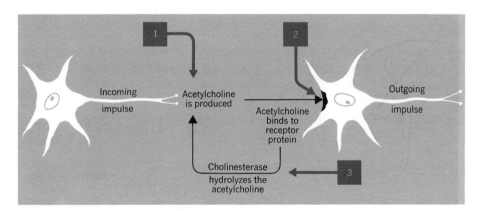

Figure 16.15 The nerve poisons. The formation and breakdown of acetylcholine must occur for normal nerve transmission. Acetylcholine is synthesized at the end of one nerve cell fiber and then migrates to a receptor protein on the next nerve cell, causing the signal to be sent along. Once the signal is sent, the acetylcholine is hydrolyzed by the enzyme cholinesterase, which leaves the cell ready for the next signal. Nerve poisons disrupt these processes in the following ways:
(1) Botulism toxin blocks the synthesis of acetylcholine so that nerve signals are not sent. Paralysis occurs, and death is caused by respiratory failure.
(2) Nicotine, curare, atropine, morphine, codeine, cocaine, and local anesthetics such as procaine combine with the receptor protein, blocking its reaction with acetylcholine. Therefore, the second cell does not receive the impulse and it is not sent on.
(3) Anticholinesterases are poisons such as nerve gases, organophosphate insecticides, and some mushroom toxins that inhibit cholinesterase, causing overstimulation of nerve cells by acetylcholine. This results in irregular heart rhythms, convulsions, and death.

ions, preventing them from combining with apoenzymes. Cyanides inhibit the enzyme cytochrome oxidase, one of the enzymes of the electron transport chain; this enzyme requires iron as an activator. Ions of heavy metals such as mercury and lead are extremely toxic because they bind with the —SH functional group on enzymes, permanently denaturing them. Arsenic compounds are poisonous because they contain the arsenate ion (AsO_4^{3-}), which can be mistaken in enzyme synthesis for the phosphate ion (PO_4^{3-}). Enzymes with arsenate ions in their structure will not function properly.

16.16 Reversible Inhibition

Reversible inhibition of enzyme activity takes two forms: competitive and noncompetitive inhibition. **Competitive inhibition** results when a compound whose structure is very similar to the substrate competes with the substrate for the active site on the enzyme. When inhibitors become bound to active sites, there are fewer enzyme molecules available to the substrate, so that enzyme activity decreases.

$$[\quad \underset{\text{Enzyme}}{E} \; + \; \underset{\text{Inhibitor}}{I} \; \rightleftharpoons EI \quad] \text{ competes with } [\quad \underset{\text{Enzyme}}{E} \; + \; \underset{\text{Substrate}}{S} \; \rightleftharpoons ES \quad]$$

For example, succinic acid and malonic acid have very similar structures. Succinate dehydrogenase catalyzes the removal of hydrogen from succinic acid, and this reaction is inhibited by malonic acid (Figure 16.16).

COOH   COOH
| |
CH_2 HC
| + E—FAD $\longrightarrow$ ‖   + E—FADH$_2$
CH_2 ↑   CH
| Succinate |
COOH   dehydrogenase COOH
Succinic   Fumaric acid
acid

COOH
|
CH_2
|
COOH
Competitive inhibitor: Malonic acid

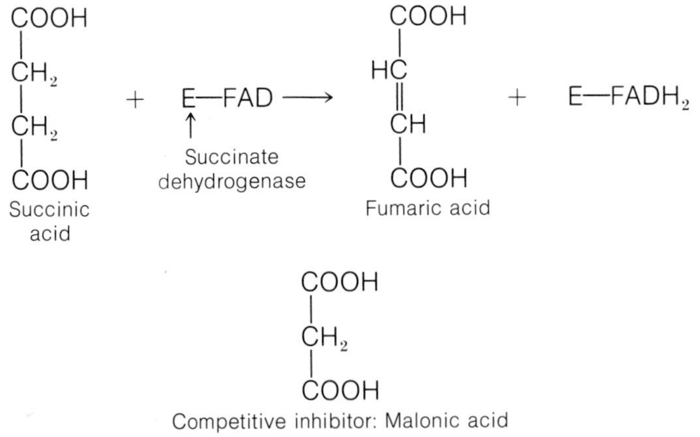

Figure 16.16 Competitive inhibition. Succinate dehydrogenase is the enzyme that catalyzes the removal of two hydrogens from succinic acid. Malonic acid is a competitive inhibitor of this reaction, competing with the succinic acid for the active site on the enzyme. (This reaction involves the hydrogen acceptor FAD, flavin adenine dinucleotide, which is covalently bonded to the enzyme and which accepts the two hydrogens from succinic acid.)

Noncompetitive inhibition results when the inhibitor doesn't combine with the active site on the enzyme, but rather combines with some other portion of the enzyme molecule that is essential to its function. A common type of noncompetitive inhibitor combines reversibly with —SH groups that are not located on the active site, but are still essential to the enzyme's activity. Iodoacetic acid inhibits the conversion of glucose to lactic acid in muscles by reacting with the —SH groups found on the enzyme glyceraldehyde phosphate dehydrogenase (Figure 16.17).

$$\text{E—SH} + \text{ICH}_2\text{COOH} \rightleftarrows \text{E—S—CH}_2\text{COOH} + \text{HI}$$

| Active enzyme | Iodoacetic acid | | Inactive enzyme |

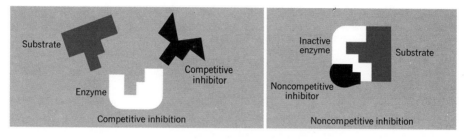

Figure 16.17 In competitive inhibition, the inhibitor and the substrate compete for the active site on the enzyme. In noncompetitive inhibition, the inhibitor combines with some other portion of the enzyme molecule that is essential for the activity of the enzyme.

16.17 Antibiotics

Antibiotics are chemicals that are extracted from organisms such as molds, bacteria, and yeasts, and that inhibit growth or destroy other microorganisms. They belong to a large class of chemicals called **antimetabolites,** which inhibit enzyme function. Antibiotics are used to treat many diseases in man and animals, and in small amounts are used to promote the growth of poultry and livestock.

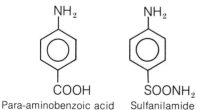

Para-aminobenzoic acid   Sulfanilamide

Figure 16.18 The sulfa drug sulfanilamide prevents bacterial growth by competitive inhibition. Bacteria use para-aminobenzoic acid in the synthesis of the vitamin folic acid. The structure of sulfanilamide closely resembles that of para-aminobenzoic acid, and sulfanilamide molecules compete for the active site on the bacterial enzyme, inhibiting folic acid synthesis.

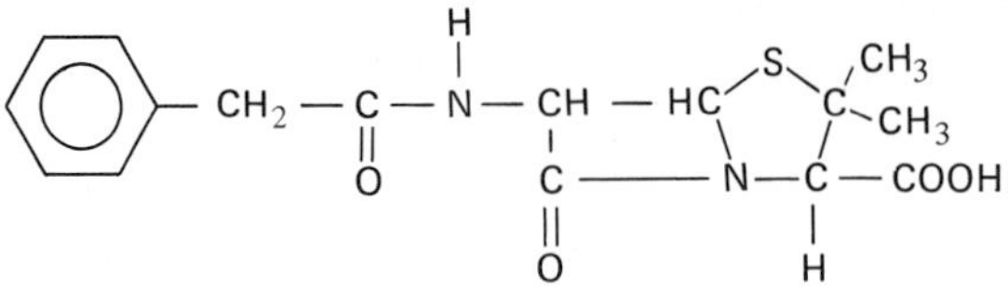

Penicillin G

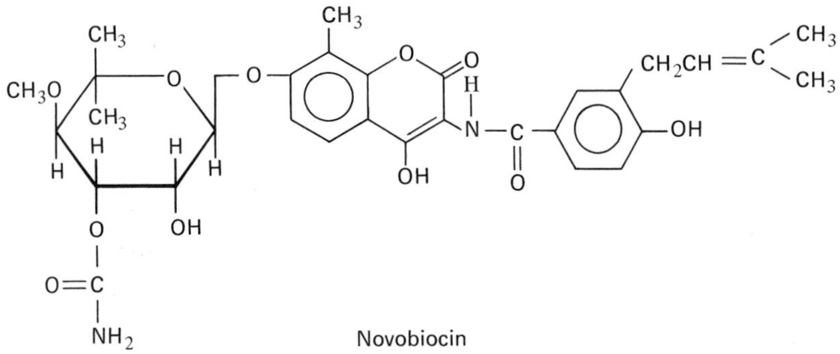

Novobiocin

Figure 16.19 The two antibiotics penicillin G and novobiocin affect bacterial growth by inhibiting bacterial cell wall synthesis.

Antibiotics can function in many ways to prevent growth or to destroy a disease-causing organism (Figure 16.18). One way in which antibiotics function is by impairing the structure or obstructing the synthesis of bacteria cell walls. Penicillin G works by inhibiting the reaction that is the final step in the formation of the cell wall, thus causing the bacterial cell to burst (Figure 16.19). The inhibitory action of novobiocin and bacitracin is very similar to that of penicillin G. Polypeptide antibiotics such as polymyxin and tyrocidine act on the cell membranes of microorganisms, increasing the permeability of the membrane and releasing vital nutrients from the cell. A third way in which antibiotics work is by interfering with protein synthesis within the cell. Chloramphenicol and the tetracyclines alter the nature of the RNA molecules synthesized by the cell, thus disrupting protein synthesis.

Additional Reading

Articles

1. E. V. Jensen and E. R. DeSombre, "Estrogen-Receptor Interaction," *Science,* October 12, 1973, page 126.

2. D. E. Koshland, Jr., "Protein Shape and Biological Control," *Scientific American,* October 1973, page 52.

3. J. Z. Majtenyi, "Antibiotics—Drugs from the Soil," *Chemistry,* January 1975, page 6 and March 1975, page 15.

4. I. Pastan, "Cyclic AMP," *Scientific American,* August 1972, page 97.

5. J. E. Pike, "Prostaglandins," *Scientific American,* November 1971, page 84.

6. I. Raw, "Enzymes, How They Operate," *Chemistry,* June 1967, page 8.

7. M. J. Schneider, "Vitamin C: How Much Do We Need?" *The Sciences,* January–February 1975, page 11.

Questions and Problems

1. Define the following terms:

   (a) Enzyme
   (b) Apoenzyme
   (c) Cofactor
   (d) Activator
   (e) Coenzyme
   (f) Prosthetic group
   (g) Zymogen
   (h) Substrate
   (i) Active site
   (j) Turnover number
   (k) Lock-and-key theory
   (l) Induced-fit theory
   (m) Vitamin
   (n) Multienzyme system
   (o) Allosteric enzyme
   (p) Allosteric site
   (q) Hormone
   (r) Endocrine glands
   (s) Irreversible inhibition
   (t) Reversible inhibition
   (u) Competitive inhibition
   (v) Noncompetitive inhibition

2. In general terms, describe the similarities and differences between the action of enzymes and inorganic catalysts.

3. What is the difference between an apoenzyme and a coenzyme?

4. Phosphoglucomutase catalyzes the conversion of glucose-1-phosphate to glucose-6-phosphate.

$$\text{Glucose-1-phosphate} \underset{}{\overset{\text{Phosphoglucomutase}}{\rightleftharpoons}} \text{Glucose-6-phosphate}$$

(a) In general terms, describe the four steps by which phosphoglucomutase accomplishes this conversion.

(b) Phosphoglucomutase has a turnover number of 1×10^3. How many molecules of glucose-1-phosphate would be converted in 1 hour by 1 molecule of phosphoglucomutase? How many moles?

5. (a) Explain how the lock-and-key analogy describes the method of enzyme action.

(b) How does the induced-fit theory modify the lock-and-key analogy?

6. Why doesn't pepsin, the protein-cleaving enzyme released in the stomach, continue to function in the small intestine?

7. Why is boiling water an effective sterilizing agent?

8. Under normal conditions a human sperm cell can live 24 to 36 hours. However, several doctors have reported success in artificial insemination using human sperm that have been frozen up to six months. In one case, pregnancy resulted from the use of sperm that had been frozen at $-196.5°C$ for six months. Explain how it is possible for sperm cells to survive for six months and then resume normal activity.

9. Excess vitamins A and D are stored in the body, but excess vitamins C and B_1 are readily excreted.

(a) What property allows vitamins A and D to be stored in the body while vitamins C and B_1 are excreted?

(b) Explain the statement that an excess of vitamin D in the diet is potentially as dangerous as a deficiency.

10. (a) What is an allosteric enzyme?

(b) Why are allosteric enzymes necessary for the normal operation of the cell?

(c) Explain the action of the regulatory molecule on the allosteric enzyme.

11. Describe three ways in which reaction rates are controlled within the human body.

12. What are the differences between vitamins and hormones? (Include both their sources and functions in your discussion.)

13. Prostaglandins are among the most potent of biological agents. Describe some of their varied physiological effects. In what way do they function like cyclic AMP?

14. The insecticide Parathion must be handled with great care since it is converted by the liver to a molecule that resembles nerve gas in its action. Explain how Parathion acts to poison human beings.

15. Public health officials estimate that more than 200,000 children become ill from lead poisoning each year, many of them from eating chips of lead-based paints. What is the chemical action of lead that causes this illness?

16. Both ethanol and methanol are oxidized in the body: ethanol to acetic acid, and methanol to formic acid. It is the formic acid that causes the acidosis of methanol poisoning. The first step in the conversion of both ethanol and methanol involves the enzyme alcohol dehydrogenase. One therapy for methanol poisoning is the use of a nearly intoxicating dose of ethanol. In the light of your knowledge of enzyme inhibition, explain why this therapy is effective.

17. What is an antimetabolite?

18. Explain three ways in which antibiotics destroy disease-causing organisms.

19. Antibiotics are often added in small amounts to foods fed poultry to promote growth. Suggest some reasons why the concentration of antibiotic used must be carefully monitored.

chapter 17

Nucleic Acids

Learning Objectives

By the time you have finished this chapter, you should be able to:

1. List the three components of nucleotides.

2. Name the bases and sugars found in DNA and RNA.

3. Describe the structure of a DNA molecule.

4. List the three types of RNA and describe their functions.

5. Describe the genetic code and its relationship to amino acids and polypeptide chains.

6. Describe the steps in protein synthesis.

7. State two ways in which cells can regulate protein synthesis.

8. Define "mutation," and indicate several ways in which mutations can disrupt the normal functions of an organism.

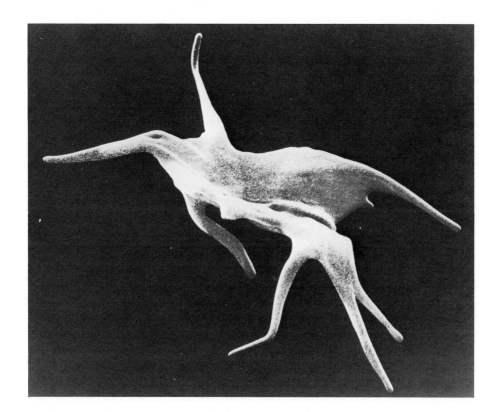

Mrs. Smith watched her son George walk slowly down the street to the school bus stop. She was puzzled and concerned. He'd been complaining on and off for a year about vague pains in his joints and abdomen. Twice he'd run a low fever and suffered from vomiting, and she had kept him home from school. He was the youngest of her four children, and the only sickly one among them.

Early in the afternoon Mrs. Smith received a call from the school nurse. She had just taken George to the hospital—he had collapsed during a vigorous game of kickball, screaming over a pain in his stomach. When Mrs. Smith reached the hospital she found George under sedation and receiving liquids intravenously. The doctors first thought that George might have appendicitis; but after talking with Mrs. Smith and learning of George's symptoms over the last year, they decided to run several blood tests. As they had suspected, the doctors found that George had sickle cell anemia, and was suffering from a sickle cell crisis. The pain in George's abdomen and the swelling in his joints slowly went away, and a week later he was able to return to school.

A few weeks later George and his mother went to a neighboring town for consultation with doctors at a large clinic specializing in the diagnosis and treatment of sickle cell anemia. There they learned that sickle cell anemia is an inherited disease that affects red blood cells. They were told that a mutation had occurred in the evolution of the African black population, causing an error in the gene for hemoglobin, the molecule in red blood cells that carries the oxygen to the body. At present, one out of every two African blacks carries this defective gene, and one out of every 10 American blacks carries this trait. Children who inherit two defective genes from their parents will suffer from sickle cell anemia, and almost all their hemoglobin will be defective. They will experience periodic sickle cell crises, resulting from the clogging of their capillaries by abnormal red blood cells, and causing severe pain, fever, and swelling of the joints. Such crises can be set off by an infection, cold weather, trauma, strenuous exercise, or emotional stress. Children having sickle cell anemia are anemic, often jaundiced, susceptible to disease, and will usually die in childhood.

Those individuals having only one defective gene will carry the sickle cell trait. About one-half of their hemoglobin is defective, but under normal conditions they will show no clinical signs of the disease. However, unusual stress can bring on a crisis in these individuals also.

Sickle cell anemia was first described clinically in 1910, but it was not until 1949 that Linus Pauling demonstrated the molecular basis of the disease. He showed that sickle cell hemoglobin (abbreviated HbS) exhibits different mobility in an electric field than does normal hemoglobin (HbA) (Figure 15.15b). Hemoglobin (molecular weight 64,458) is a molecule consisting of a protein part, called globin, and a nonprotein part, called heme. The globin part consists of four polypeptide chains: two alpha chains containing 141 amino acid units, and two beta chains containing 146 amino acid units. Each chain is wrapped around a heme group containing an Fe^{2+} atom; this is the group that actually carries the oxygen. (See Chapter 12, Figure 12.12 and Chapter 15, Figure 15.11. Therefore, each hemoglobin molecule can carry four molecules of oxygen.

In 1956, Ingram showed that the entire difference between HbS and HbA was in one amino acid on the beta chain. In HbS, a valine molecule is erroneously substituted for a glutamic acid molecule in position 6 on the beta chain (See page 441).

Valine   Glutamic acid

At pH 7, the glutamic acid will have a negative charge, while the valine will have no charge. This accounts for the difference in mobility of the two hemoglobin molecules in an electric field.

(a)

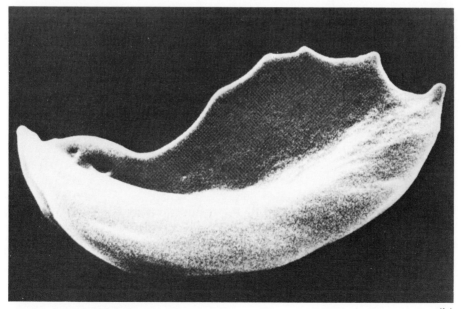

(b)

Figure 17.1   (a) Normal red blood cells. (b) Red
blood cells from a person with sickle cell anemia.
[(a) Courtesy Francois Morel. From *J. Cell Biol.*
48:91-100, 1971. (b) Courtesy Springer-Verlag,
Publishers, from CORPUSCLES by Marcel Bessis.]

At low oxygen concentrations, red blood cells containing HbS will assume abnormal shapes, some resembling sickles (Figure 17.1). This sickling results from the substitution of valine (which contains a nonpolar R-group), for glutamic acid (which contains a polar R-group). At low oxygen concentrations the nonpolar valine on one HbS molecule is attracted to a "hole" or region of nonpolar R-groups on a neighboring molecule. This strong attraction causes the HbS molecules to form into long filaments, distorting the red blood cells into bizarre shapes. The abnormal shape will correct itself in most red blood cells as the oxygen concentration is increased, but some cells will remain irreversibly sickled.

The sickling activity occurs in the capillaries, where the red blood cells release their oxygen. The sickled cells will increase the viscosity of the blood and can block the capillaries, depriving tissues of blood-carried oxygen. Such blockage brings about the clinical symptoms of a sickle cell crisis. The anemia suffered by these individuals results from the decreased life span of their red blood cells.

Individuals suffering from sickle cell anemia, and carriers of the trait, can be identified through blood screening tests, but at present there is no effective treatment for the disease. As one approach to this problem, drugs are currently being tested that may help prevent the hydrophobic bonding of HbS at low oxygen concentrations.

Molecular Basis of Heredity

Sickle cell anemia is just one of more than 2000 diseases in man that are known to be caused by disorders in genes. Genes are specific segments of molecules called DNA. Each gene contains the information necessary to make one polypeptide chain. To understand how this information carried by genes directs the formation of a polypeptide chain, we must first become familiar with the molecular nature of the genetic material, DNA.

DNA, Deoxyribonucleic Acid

17.1 Nucleotides

DNA, deoxyribonucleic acid, and RNA, ribonucleic acid, belong to a class of polymers called **nucleic acids.** Nucleic acids are polymers of monomer units called **nucleotides.** But unlike the monomer units we have studied previously, nucleotides can be further hydrolyzed to yield three components:

1. A nitrogen-containing base.

2. A five-carbon sugar.

3. Phosphoric acid.

The Nitrogen Bases

There are two classes of nitrogen-containing bases found in nucleotides: **pyrimidines** and **purines** (Figure 17.2). The bases derived from pyrimidine are uracil (U), thymine (T), and cytosine (C); those derived from purine are adenine (A) and guanine (G). The base uracil is found only in nucleotides of RNA, and the base thymine is found only in nucleotides of DNA.

The Pyrimidines

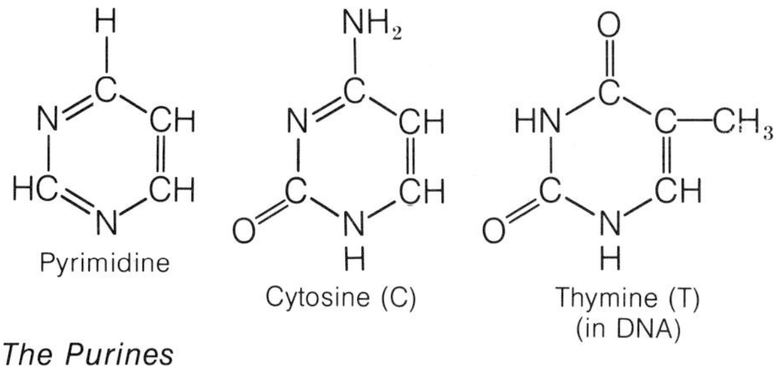

Pyrimidine

Cytosine (C)

Thymine (T)
(in DNA)

Uracil (U)
(in RNA)

The Purines

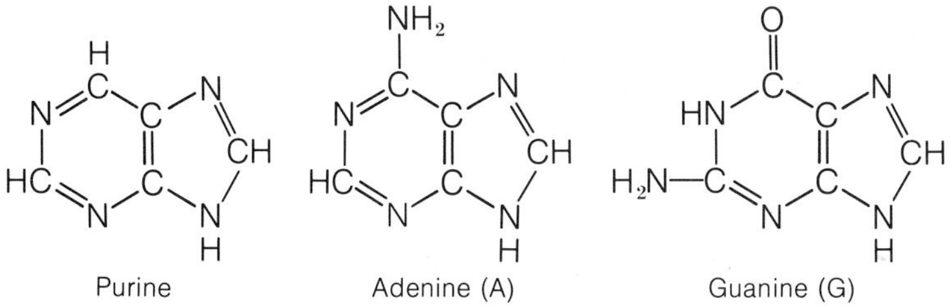

Purine

Adenine (A)

Guanine (G)

Figure 17.2   The bases found in nucleotides.

The Pentose Sugars

The second component of nucleotides is the pentose sugar. RNA contains the sugar ribose, and DNA contains a derivative of ribose, 2-deoxyribose (Figure 17.3). The structure of one nucleotide containing the base adenine

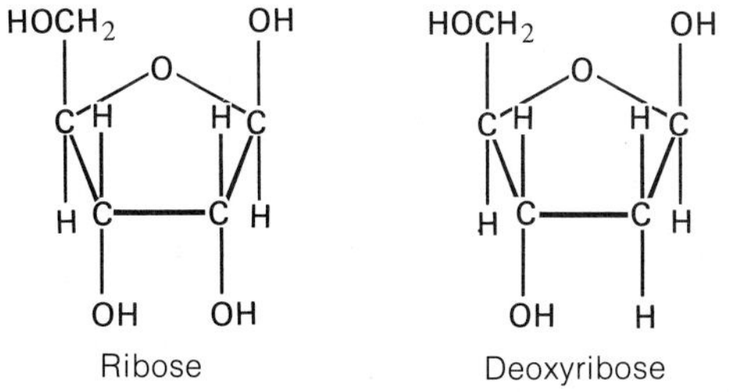

Ribose

Deoxyribose

Figure 17.3   The pentose sugars found in nucleotides.

and the sugar ribose is shown in Figure 17.4. Other ribonucleotides will have similar structures, but different bases. The deoxyribonucleotides will contain the sugar deoxyribose instead of ribose.

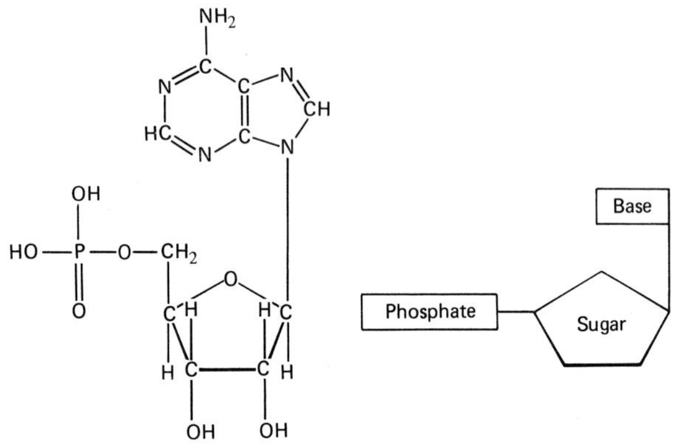

Adenylic acid —— AMP

A nucleotide

Figure 17.4   A nucleotide. Adenylic acid (adenosine monophosphate or AMP) is the nucleotide containing the sugar ribose and the base adenine.

Free nucleotides occur in significant numbers in the cell, and perform many functions. In addition to being the basic structural unit of nucleic acids, they also participate in biosynthetic reactions, serve as coenzymes, and are important in the transport of energy from energy-releasing reactions to energy-requiring reactions. ATP, for example, is an energy-carrying mononucleotide.

DNA, and proteins that are bound to the DNA molecule, make up the chromosomes in the nucleus of animal cells. The way in which nucleotides are arranged in the DNA molecule determines the genetic information that is carried by that molecule.

17.2 The Structure of DNA

The nucleic acid DNA was first isolated in 1868 by Friedrich Miescher, but the structure of the molecule was not determined until 1953 when J. D. Watson and F. H. C. Crick proposed a structure that explained the physical and chemical properties of DNA. The configuration proposed by Watson and Crick consists of two helical polynucleotide chains coiled around the same axis, forming a double helix (Figure 17.5). The hydrophilic (attracted to water) sugar and phosphate components of the nucleotides are found on the outside of the helix, and the hydrophobic bases are found on the inside.

The Backbone of the Helix

The nucleotides making up each strand of DNA are connected by ester bonds between the phosphate group and the deoxyribose sugar (Figure 17.6). This forms the "backbone" of each DNA strand, from which the bases extend. The bases of one strand of DNA will pair with bases on the other strand by means of hydrogen bonding. This hydrogen bonding is very specific: adenine's structure permits it to hydrogen bond only with thymine,

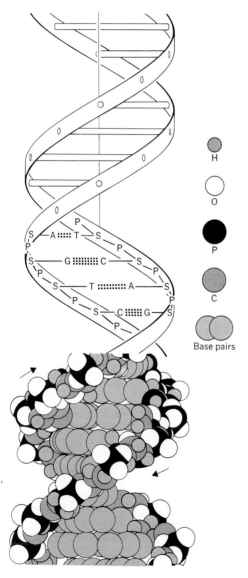

Figure 17.5 The double helix of the DNA molecule.

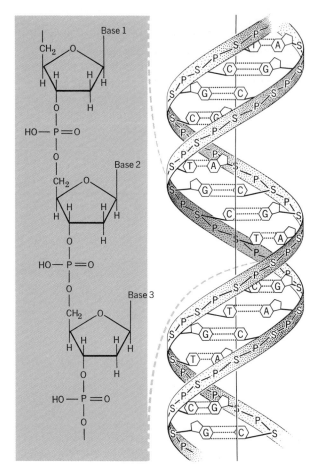

Figure 17.6 The backbone of the DNA molecule consists of the sugar and phosphate groups of the nucleotides. The deoxyribose sugar of one nucleotide is bonded to the phosphate group on the next nucleotide by an ester bond.

and guanine will bond only with cytosine (Figure 17.7). As a result, the two strands of DNA are not identical, but rather are complementary; where thymine appears on one strand, adenine will appear on the other.

17.3 Replication of DNA

When a cell divides, the DNA molecules must **replicate** (that is, must make exact copies of themselves) so that each daughter cell will have DNA identical to the parent cell. The replication of DNA is catalyzed by the enzyme DNA polymerase. In this process, the two strands of the helix unwind, and each strand serves as a template or pattern for the synthesis

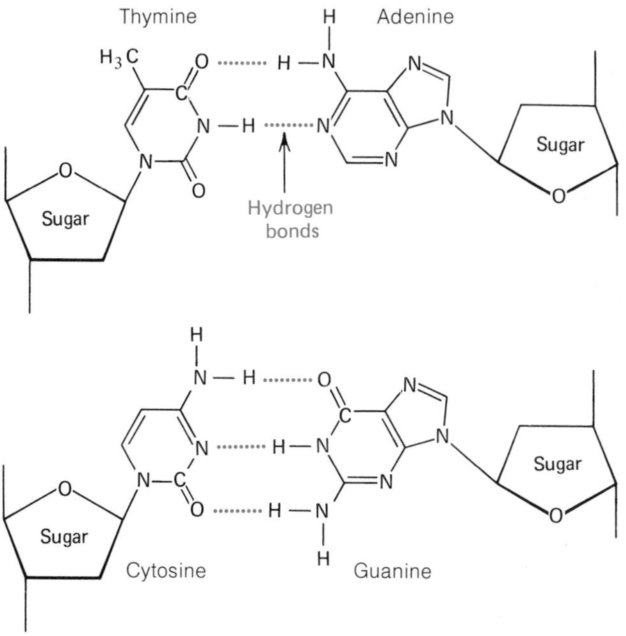

Figure 17.7 Hydrogen bonding in DNA. The bases on the nucleotides of DNA extend toward the inside of the helix, and the two strands of the helix are held together by hydrogen bonding between the bases. This hydrogen bonding is very specific: thymine will form hydrogen bonds only with adenine, and cytosine only with guanine.

of a new strand of DNA (Figure 17.8). Each of the two daughter helixes will contain one original DNA strand and one newly made strand.

Genetic Code
The genetic information for the cell is contained in the sequence of the bases A, T, C, and G in the DNA molecule. Anything that alters the order of the sequence will cause a change, or **mutation,** in the genes of the cell.

RNA, Ribonucleic Acid

Molecules of RNA make up 5 to 10% of the total weight of the cell. Unlike DNA, RNA is not double-stranded; it consists of a single strand of nucleic acid. There are three distinct types of RNA found in the cell: messenger RNA (*m*RNA), ribosomal RNA (*r*RNA), and transfer RNA (*t*RNA).

17.4 Messenger RNA (*m*RNA)

*m*RNA is synthesized in the nucleus of the cell, and contains only the four bases adenine, cytosine, guanine, and uracil. It is synthesized on one

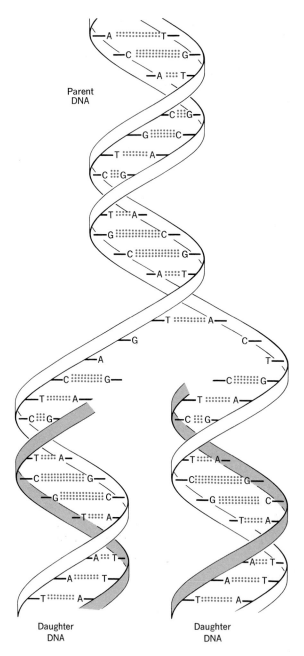

Parent
DNA

Daughter
DNA

Daughter
DNA

Figure 17.8 Replication of DNA. In one theory of
the replication of DNA, the parent strand is forced
apart and new strands are built on each of the old
strands. Each daughter DNA will contain one old
strand and one new strand.

strand of a DNA helix, so it will have a sequence of bases complementary to that of the DNA (Figure 17.9). mRNA is not a stable molecule, and is synthesized by the cell whenever it is needed. After being synthesized, a mRNA molecule will migrate to the cytoplasm of the animal cell. There it serves as a template or pattern for the sequencing of amino acids in the synthesis of proteins in the ribosomes.

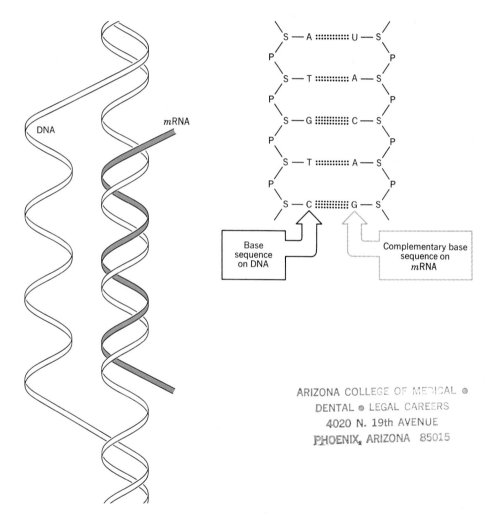

Figure 17.9 The transcription of DNA to mRNA occurs in the nucleus of the cell. The double helix of DNA opens up, and the mRNA is synthesized on one strand of the DNA. The sequence of bases on mRNA is complementary to the sequence on the DNA. However, in making the mRNA the cell uses the base uracil instead of thymine.

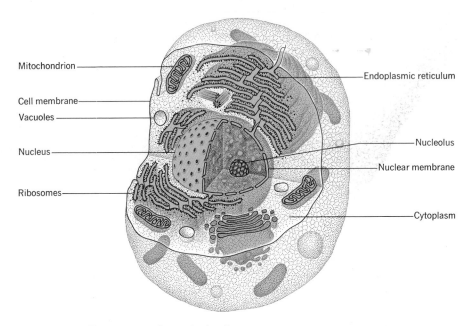

Figure 17.10 The structure of an animal cell.

17.5 Ribosomal RNA (rRNA)

rRNA is also synthesized in the nucleus of the cell from a DNA template. It migrates to the cytoplasm of the cell where, with proteins, it forms the ribosomes. The **ribosomes** are the sites of protein synthesis, and are located on the endoplasmic reticulum (Figure 17.10). The ribosomes are composed of two subunits that combine with mRNA to form the "factory" for the production of proteins (Figure 17.11).

17.6 Transfer RNA (tRNA)

tRNA is the smallest of the RNA molecules; it is water-soluble and highly mobile within the cell. tRNA molecules are synthesized in the nucleus of the cell, and each is specifically designed for a particular amino acid. A tRNA molecule becomes "charged" when a specific amino acid is joined to the terminal adenine nucleotide present on each tRNA polynucleotide chain. The tRNA molecule then carries this amino acid to the ribosomes, where the amino acid is used in protein synthesis (Figure 17.12).

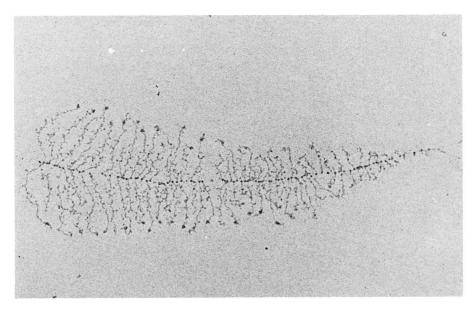

Figure 17.11 Transcription of RNA on a single gene. This electron micrograph shows the formation of fibrils of RNA on the central DNA strand. The shortest fibrils are just beginning to be formed, while the longest fibrils are almost completed. The small dots at the base of each RNA fibril on the DNA strand are molecules of the enzyme RNA polymerase. (Courtesy O. L. Miller, Jr. and Barbara R. Beatty, Biology Division, Oak Ridge National Laboratory)

Protein Synthesis

17.7 Codons and Anticodons

A **gene** is a region on the DNA molecule containing a sequence of bases that will translate into the specific sequence of amino acids making up a protein. The **genetic code** is the general term used to denote the sequence of bases found in the gene. In the early 1960s the "language" of the genetic code was deciphered through extensive studies using artificially synthesized *m*RNA molecules. It was discovered that a three-base sequence is necessary to code for each amino acid. This three-base sequence on *m*RNA is called a **codon.** There is more than one codon for most amino acids, as is shown in Table 17.1.

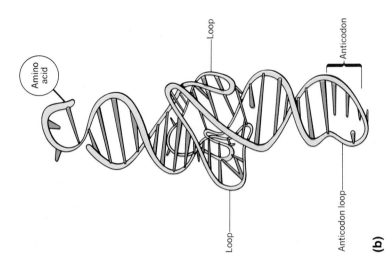

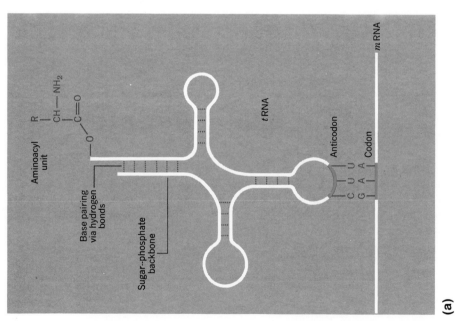

Figure 17.12 The structure of *t*RNA. (a) This is a schematic drawing of a *t*RNA. The three bases in the anticodon region are complementary to a codon on the *m*RNA. (b) This figure illustrates the three-dimensional structure of a *t*RNA.

For example, GGU, GGC, GGA, and GGG are all codons for the amino acid glycine; UUU and UUC are codons for the amino acid phenylalanine. You will note from the table that the first two bases in the codon for a given amino acid are usually the same. The third base can vary, and is less important in determining the amino acid being specified. The genetic code on the messenger RNA is not punctuated; that is, the code is read one triplet after another—without a break from one end of the mRNA molecule to the other.

mRNA: G G U C A G U G C U C C . . .

Amino acid: Gly - Gln - Cys - Ser -. . .

There are three codons (UAA, UAG, UGA) that do not code for any amino acid, and they were originally called "nonsense" codons. It is now thought, however, that these codons code for the end of the polypeptide chain of the protein and are, therefore, **terminal codons.**

An **anticodon** is the three-base sequence that is complementary to the codon on the messenger RNA molecule. Each transfer RNA molecule contains a region carrying the anticodon complementary to the codon on the mRNA molecule (Figure 17.12).

Table 17.1   The Codons for the Amino Acids

UUU Phe	UCU Ser	UAU Tyr	UGU Cys
UUC Phe	UCC Ser	UAC Tyr	UGC Cys
UUA Leu	UCA Ser	UAA	UGA
UUG Leu	UCG Ser	UAG	UGG Trp
CUU Leu	CCU Pro	CAU His	CGU Arg
CUC Leu	CCC Pro	CAC His	CGC Arg
CUA Leu	CCA Pro	CAA Gln	CGA Arg
CUG Leu	CCG Pro	CAG Gln	CGG Arg
AUU Ile	ACU Thr	AAU Asn	AGU Ser
AUC Ile	ACC Thr	AAC Asn	AGC Ser
AUA Ile	ACA Thr	AAA Lys	AGA Arg
AUG Met	ACG Thr	AAG Lys	AGG Arg
GUU Val	GCU Ala	GAU Asp	GGU Gly
GUC Val	GCC Ala	GAC Asp	GGC Gly
GUA Val	GCA Ala	GAA Glu	GGA Gly
GUG Val	GCG Ala	GAG Glu	GGG Gly

17.8 The Steps in Protein Synthesis

The steps involved in the **transcription** of DNA to *m*RNA (that is, the passage of genetic information from DNA to *m*RNA), and then in the **translation** of *m*RNA to protein (that is, the expression of the genetic information in the amino acid sequence of the protein), are identical in all living cells. Many ribosomes can be producing proteins off of the same strand of *m*RNA. When viewed under a microscope these groups of ribosomes, called polyribosomes or **polysomes,** appear as a series of dots (Figure 17.13). The following step-by-step description refers to the diagrams in Figures 17.14 and 17.15.

1. *Synthesis of mRNA. mRNA* is synthesized in the nucleus of the animal cell from a template formed by one strand of the DNA molecule. The *m*RNA then migrates from the nucleus to the cytoplasm.

2. *Attaching Amino Acids to tRNA.* Amino acids are attached to molecules of *t*RNA by enzymes that are specific for each amino acid and each *t*RNA. Each such enzyme has two active sites: one that recognizes the amino acid and one that recognizes the anticodon region of the *t*RNA. The attachment of the amino acid to the *t*RNA (that is, the activation or "charging" of the *t*RNA) is a two-step process requiring the presence of ATP.

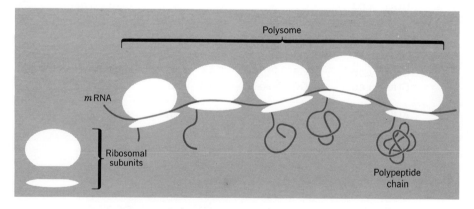

Figure 17.13 Ribosomes consist of two submits that form a complex with the *m*RNA. A polysome is a cluster of ribosomes all making polypeptide chains on the same strand of *m*RNA.

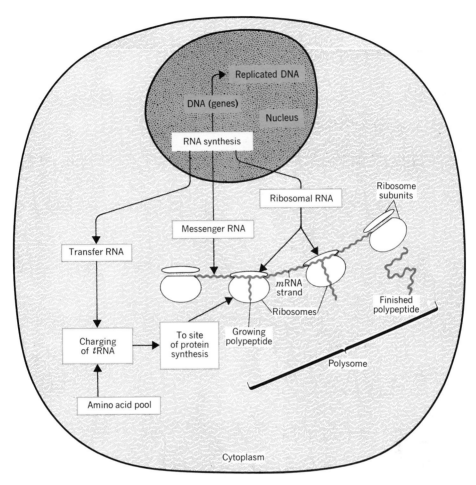

Figure 17.14 The relationship between DNA, the three RNA's, and protein synthesis. (Adapted from J. R. Holum, *Elements of General and Biological Chemistry,* 1975, John Wiley and Sons, Inc. New York. Used by permission.)

3. *Initiating the Polypeptide Chain.* To begin the synthesis of a protein, the ribosome breaks into its two subunits. The smaller unit attaches to the *m*RNA and to an initiating amino acid that is a derivative of N-formylmethionine (symbol, fMet). This complex then reforms with the larger subunit of the ribosome. This procedure insures that the synthesis of the protein will not accidentally begin in the middle of the *m*RNA molecule.

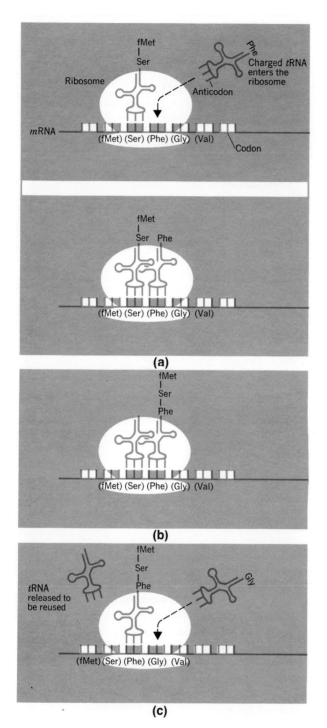

Figure 17.15 Growth or elongation of the peptide chain. The individual steps are discussed in Section 17.8.

4. *Adding Amino Acids to the Chain.* The growth or elongation of the polypeptide chain involves a series of repeated steps that take place between *t*RNA and *m*RNA inside the ribosome as it moves along the *m*RNA strand (Figure 17.15):

   a. A "charged" *t*RNA binds to the next codon on the *m*RNA.

   b. A peptide bond forms between the amino group on the newly attached amino acid and the carboxyl group on the previously bound amino acid, displacing the *t*RNA molecule that had been attached to the previously bound amino acid.

   c. This displaced *t*RNA, now empty, is bumped from the ribosome and is free to become charged once again.

5. *Termination of the Polypeptide Chain.* The elongation of the chain stops when a terminal codon is reached on the *m*RNA molecule. The polypeptide chain separates from the ribosome, and the last *t*RNA drops off the chain. The initiating amino acid is now cleaved from the protein if it is not part of the molecule. As the polypeptide chain has been growing, it has been spontaneously bending and folding into its native shape. Therefore, when it is released from the ribosome, the protein is in its active form.

17.9 Regulation of Protein Synthesis

The control of protein synthesis is important in the differentiation of cells. Each cell in the body contains all the genetic information necessary to produce an entire organism, yet less than 10% of this information will be expressed in any one cell. The genes that are active in a heart muscle cell, for example, are different from those active in a liver cell. As we have seen, cellular processes are regulated by a complex set of biological controls. The "coarse" control mechanism is the synthesis of enzymes by the cell, while the "fine" control involves the allosteric or regulatory enzymes.

Very little is known about the exact mechanisms of the control of enzyme synthesis in higher animals (that is, the regulation of gene transcription to *mRNA*). What we do know about gene regulation comes principally from studies on *Escherichia coli* (*E. coli*), a single-cell organism that occurs naturally in our digestive tract.

These studies found that different types of genes can be distinguished in the control of enzyme synthesis. A region of DNA containing the code for a specific protein is called a **structural gene.** A group of structural genes that code for the enzymes catalyzing a multienzyme system is called an **operon;** these genes will be located next to one another on the chromosome. The synthesis of *mRNA* on a structural gene is catalyzed by the enzyme RNA polymerase. RNA polymerase can function only if a strip of DNA next to the operon, called the **operator gene,** says it is okay to go ahead. This operator gene is, in turn, controlled by a protein molecule called the **repressor.** When the repressor molecule is bound to the operator gene, no synthesis of *mRNA* can take place. The removal of the repressor gene from the operator gene requires that a molecule called the **inducer** bonds to the repressor. This union changes the shape of the repressor, and it will no longer fit on the operator gene. With the repressor so removed, the operator gene is "turned on," and RNA polymerase is free to catalyze the synthesis of *mRNA* on the operon.

Let's look at the best understood example of this control mechanism: the lac operon of *E. coli* (Figure 17.16). *E. coli* is able to live on the sugar lactose as its sole source of carbon. The lactose cannot be used, however, until this disaccharide has been broken down into glucose and galactose by the enzyme *β*-galactosidase. The structural gene for *β*-galactosidase does not continuously produce *mRNA* for this enzyme, and there is a repressor molecule on the operator gene. When *E. coli* is placed in a medium containing lactose, the lactose molecule itself acts as the inducer. It attaches to the repressor, causing this molecule to move off the operator gene. This permits the production of the *mRNA* needed for the synthesis of *β*-galactosidase by the ribosomes. The transcription of *mRNA* will continue until the lactose has all been broken down. At this point the repressor molecule will then again be free, and will return to the chromosomes to repress the operator gene. *β*-Galactosidase production will then stop.

Figure 17.16 *(on facing page)* Control of protein synthesis in the lac operon of *E. coli.* (1) *mRNA* coded for the repressor molecule is synthesized on the repressor gene. The repressor molecule is produced from the *mRNA* template in the ribosomes. (2) When the repressor molecule is attached to the operator gene, the transcription of *mRNA* from the structural gene for *β*-galactosidase is halted. (3) When the inducer molecule (in this case, lactose) attaches to the repressor molecule, the repressor is no longer able to bind to the operator gene. Transcription of the structural gene is no longer blocked, and *mRNA* coded for *β*-galactosidase is synthesized. This transcription will terminate when all the lactose has been hydrolyzed, leaving the repressor molecule again free to bind to the operator gene.

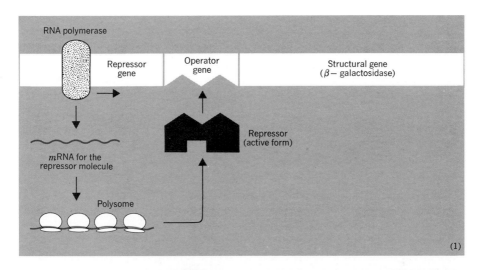

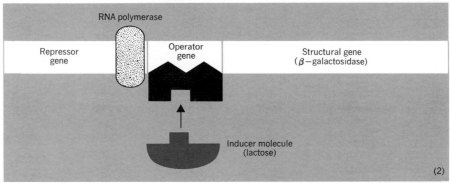

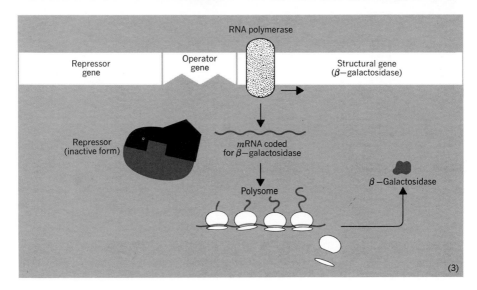

17.10 "Central Dogma" of Molecular Biology

In 1958 Crick proposed a theory of gene action that has come to be known as the **central dogma of molecular biology.** This theory states that genetic information can flow in only one direction—from **DNA** to **RNA** to **protein.** Discoveries made by H. M. Temin and D. Baltimore in the early 1970s have modified this dogma. They discovered that an enzyme called RNA-directed DNA polymerase (or reverse transcriptase), which is present in some RNA tumor viruses, is capable of synthesizing short segments of double-stranded DNA from viral RNA (Figure 17.17). The ability of the RNA virus to synthesize DNA may explain the process by which a virus can disappear after infecting an organism, only to reappear months or years later. The viral-produced DNA may be incorporated into the cellular DNA, and may be passed on from cell to cell until it is again activated in some unknown way. Researchers are actively studying this mechanism as a possible explanation of the process by which a normal cell changes into a cancerous cell.

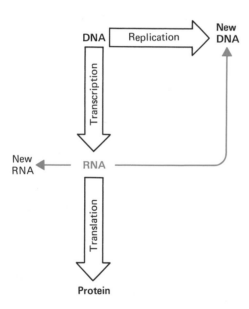

Figure 17.17 The flow of genetic information. The central dogma of molecular biology, which states that genetic information flows from DNA to RNA to protein, has been modified to include evidence indicating that RNA can produce new RNA and, in some cases, DNA. However, it has yet to be shown that the amino acid sequence of a protein can code for a nucleotide sequence in a nucleic acid.

17.11 Mutations

A **mutation** is any chemical or physical change in the DNA molecule that results in the synthesis of a protein having an altered amino acid sequence. There are many ways by which a gene can be changed or mutated. The sequence of bases in the DNA molecule is critical and very specific. Replacing a base in the sequence, adding a base, or deleting a base will throw the code off and may change the amino acid sequence of the resulting protein. Such changes can occur spontaneously, or may be caused by radiation or chemical agents.

Often the defective proteins produced through a mutation lack biological activity, resulting in the death of the cell. In other cases, the defective gene may cause abnormalities or disease. We saw at the beginning of this chapter that a mutation in the structural gene for hemoglobin resulted in an error in one amino acid on the beta polypeptide chain. This causes the production of the abnormal hemoglobin characteristic of sickle cell anemia. More than 2000 diseases in man result from mutations in genes; included are such diseases as cystic fibrosis, albinism, hemophilia, galactosemia, color blindness, and PKU (Table 17.2).

Table 17.2   Some Diseases of Fat Metabolism That Are Hereditary*

Disease	Symptoms
Fabry's disease	Reddish-purple skin rash, kidney failure, and pain in legs.
Gaucher's disease	Spleen and liver enlargement, mental retardation in infantile form, and erosion of long bones and pelvis.
Generalized gangliosidosis	Mental retardation, liver enlargement, skeletal deformities, and red spot in retina in about 50% of the cases.
Niemann-Pick disease	Mental retardation, liver and spleen enlargement, and red spot in retina in about 30% of the cases.
Tay-Sachs disease	Mental retardation, red spot in retina, blindness, and muscular weakness.

* In these diseases, sphingolipids accumulate in tissues because the enzymes that normally catalyze the cleavage of these lipids are defective.

17.12 PKU — A Final Look

We began this book with the story of Billy, a child suffering from the disease phenylketonuria. PKU is a disease resulting from a mutation in the structural gene that codes for the liver enzyme phenylalanine hydroxylase. This defective gene is carried by 2% of the population; when a child inherits the defective gene from both of his parents, he will suffer from PKU. In this disease the defective phenylalanine hydroxylase produced in the liver will not catalyze the conversion of phenylalanine to tyrosine (Figure 17.18). This results in the buildup in the body of phenylalanine and PKU metabolites — substances produced by the metabolism of the excess phenylalanine. It is this chemical imbalance that causes the abnormalities observed in an untreated PKU child (Figure 17.19).

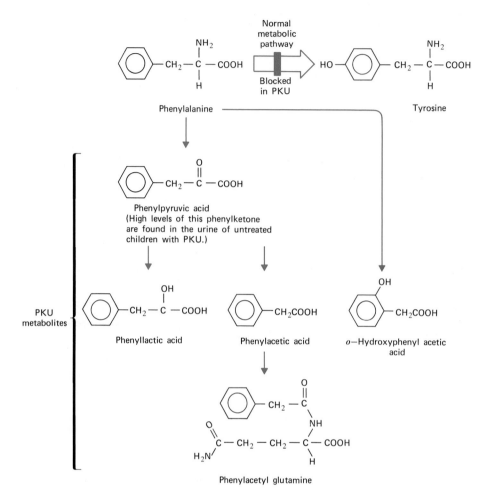

Figure 17.18 The blockage of a metabolic pathway in PKU results in the production of PKU metabolites as the body tries to metabolize the phenylalanine by alternate chemical routes.

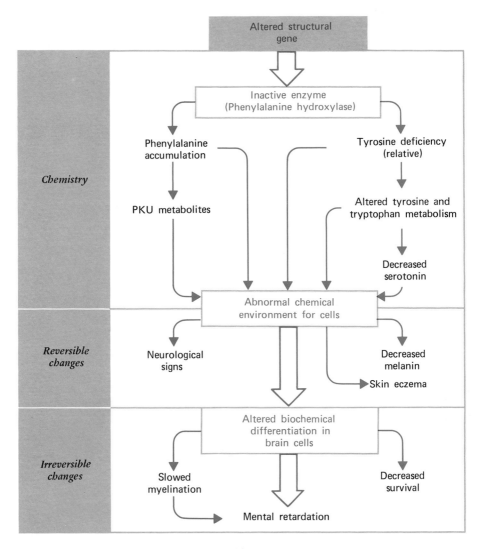

Figure 17.19 The metabolic changes caused by the mutation of PKU. (Adapted from *The Metabolic Basis of Inherited Disease* by Stanbury et al. Copyright © 1972 by McGraw-Hill Inc. Used with permission of McGraw-Hill Book Company.)

Additional Reading

Books

1. H. Abramson, J. Bertles, and D. Wethers, ed., *Sickle Cell Disease,* C. V. Mosby, St. Louis, 1973.

2. J. S. Lin-Fu, *Sickle Cell Anemia,* U. S. Department of Health, Education, and Welfare, Rockville, Maryland, 1972.

3. A. Sayre, *Rosalind Franklin and DNA,* Norton, New York, 1975.

4. J. D. Watson, *Double Helix,* Atheneum, New York, 1968.

Articles

1. J. Arehart-Treichel, "How Mammals Get the Message: *m*RNA," *Science News,* March 23, 1974, page 197.

2. _____ "Sickle Cells: The Molecular Attack," *Science News,* February 16, 1974, page 104.

3. T. Friedmann, "Prenatal Diagnosis of Genetic Disease," *Scientific American,* November 1971, page 34.

4. S. Gilmore, "The Structure of Chromatin," *New Scientist,* February 13, 1975, page 390.

5. A. Kornberg, "The Synthesis of DNA," *Scientific American,* October 1968, page 64.

6. T. Maniatis and M. Ptashne, "A DNA Operator-Repressor System," *Scientific American,* January 1976, page 64.

7. O. L. Miller, Jr., "The Visualization of Genes in Action," *Scientific American,* March 1973, page 34.

8. A. E. Mirsky, "The Discovery of DNA," *Scientific American,* June 1968, page 78.

9. M. Ptashne and W. Gilbert, "Genetic Repressors," *Scientific American,* June 1970, page 36.

10. G. S. Stein, J. S. Stein, and L. J. Kleinsmith, "Chromosomal Proteins and Gene Regulation," *Scientific American,* February 1975, page 47.

11. H. M. Temin, "RNA-directed DNA Synthesis," *Scientific American,* January 1972, page 24.

12. The Genetic Code
    F. H. C. Crick, *Scientific American,* October 1962.
    M. W. Nirenberg, *Scientific American,* March 1963.
    F. H. C. Crick, *Scientific American,* October 1966.
    C. Yanofsky, *Scientific American,* May 1967.

13. Carcinogenesis
    T. H. Maugh, "Chemical Carcinogenesis: Genetic Mechanism," *Science,* March 8, 1974, page 941.
    J. L. Marx, "Viral Carcinogenesis: Role of DNA Viruses," *Science,* March 15, 1974, page 1066.
    T. H. Maugh, "RNA Viruses: The Age of Innocence Ends," *Science,* March 22, 1974, page 1181.

Questions and Problems

1. Define the following terms:

   (a) Hemoglobin
   (b) Sickle cell anemia
   (c) Nucleotide
   (d) Nucleic acid
   (e) Pyrimidine
   (f) Purine
   (g) Double helix
   (h) *m*RNA
   (i) *t*RNA
   (j) *r*RNA
   (k) Ribosome
   (l) Gene
   (m) Genetic code

   (n) Codon
   (o) Terminal codon
   (p) Anticodon
   (q) Polysomes
   (r) Structural gene
   (s) Operon
   (t) Operator gene
   (u) Repressor
   (v) Inducer
   (w) Replication
   (x) Transcription
   (y) Translation
   (z) Mutation

2. What are the three components comprising a nucleotide?

3. What is the difference between a nucleic acid and a nucleotide?

4. Which nitrogen bases found in DNA are pyrimidines? Which are purines?

5. Look at Figure 17.3. From the structure of deoxyribose, explain the meaning of the prefix *deoxy-*.

6. Write the structure of the nucleotide containing the sugar deoxyribose and the base guanine.

7. How are the two polynucleotide chains held together in the double helix?

8. "The two strands of DNA are not identical, but rather are complementary." Explain this statement.

9. In general terms describe the replication of a DNA molecule.

10. What are the three main structural differences between a molecule of DNA and a molecule of RNA?

11. What are the three types of RNA found in the cell? Describe the function of each type.

12. How is genetic information carried on the DNA molecule?

13. Describe the mechanism by which the synthesis of β-galactosidase in E. coli is stopped when the enzyme is no longer needed.

14. What is the "central dogma" of molecular biology? How has this dogma been modified in the light of recent experimental data?

15. The disease sickle cell anemia results from a mutation of a specific region of DNA that codes for the polypeptide chain of a hemoglobin molecule.

   (a) Suggest three possible causes of this mutation.
   (b) The mRNA for the β-chain of normal hemoglobin has the base triplet GAA or GAG in the sixth position. What is the change in this base sequence that results in the production of the β-chain of sickle cell hemoglobin?
   (c) In general terms describe the steps involved in the production of one normal β-chain.
   (d) Account for the difference in mobility of normal and sickle cell hemoglobin in an electric field.
   (e) Explain how the substitution of one amino acid in the β-chain of hemoglobin can result in the symptoms of sickle cell anemia.

16. One of the arguments for eliminating the use of fluorocarbons in aerosol sprays is that fluorocarbons might destroy the ozone layer in the upper atmosphere. It is this layer that absorbs ultraviolet radiation coming in from outer space. The destruction of this ozone layer could result in an increase in the number of cases of skin cancer. Suggest a possible mechanism for the increase in skin cancer that might be caused by a reduction in the ozone layer.

appendix 1
Numbers in Exponential Form

Whenever we are working with very large or very small numbers, it is convenient to write them in **exponential form** (also called scientific notation). Numbers written in exponential form are expressed as the product of a number between one and ten multiplied by the number ten raised to some power. The **exponent** is the power to which the number 10 is raised, and is shown as a superscript to the right of the number 10.

$$4.75 \times 10^3 \longleftarrow \text{Exponent}$$

If the exponent is a positive number, it indicates how many times the number is to be multiplied by the number 10. For example,

$$10^3 = 1 \times 10^3 = 1 \times 10 \times 10 \times 10 = 1000$$

$$4.5 \times 10^6 = 4.5 \times 10 \times 10 \times 10 \times 10 \times 10 \times 10 = 4{,}500{,}000$$

If the exponent is a negative number, it tells us how many times the number is to be divided by the number 10. For example,

$$10^{-3} = 1 \times 10^{-3} = \frac{1}{10 \times 10 \times 10} = 0.001$$

$$4.5 \times 10^{-6} = \frac{4.5}{10 \times 10 \times 10 \times 10 \times 10 \times 10} = 0.0000045$$

The following table illustrates what various numbers look like in exponential form:

Number	Exponential Form	Number	Exponential Form
10	1×10^1	45	4.5×10^1
100	1×10^2	356	3.56×10^2
1000	1×10^3	8400	8.4×10^3
10,000	1×10^4	24,500	2.45×10^4
100,000	1×10^5	680,000	6.8×10^5
1,000,000	1×10^6	7,450,000	7.45×10^6
0.1	1×10^{-1}	0.5	5×10^{-1}
0.01	1×10^{-2}	0.037	3.7×10^{-2}
0.001	1×10^{-3}	0.004	4×10^{-3}
0.0001	1×10^{-4}	0.00056	5.6×10^{-4}
0.00001	1×10^{-5}	0.000082	8.2×10^{-5}
0.000001	1×10^{-6}	0.0000091	9.1×10^{-6}
0.0000001	1×10^{-7}	0.0000002	2×10^{-7}

Multiplying Numbers in Exponential Form

To multiply two numbers expressed in exponential form, you:

1. *Multiply* the two numbers appearing to the left of the number 10.

2. Then, *add* the two exponents to determine the power of 10 to use in the product.

For example,

(a) $(1 \times 10^4) \times (1 \times 10^6) = 1 \times 10^{(4+6)} = 1 \times 10^{10}$
(b) $(4 \times 10^2) \times (6 \times 10^5) = (4 \times 6) \times 10^{(2+5)} = 24 \times 10^7 = 2.4 \times 10^8$
(c) $(2 \times 10^4) \times (3 \times 10^{-6}) = (2 \times 3) \times 10^{(4+(-6))} = 6 \times 10^{-2}$

[Note from example (b) that exponential form requires that the power of ten be multiplied by a number between one and ten.]

Dividing Numbers in Exponential Form

To divide two numbers in exponential form:

1. *Divide* the two numbers appearing to the left of the number 10.

2. Then, *subtract* the exponent of the denominator from the exponent of the numerator.

For example,

(a) $\dfrac{1 \times 10^6}{1 \times 10^4} = \dfrac{1}{1} \times 10^{(6-4)} = 1 \times 10^2$

(b) $\dfrac{8 \times 10^7}{2 \times 10^5} = \dfrac{8}{2} \times 10^{(7-5)} = 4 \times 10^2$

(c) $\dfrac{8 \times 10^4}{3 \times 10^{-2}} = \dfrac{8}{3} \times 10^{(4-(-2))} = 2.67 \times 10^6$

(d) $\dfrac{4 \times 10^{-3}}{8 \times 10^2} = \dfrac{4}{8} \times 10^{(-3-2)} = 0.5 \times 10^{-5} = 5 \times 10^{-6}$

appendix 2
Chemical Calculations

In performing the many calculations needed in any area of science, it is usually very helpful to keep careful track of the units of measure represented by each of the numbers being used. That is, rather than just writing down the number 15, say, you should write 15 grams, or 15 feet, or 15 gallons, or whatever units of measure the number 15 happens to represent. Doing this will not only keep you from getting confused in the middle of a calculation, but can often remind you of the steps you must perform to finish solving the problem. For example, suppose you must do a calculation to determine the speed of some object. If you know that speed is commonly measured in such units as miles per hour $\left(\dfrac{miles}{hour}\right)$, feet per second $\left(\dfrac{feet}{second}\right)$, or meters per second $\left(\dfrac{meters}{second}\right)$, this tells you that your eventual goal is to divide some measurement of distance (in the appropriate units) by a measurement of time (in the appropriate units). You will find that the calculations required in this textbook are much easier to perform if you make a point of keeping your numbers properly labeled with their units of measure.

The Role of the Number "One"

There are two properties of the number "one" with which you are very familiar, and that are crucial to the performance of chemical calculations. The first property is that any number, when multiplied by one, remains the same. More generally, any quantity remains the same when multiplied by the number one. For example, $36 \times 1 = 36$, 7 apples $\times 1 = 7$ apples, and $92\, \dfrac{miles}{hour} \times 1 = 92\, \dfrac{miles}{hour}$.

The second property of the number one is that this number can be written as $\dfrac{2}{2}$, or $\dfrac{156}{156}$, or as the quotient of any number divided by itself. More generally, the number one can be represented by any quantity divided by itself. For example, the number one can be represented by $\dfrac{5\ apples}{5\ apples}$, or $\dfrac{156\ camels}{156\ camels}$ — but cannot be represented by $\dfrac{17\ apples}{17\ oranges}$, since the numerator is not the same as the denominator. Taken one step further, the number one can be represented by $\dfrac{12\ inches}{1\ foot}$, or by $\dfrac{12\ eggs}{1\ dozen\ eggs}$, or by $\dfrac{1\ inch}{2.54\ cm}$, since the quantity in the numerator is equal to the quantity in the denominator, even though they are here expressed in different terms. Such ratios equivalent to the number one are called **unit factors,** and are the key to most of the calculations in this textbook.

To illustrate how unit factors are used in a calculation, let's look at a problem you could easily solve: how many inches are there in 2 feet? You would immediately say 24 inches, which you would have calculated by multiplying $2 \times 12 = 24$. But what did you do when you started with a distance that you called "2," and said that it is the same as a distance that you called "24"? What you actually did was to use a unit factor as follows:

$$2 \text{ feet} \times \frac{12 \text{ inches}}{1 \text{ foot}} = \frac{2 \times 12 \ (\cancel{\text{feet}}) \ (\text{inches})}{(\cancel{\text{feet}})} = 24 \text{ inches}$$

You are able to feel confident that the distance you started out with (2 feet) is the same distance that you ended up with (24 inches) because all you really did was to multiply your initial distance by a unit factor—that is, by the number one. When you make a point of keeping close track of the units of measure for each of the numbers in the problem (as we just did), you see that similar units in the numerator and denominator "cancel," leaving you with an answer in the desired units of measure.

Working with the Metric System

Most of the problems encountered in this textbook can be solved through the use of appropriate unit factors. Let's look at some sample problems using the metric system to see how this is done.

Example 1: _____
How many centimeters are there in 4 meters? We might rephrase this problem as follows:

$$4 \text{ meters} = (?) \text{ centimeters}$$

Since the problem requires us to change meters into centimeters, we must look for a unit factor that shows the relationship between these two units of measure. Using the equality 1 meter = 100 centimeters, we can construct two unit factors that might be appropriate:

$$\frac{1 \text{ meter}}{100 \text{ centimeters}} \quad \text{or} \quad \frac{100 \text{ centimeters}}{1 \text{ meter}}$$

The second unit factor is the correct one to choose since it will allow units of measure to cancel so as to give us an answer in centimeters.

$$4 \ \cancel{\text{meters}} \times \frac{100 \text{ centimeters}}{1 \ \cancel{\text{meter}}} = \frac{4 \times 100}{1} \text{ centimeters} = 400 \text{ centimeters}$$

Example 2: _____
How many milligrams are there in 0.024 grams? We know that 1 gram = 1000 milligrams, so from this equality we can determine two possible unit factors showing the relationship between grams and milligrams:

$$\frac{1 \text{ gram}}{1000 \text{ milligrams}} \quad \text{or} \quad \frac{1000 \text{ milligrams}}{1 \text{ gram}}$$

Since our problem asks 0.024 g = (?) mg, the second factor is the correct one to use.

$$0.024 \ \cancel{\text{g}} \times \frac{1000 \text{ mg}}{1 \ \cancel{\text{g}}} = 0.024 \times 1000 \text{ mg} = 24 \text{ mg}$$

Calculations Involving the Metric System and English System

The following table of conversion factors between the metric and English systems will be helpful in setting up the unit factors for the next set of examples.

1 m = 39.37 in	1 kg = 2.2 lb	1 l = 1.06 qt
1.6 km = 1 mi	908 kg = 1 ton	3.78 l = 1 gal
2.54 cm = 1 in	454 g = 1 lb	(1 ml = 1 cc)

Example 3: _____
The distance between Seattle and San Francisco is 820 miles. How many kilometers is this?

Since we want to change miles into kilometers, we need a unit factor showing the relationship between these two units. From the above table we have a choice of

$$\frac{1.6 \text{ km}}{1 \text{ mi}} \quad \text{or} \quad \frac{1 \text{ mi}}{1.6 \text{ km}}$$

Our problem asks 820 mi = (?) km, so the first unit factor is the correct one to use.

$$820 \text{ mi} \times \frac{1.6 \text{ km}}{1 \text{ mi}} = 820 \times 1.6 \text{ km} = 1312 \text{ km}$$

Example 4: _____
How many milliliters are there in one pint?

This problem will require us to use several different unit factors to make the transition from pints to milliliters. Our choices are as follows:

$$\frac{2 \text{ pints}}{1 \text{ quart}} \quad \text{or} \quad \frac{1 \text{ quart}}{2 \text{ pints}}$$

$$\frac{1 \text{ liter}}{1.06 \text{ quarts}} \quad \text{or} \quad \frac{1.06 \text{ quarts}}{1 \text{ liter}}$$

$$\frac{1 \text{ liter}}{1000 \text{ ml}} \quad \text{or} \quad \frac{1000 \text{ ml}}{1 \text{ liter}}$$

We can solve the entire problem in one step if we carefully arrange unit factors so that all the units of measure cancel, leaving us with an answer expressed in milliliters.

$$1 \text{ pint} \times \frac{1 \text{ quart}}{2 \text{ pints}} \times \frac{1 \text{ liter}}{1.06 \text{ quarts}} \times \frac{1000 \text{ ml}}{1 \text{ liter}} = \frac{1000}{2 \times 1.06} \text{ ml}$$

$$= 471.7 \text{ ml}$$

Chapter Opening Photo Credits

Chapter 1, page 1:	Fredrik D. Bodin/Stock, Boston
Chapter 2, page 11:	Fritz Henle/Photo Researchers
Chapter 3, page 58:	Courtesy U.S. Air Force
Chapter 4, page 83:	Courtesy D. Bruce Sodee, M.D., FACP, DABNM, Director, Nuclear Medicine Department, Hillcrest Hospital, Mayfield Heights, Ohio
Chapter 5, page 110:	Courtesy Argonne National Laboratory
Chapter 6, page 139:	Wide World Photos
Chapter 7, page 169:	Wide World Photos
Chapter 8, page 191:	Ron Nelson
Chapter 9, page 227:	From Alexander I. Oparin, CHEMICAL ORIGIN OF LIFE, translated by Ann Synge, 1964. Courtesy Charles C. Thomas, Publisher
Chapter 10, page 277:	Hugh Rogers/Monkmeyer
Chapter 11, page 311:	Mimi Forsyth/Monkmeyer
Chapter 12, page 335:	Culver Pictures
Chapter 13, page 363:	Michal Heron/Monkmeyer
Chapter 14, page 405:	Dan Budnick/Woodfin Camp
Chapter 15, page 430:	World Health Organization
Chapter 16, page 463:	Phyllis Lefohn
Chapter 17, page 493:	From Marcel Bessis, CORPUSCLES: Atlas of Red Blood Cell Shapes, 1973. Courtesy Springer-Verlag New York, Inc.

Index*

for Chemistry and The Living Organism

---

* The page numbers in italics indicate pages containing tables.